Issues in Toxicology

Series Editors:
Professor Diana Anderson, *University of Bradford, UK*
Dr Michael D Waters, *Integrated Laboratory Systems, Inc, N Carolina, USA*
Dr Martin F Wilks, *University of Basel, Switzerland*
Dr Timothy C Marrs, *Edentox Associates, Kent, UK*

Titles in the Series:

How to obtain future titles on publication:
A standing order plan is available for this series. A standing order will bring delivery of each new volume immediately on publication.

For further information please contact:
Book Sales Department, Royal Society of Chemistry, Thomas Graham House, Science Park, Milton Road, Cambridge, CB4 0WF, UK
Telephone: +44 (0)1223 420066, Fax: +44 (0)1223 420247
Email: booksales@rsc.org
Visit our website at www.rsc.org/books

Contents

Issues in Toxicology No. 19
Reducing, Refining and Replacing the Use of Animals in Toxicity Testing
Edited by David G. Allen and Michael D. Waters
© The Royal Society of Chemistry 2014
Published by the Royal Society of Chemistry, www.rsc.org

Chapter 5   Computers Instead of Cells: Computational Modeling of Chemical Toxicity   163

Hao Zhu, Marlene Kim, Liying Zhang and Alexander Sedykh

CHAPTER 1

History of the 3Rs in Toxicity Testing: From Russell and Burch to 21st Century Toxicology

MARTIN L. STEPHENS*[a] AND NINA S. MAK[b]

[a] Center for Alternatives to Animal Testing, Johns Hopkins University, 615 N. Wolfe Street, W7032, Baltimore, MD 21205, USA; [b] Alternatives Research & Development Foundation, 801 Old York Road, #316, Jenkintown, PA 19046, USA, Email: nmak@ardf-online.org
*Email: mstephen@jhsph.edu

1.1 Introduction

Toxicity testing is an important part of the process of assessing the hazards, safety, or risk that chemicals and other substances pose to humans, animals, or the environment. The early toxicity tests that went on to enter routine use were developed in the first half of the 20th century. These included the LD50 test for acute systemic toxicity[1] and the Draize test for eye irritancy.[2] These procedures used vertebrate animals as test subjects – typically rodents in the LD50 test and rabbits in the Draize test.

As the new science of toxicology progressed, it continued to rely heavily on animals as test subjects. This was due largely to the rise of animal (mostly rodent) breeding for science in general, the virtual absence of more sophisticated ways of assessing toxicity, and the low status of animals in society.

While animal use in toxicology and the life sciences in general rose over the course of the early and mid-20th century, two parallel trends emerged. Concern

Issues in Toxicology No. 19
Reducing, Refining and Replacing the Use of Animals in Toxicity Testing
Edited by David G. Allen and Michael D. Waters
© The Royal Society of Chemistry 2014
Published by the Royal Society of Chemistry, www.rsc.org

for animal welfare was growing, and the life sciences were flourishing, with dramatic advances in knowledge and technique. It was in this context that the Universities Federation for Animal Welfare (UFAW), founded in England in 1926, made the fateful decision in the mid-1950s to undertake an ambitious survey of humane experimental techniques in animal-based experimentation throughout the life sciences.[3] The project culminated in a pioneering book, *The Principles of Humane Experimental Technique* (hereinafter "*The Principles*").[4]

The book's authors – scientists William Russell and Rex Burch – proposed the 3Rs framework for making progress on both scientific and animal welfare fronts. Specifically, they advocated using scientific ingenuity to replace, reduce, and refine the use of animals wherever feasible without compromising scientific rigor. Russell and Burch's extensive discussion of each of the 3Rs included numerous and diverse examples of each approach, drawing on their broad knowledge of the life sciences (and that of the experts they consulted).

"Refinement" referred to modifications in procedures that resulted in the animals experiencing less pain, distress, or discomfort. In discussing refinement, Russell and Burch considered a wide range of issues, including anesthesia, analgesia, euthanasia technique, injection sites, use of less sentient species, and adoption of less intense experimental procedures to induce stress. The scope of refinement eventually expanded beyond limiting negative effects and came to include enhancing animal welfare, such as through housing social animals in groups rather than individually, or enriching their cage environment with objects such as nesting material.[5]

"Reduction" referred to careful design and analysis of animal-based experiments so that fewer animals could be used. In this context, Russell and Burch discussed a variety of approaches, such as calculating the minimum group size(s) needed for a particular experiment, conducting testing sequentially rather than concurrently to exploit information learned in prior stages, and employing advanced experimental designs (e.g., blocking) that increased statistical power while using fewer animals. They also called for increased use of genetically uniform animals, or the offspring of crosses between two different in-bred lines, as a means of controlling inter-individual variation. And more generally, they argued that it is contrary to the spirit of reduction to waste animals on experiments that are poorly conceived, designed, or statistically analyzed.

Finally, "replacement" referred to ways to avoid using whole, sentient animals, by the use of: (i) non-animal approaches such as *in vitro* methods, microorganisms, ethical human studies, and computer simulation, (ii) experiments using invertebrates, or early stage vertebrate embryos, and (iii) anesthetized vertebrates. Over time, use of anesthetized vertebrates has come to be viewed as refinement rather than replacement.

In the 1960s and 1970s, animal protection organizations began to use the term "alternatives" for the 3Rs, especially replacement, as part of their challenge to the status quo. Thus the "alternatives" label proved to be more politically charged than did the "3Rs."

The forceful championing of alternatives left some scientists uneasy.[6] In response, they began to push back. In 1985, for example, Swiss scientist

W. H. Wiehe argued that "alternative methods are a fallacy" and that *in vivo* experiments are "irreplaceable."[7] Ironically, a 1978 book by the Research Defence Society (in England) had explicitly and approvingly used this term "alternatives" to refer to each of the 3Rs in a book entitled "Alternatives to Animal Experiments."[8]

Yet a little more than a generation after Wiehe's dismissive remark, following the 2007 publication of a US National Research Council (NRC) report on "Toxicity Testing in the 21st Century, A Vision and a Strategy,"[9] prominent scientists were predicting the near elimination – if not the total replacement – of animal use in toxicity testing through the development of "21st Century Toxicology." Melvin Andersen and Daniel Krewski (members of the committee that drafted the NRC report) noted that the report "envisions a not-so-distant future in which virtually all routine toxicity testing would be conducted in human cells or cell lines *in vitro*."[10] Key government scientists in the United States, led by Francis Collins, currently director of the National Institutes of Health (NIH), wrote in a 2008 policy forum of the prestigious journal *Science* that federal initiatives are "promoting the evolution of toxicology from a predominantly observational science at the level of disease-specific models *in vivo* to a predominantly predictive science focused on broad inclusion of target-specific, mechanism-based, biological observations *in vitro*."[11] That same year Collins' predecessor, Elias Zerhouni, had referred to these initiatives as the beginning of the end of animal testing.[12] In an 2011 editorial about the efforts of the US Food and Drug Administration (FDA) to modernize regulatory science, FDA Commissioner Margaret Hamburg wrote in *Science* that the agency is "working to eventually replace animal testing with a combination of *in silico* and *in vitro* approaches."[13]

How have we gotten from Russell and Burch to the beginnings of 21st Century Toxicology? That journey will be the focus of this chapter.

1.1.1 Measuring 3Rs Activity: Our Approach

In our effort to trace the path from Russell and Burch to 21st Century Toxicology, we compile information on a wide range of 3Rs activities in the field of toxicology from 1959, when *The Principles* was published, to the present. In doing so, we want to supplement a conventional narrative approach to a historical review, which can be subjective in the choice of events and papers considered – and therefore in the interpretations offered – with more objective measures.

To that end, we begin with comprehensive citation and literature searches to trace the influence of Russell and Burch's 3Rs framework and the prevalence of 3Rs-related research in toxicology over time, as revealed by patterns in the toxicological literature. When did Russell and Burch's framework start influencing the field of toxicology? How relevant have the 3Rs been to toxicology research? These are the kind of questions we sought to answer.

We then present timelines of various 3Rs activities to inform our historical analysis. These activities include the founding of 3Rs organizations and centers, the establishment of 3Rs funding sources, the enactment of animal welfare/ alternatives laws, the founding of 3Rs journals and websites, the occurrence of 3Rs workshops and conferences, and other milestones.

Following this, we integrate the findings from the literature searches and timelines to briefly tell the story of the 3Rs in toxicology, framing the narrative around what we regard as four phases of activity:

- incubation (1959–1979),
- increasing acceptance and spread (1980–early 1990s),
- maturation (early 1990s–2007), and
- paradigm shift (2007–present).

We then look at measures of the impact of this 3Rs activity on toxicity testing, focusing on alternative methods that have been successfully validated and accepted for regulatory use and any impact of these methods on trends in animal use. We conclude with a section on remaining challenges to replacing animal use in toxicology.

A number of narrative histories of the 3Rs in toxicology or in the life sciences generally have been written previously.[14–17] Similarly, although not necessarily focused on toxicology, a number of reviews have been written regarding refinement,[18–20] reduction,[21–23] and replacement,[24–27] separately or in combinations.[28,29] In our survey, we bring some of these earlier reviews up-to-date, focusing exclusively on toxicology and treating the 3Rs as a holistic framework, rather than covering each R thematically. We also explore literature searches as a source of additional historical insight and attempt to objectively assess the impact of the past 50-plus years of 3Rs activity in toxicology.

1.2   3Rs-Related Trends in the Toxicological Literature

Citation and literature searches can reveal historical patterns in the uptake and prevalence of 3Rs-related research in toxicology. To our knowledge, this approach has not been thoroughly investigated before. Some noteworthy findings are discussed here, and are incorporated in our discussion of historical phases of 3Rs activity (Section 1.4).

Our searches were designed specifically to capture toxicology-related papers that: (i) cite the book that launched the 3Rs framework (*The Principles*, both the 1959 original and a 1992 reprinting[30]), (ii) cite the NRC report that proposed a paradigm shift to non-animal-based toxicity testing (*Toxicity Testing in the 21st Century*), and (iii) explicitly address one or more of the 3Rs or alternatives, based on the presence of selected 3Rs terminology, synonymous phrases, or database indexing terms. This last set of papers was further analyzed to explore the prevalence of key 3Rs topics, such as validation of new methods, based on the inclusion of relevant terminology in the paper's title, abstract, or keywords. (See Appendices A and B for further details of our search strategies.)

An analysis of citations in the toxicological literature to Russell and Burch's pioneering book and the NRC's seminal report is perhaps the most direct assessment of the influence of these works over time and the extent of discussion about the ideas they espoused. The literature searches for 3Rs-related publications, meanwhile, give a sense of the prevalence of "alternatives" work devoted to investigating, considering, or applying 3Rs concepts within the field of toxicology over the past 50-plus years, irrespective of whether or not Russell and Burch or *Toxicity Testing in the 21st Century* were explicitly cited.

According to our results, from 1959, when Russell and Burch's *The Principles* was first published, to 2011, the last year for which we have complete data, 438 publications in the toxicology/pharmacology literature cited their work.[a] The search for toxicological papers addressing one or more of the 3Rs yielded nearly 3200 publications.

These findings indicate that far more papers discussed or applied replacement, reduction, refinement, or alternatives concepts than cited Russell and Burch's pioneering book. Perhaps one reason for this discrepancy, at least in recent years, is that authors take the 3Rs framework as a given, without the perceived need to reference its origin. As the 3Rs framework becomes increasingly integrated into toxicity testing programs and animal research policies (e.g., via the REACH program in Europe[31]), these efforts may act as other, more recent drivers for authors considering the question.

Notwithstanding the discrepancy in absolute numbers, the historical trend in toxicological publications citing *The Principles* (Figure 1.1) is similar to those addressing one or more of the 3Rs (Figure 1.2). There was a clear initial period of relative dormancy following publication of *The Principles* in 1959, lasting until the early 1980s, as revealed by the relative dearth of publications citing Russell and Burch or addressing the 3Rs. Citations and 3Rs papers began appearing regularly around 1980, with a fairly steady growth starting in the 1990s and a sharp uptick after 2007.

The dip in citations to *The Principles* starting in 2000 (Figure 1.1) is possibly an artifact of our search strategy, as we had augmented our search with two additional databases for the years prior to 2000 (see Appendix A for details). Removing the citations unique to those extra databases produces a more consistent upward trend (data not shown).

Parsing our universe of 3Rs-related papers from Figure 1.2 reveals finer-grained trends in the 3Rs literature. Figure 1.3, for example, shows that in the earlier years, a far greater percentage of 3Rs papers mentioned reduction (in the title, abstract, or keywords) than either refinement or replacement. This is surprising in that much of the reduction literature is generic – not specifically focused on toxicology or any other particular discipline. We suspect that some of these early toxicology papers were using "reduction" loosely to refer to diminishing animal use through either replacement or reduction.

---

[a]We emphasize that this represents just some of all the citations to Russell and Burch's book throughout the scientific literature during this period.

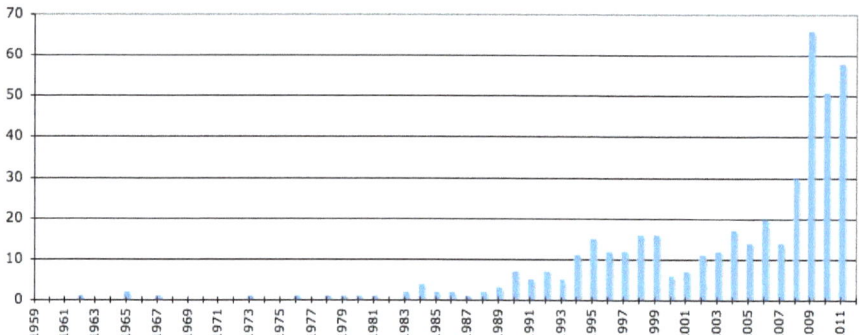

Figure 1.1 Citations to Russell and Burch's *The Principles of Humane Experimental Technique* (1959 and 1992 editions) by year (1959–2011), as identified by citation searches of Web of Science, BIOSIS, and SCOPUS databases through 2011, as well as SciSearch and Google Scholar databases through 1999, limited to the toxicological or pharmacological literature.

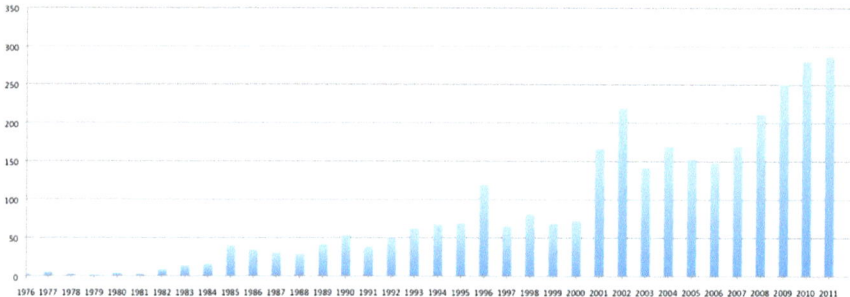

Figure 1.2 Publications related to the 3Rs and alternatives by year (1976–2011), as identified by literature searches using Embase and Ovid Medline databases and limited to the toxicological literature.

Over time, however, refinement and replacement papers became more prevalent, narrowing the gap between reduction papers on the one hand and refinement and replacement papers on the other. By 2011, approximately 40% of the papers we examined were related to reduction, 30% to replacement, and 20% to refinement. Removing those papers that mention all three Rs from the analysis does little to change this trend (data not shown). Similarly, inclusion of those papers found in the database searches using synonyms of "reduction" or "refinement" (e.g., "decreased use of animals," "lessen pain," etc.) does not substantively change the results, probably because these represent a very small percentage of papers in a given year (data not shown). (No searches were conducted using synonyms of "replacement.")

Interestingly, papers that did not explicitly mention any of the 3Rs (replacement, reduction, or refinement) typically constituted the majority until the early 2000s. These papers were likely identified during our literature search

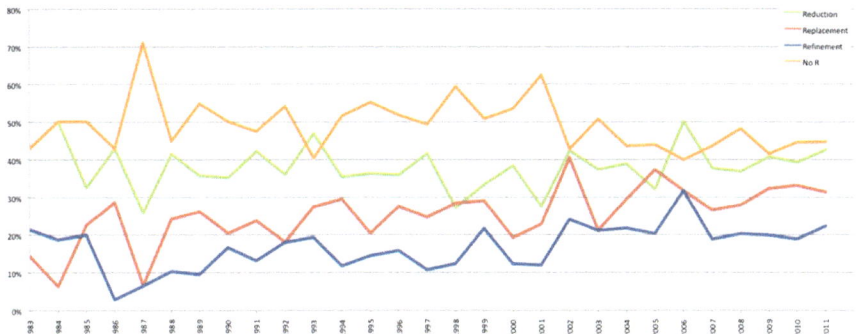

Figure 1.3   Percentage of 3Rs-related publications of those identified in Figure 1.2
mentioning "replac" (Replacement), "reduc" (Reduction), or "refin" (Refinement) in their title, abstract, or keywords (1983–2011). Some papers fall
under multiple categories if they have multiple R terms. "No R" refers to
those papers that have none of the R terms in the title, abstract, or keywords (e.g., papers that use the term "alternative" or "3Rs" instead).

owing to their use of the more general term "alternative" or "3Rs," rather than
a specific R term. During the 2000s, papers not mentioning any R became less
frequent, and these papers made up an almost equal percentage as reduction
papers by 2011.

Looking at our universe of 3Rs-related toxicology papers from Figure 1.2 in
a different way, we determined the frequency with which certain key topics
appeared in this collection. We made no attempt to be exhaustive, although we
did choose a representative sampling of alternatives-related topics. These were
as follows:

- "*In vitro*," "cell culture," and/or "tissue culture": these terms broadly
  represent the most commonly used type of non-animal testing.
- "*In silico*" and/or "SAR" (Structure–Activity Relationships): these
  computer-based methods, including (Quantitative) Structure–Activity
  Relationships, generate toxicological predictions based on chemical
  properties.
- "Validation": this is the process by which the relevance and reliability of
  methods are assessed for a particular purpose (see also Section 1.5.1).
  These are formal assessments of whether test methods (often new alternative methods) are fit for purpose and ready for consideration to be included in regulatory toxicology.
- "Testing strategies": these are frameworks incorporating two or more
  types of testing. Commonly known as "integrated" or "intelligent" testing
  strategies, these efforts are often employed as a means of increasing testing
  efficiency and thereby limiting animal testing.
- "Humane endpoint": a type of refinement in which an experiment is terminated at an earlier point, sparing animals unnecessary suffering without
  loss of experimental information.[32]

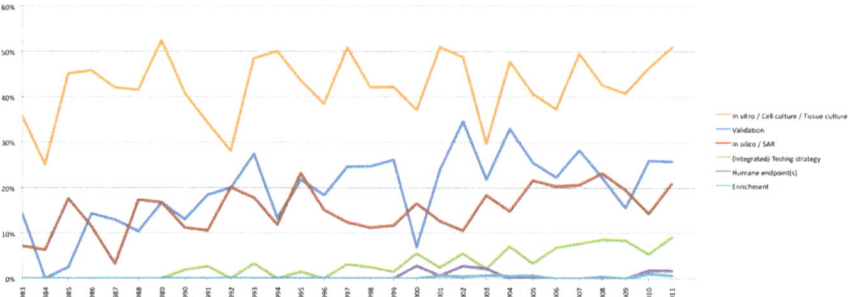

Figure 1.4   Percentage of 3Rs-related publications of those identified in Figure 1.2 mentioning select key topic phrases in their title, abstract, or keywords (1983–2011). Variants of key phrases were included when applicable, e.g., "validat" for Validation; "SAR," "structure activity relation," or "structure-activity relation" for (Q)SAR; and "testing strateg" for (Integrated) testing strategy.

- "Enrichment": a type of refinement in which an animal's living situation is enhanced through various means, including the provision of a social partner, nesting material, or food puzzles.[33]

Papers mentioning "*in vitro*," "cell culture," and/or "tissue culture" in the title, abstract, or keywords consistently constituted 30–50% of the 3Rs papers over time (Figure 1.4). Other topics, such as "*in silico*" and/or "SAR" and, to a greater degree, "validation," rose in prominence over the years. "Validation" represented a quarter to a third of the 3Rs papers most years from 1993–2011. The concept of (integrated) testing strategies first appeared in this literature in 1990 and increased noticeably in the 2000s. Other topics, such as "humane endpoint" and "enrichment" have a recent but minor presence in the 3Rs literature in toxicology.

Another important topic relevant to the history of the 3Rs in toxicology is "21st century toxicology" as exemplified in the 2007 NRC report on *Toxicity Testing in the 21st Century* (see also Sections 1.1, 1.3, and 1.4.4). This report has considerable relevance to replacement in toxicology. Interestingly, it was not the output of the 3Rs community per se but it was clearly the product of an era informed by Russell and Burch and the 3Rs approach, as well as by a broader concern for animal welfare and an appreciation for how far science and technology had advanced since the early days of animal testing.[34] We conducted a separate citation analysis on this report. From 2007, when *Toxicity Testing in the 21st Century* was published, to 2011, we identified 216 citations to it, indicating that the report has generated extensive discussion and implementation during the few years since its publication. Perhaps tellingly, citations to the report in the last two years have surpassed those to *The Principles* (Figure 1.5), and preliminary data indicate that the same is likely to be true in 2012.

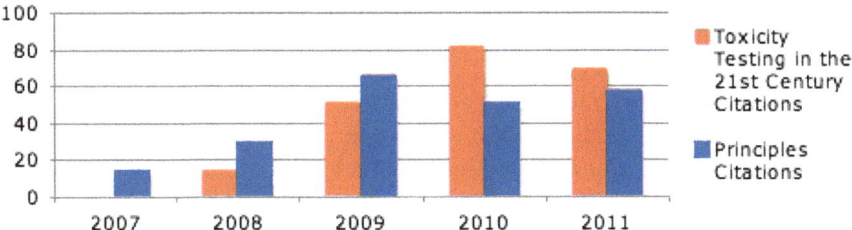

Figure 1.5 Citations to the National Research Council's *Toxicity Testing in the 21st Century: A Vision and a Strategy* (2007) by year (2007–2011), as identified by citation searches of Web of Science, BIOSIS, and SCOPUS databases through 2011 and limited to the toxicological and pharmacological literature, alongside citations to Russell and Burch's *The Principles of Humane Experimental Technique* from 2007–2011 (see Figure 1.1).

Our citation and literature searches provide insights regarding the influence of Russell and Burch and their 3Rs framework in toxicology, but they cannot be expected to reveal the whole story. For one thing, while our searches were comprehensive, they were not exhaustive. This, combined with other limitations in database searching, means we have likely underestimated the absolute number of citations and 3Rs papers. (For further discussion of the limitations of our approach, see Appendices A and B.)

In addition, our searches do not capture the 3Rs work done outside the field of toxicology but that has nevertheless influenced the work of toxicologists. During the early 1980s, for example, the work of Russell and Burch figured into more general questions of animal use, when papers related to the ethics and necessity of animal experimentation cropped up as the U.S. Congress debated and ultimately passed significant amendments to the Animal Welfare Act dealing with animals in research.

Many refinement and reduction papers have also been framed generally, rather than narrowly tailored to toxicology. "Humane endpoints" has been a hot topic in refinement for years, but is barely present in the toxicological literature. Toxicologists likely reference and make use of these sorts of papers, even if they are not indexed as toxicology papers.

1.3 Timelines of 3Rs Activities in Toxicology

Timelines can provide an event-based view of the development of the 3Rs in toxicology over time. We compiled 3Rs timelines that update those published previously[14–16] and focus them primarily on toxicology. We also group events according to activity type rather than by time blocks to allow for an additional level of analysis. Unlike the literature searches in Section 1.2, the timelines are largely self-explanatory, so we simply provide an annotated list of them here as a prelude to weaving them into our discussion of historical phases of 3Rs activity in Section 1.4.

Our timelines are organized into the following activity types:

- *Early funding sources* (Table 1.1): the establishment of funding sources for "humane research," which began in the 1960s, was one of the earliest manifestations of the impulse to promote alternatives research.
- *Early animal welfare/alternative laws* (Table 1.2): countries incorporated the 3Rs in their legislation beginning in the late 1970s.

Table 1.1   The establishment of some early funding sources for 3Rs research.

Year	Event	Country	Note	Source
1961	Humane Research Trust (initially the Lawson Tait Trust)	UK	The first research fund to support the scientific development of alternatives	http://www.humaneresearch.org.uk
1964	Swedish Fund for Research Without Animal Experiments	Sweden	Founded by Swedish Society Against Painful Experiments on Animals. First research grants awarded in 1971	http://www.forskautandjurforsok.se/in-english/
1973	Lord Dowding Fund for Humane Research	UK	A program of the National Anti-Vivisection Society	http://www.ldf.org.uk
1972	Felix Wankel Prize	Germany	A biennial award of up to 30,000 Euros	http://www.felix-wankel-forschungspreis.de
1979	The first government funding for alternatives	Sweden		Ref. 16
1981	Center for Alternatives to Animal Testing grants program	US		http://caat.jhsph.edu/programs/grants/
1985	International Foundation for Ethical Research	US	Launched by the New England Anti-Vivisection Society, currently affiliated with the National Anti-Vivisection Society	http://www.ifer.org
1989	Alternatives Research & Development Foundation	US	Affiliated with the American Anti-Vivisection Society	http://www.ardf-online.org

Table 1.2 The enactment of representative alternatives-related laws.

Year	Event	Country	Note
1977	Animal Protection Law includes a section on alternatives	The Netherlands	Has grown into a program that provides funding for alternatives research
1981	Legislation requires consideration of alternatives	Switzerland	
1985	Animal Welfare Act amendments call for consideration of alternatives	United States	
1986	European Community Directive 86/609	European Community	Requires member countries to develop legislation promoting the Three Rs
1986	Laws requiring consideration of alternatives in animal research	Germany	
1992	Cosmetic Directive bans marketing of cosmetics tested on animals after Jan. 1, 1998	European Union	A decision on the ban is later postponed until June 30, 2000
1994	The 1993 National Institutes of Health Revitalization Act leads to the establishment of ICCVAM	United States	
2000	ICCVAM Authorization Act	United States	Strengthened ICCVAM's status and mandate
2006	REACH mandates that chemical manufacturers or importers submit safety information	European Union	Articles 1 and 13 have pro-alternatives language

Abbreviations: ICCVAM: Interagency Coordinating Committee on the Validation of Alternative Methods; REACH: Registration, Evaluation, Authorisation and Restriction of Chemicals. Source: Ref. 16.

- *Journals and websites* (Table 1.3): several journals and websites devoted to alternative methods were founded, beginning in the early 1970s.
- *Early workshops and conferences* (Table 1.4): several conferences devoted to alternative methods were organized beginning in the mid-1970s, including ones that became part of a series, such as the Center for Alternatives to Animal Testing (CAAT) symposia, as well as the ongoing Linz conferences and World Congresses on Alternatives and Animal Use in the Life Sciences.
- *Alternatives organizations and centers* (Table 1.5): an impressive number of organizations and centers dedicated to the 3Rs were founded over the years. Among the most prominent are the Fund for the Replacement of Animals in Medical Experiments or FRAME (1969), CAAT (1981),

Table 1.3   The founding of some representative journals and websites devoted to the 3Rs or non-animal methods.

Year	Journal/Website	Note	Source
1973	The journal *Alternatives to Laboratory Animals (ATLA)*	Founded by Fund for the Replacement of Animals in Medical Experiments	http://www.frame.org.uk
1984	The journal *ALTEX*, devoted to the 3Rs of animal experimentation	Founded by Animalfree Research, now issued by Swiss Society ALTEX Edition	http://www.animalfree-research.org
1986	The journal *Toxicology In Vitro*		http://www.journals.elsevier.com/toxicology-in-vitro/
1990	The journal *Alternatives to Animal Testing and Experimentation (AATEX)*	Published by the Japanese Society for Alternatives to Animal Experimentation	http://www.asas.or.jp
1996	*Altweb*, a website devoted to the Three Rs in the life sciences	Founded by the Center for Alternatives to Animal Testing and numerous partners	http://altweb.jhsph.edu
2007	*AltTox*, a website devoted to non-animal alternatives in toxicology	Founded by the Humane Society of the United States and the Procter and Gamble Company	http://www.alttox.org

Center of the Documentation and Evaluation of Alternative Methods to Animal Experimentation [better known by its German acronym, ZEBET] (1989), European Centre for Alternatives to Animal Methods or ECVAM (1992), the US-based Interagency Coordinating Committee on the Validation of Alternative Methods or ICCVAM (1994), and Japanese Center of the Validation of Alternative Methods or JaCVAM (2005).

- *Developments related to pathway-based testing* (Table 1.6): the publication of the US NRC report in 2007 on *Toxicity Testing in the 21st Century, A Vision and a Strategy*[9] revolutionized thinking about the future of toxicity testing (see also Sections 1.1, 1.2, and 1.4.4). Its emphasis on upstream, pathway-based testing has led to – or provided intellectual backing for – a number of important efforts with dramatic implications for replacing animal use in toxicology, including the U.S. Environmental Protection Agency's ToxCast program and the multi-agency Tox21 program.

These timelines depict representative events and are not necessarily exhaustive. Particular attention was paid to early developments; for some

Table 1.4 Representative early workshops and conferences on the 3Rs and alternatives.

Year	Event	Note	Source
1975	The National Academy of Sciences holds the US's first major scientific meeting on alternatives	Topic: "The Future of Animals, Cells, Models, and Systems in Research, Development, Education, and Testing"	Ref. 57
1978	FRAME hosts Europe's first big scientific meeting on alternatives	Topics: "Alternatives in Drug Development and Testing"	http://www.frame.org.uk/page.php?pg_id = 42
1982	CAAT's first symposium	Center for Alternatives to Animal Testing	Ref. 14
1982	FRAME hosts "Animals and Alternatives in Toxicity Testing" conference	The FRAME Toxicity Committee presents its first report	http://www.frame.org.uk/page.php?pg_id = 42
1988	The first meeting of the Industrial In Vitro Toxicology Group	Corporate toxicologists applying in vitro methods	Ref. 14
1991	First conference in Linz, Austria, later the European Congresses on Alternatives to Animal Testing	Initially organized by animal protectionists, then by MEGAT/EUSAAT and later joined by ZET (Austrian alternatives platform)	Horst Spielmann (personal communication)
1993	The first World Congress on Alternatives and Animal Use	Subsequent WCs held in 1996, 1999, 2002, 2005, 2007, 2009, and 2011.	Ref. 58
1996	OECD holds a workshop on validation and regulatory acceptance	Aim: to develop internationally harmonized criteria	Ref. 59

Abbreviations: CAAT: Center for Alternatives to Animal Testing; FRAME: Fund for the Replacement of Animals in Medical Experiments; MEGAT: Middle European Society for Alternatives to Animal Testing; EUSAAT: the European Society for Alternatives to Animal Testing; OECD: Organisation for Economic Co-operation and Development.

activities, recent developments were too numerous to be included in their entirety.

1.4 Phases in the History of the 3Rs in Toxicology

Taking together the results of our literature searches (introduced in Section 1.2) and our timelines of important events (introduced in Section 1.3), we find it helpful to view the history of the 3Rs from Russell and Burch to 21st Century Toxicology as a progression through four phases: incubation, increasing acceptance, maturation, and paradigm shift. We divide the following historical narrative into these (somewhat overlapping) phases simply as a heuristic device to aid interpretation and understanding. We do not intend the phases to be

Table 1.5 The founding of representative alternatives organizations and centers.

Year	Event	Country	Source
1969	Fund for the Replacement of Animals in Medical Experiments (FRAME)	United Kingdom	http://www.frame.org.uk
1981	Center for Alternatives to Animal Testing (CAAT)	United States	http://caat.jhsph.edu
1985	The European Research Group into Alternatives to Toxicity Testing (ERGATT)	Europe	
1986	The Dutch Alternatives to Animal Experiments Platform	Netherlands	http://oslovet.norecopa.no/platform/report/ecoplatforms.pdf
1986	Foundation for the Promotion of Alternate and Complementary Methods to Reduce Animal Testing (SET)	Germany	http://www.stiftung-set.de/index.php?id = 3&L = 1
1987	3R Research Foundation	Switzerland	http://www.forschung3r.ch/index_en.html
1989	Center for the Documentation and Evaluation of Alternative Methods to Animal Experimentation (ZEBET)	Germany	http://www.bfr.bund.de/en/zebet-58194.html
1992	European Centre for the Validation of Alternative Methods (ECVAM)	Europe	http://ec.europa.eu/dgs/jrc/downloads/jrc_press_animal_ecvam_overview.pdf
1994	Interagency Coordinating Committee on the Validation of Alternative Methods (ICCVAM)	United States	http://iccvam.niehs.nih.gov
1994	Netherlands Centre for Alternatives to Animal Use (NCA), now National Knowledge Centre on Alternatives (NKCA)	Netherlands	http://www.nca-nl.org
1996	Prince Laurent Foundation	Belgium	http://www.fondation-prince-laurent.be
1997	Institute for In Vitro Sciences (IIVS)	United States	http://www.iivs.org
1999	Spanish National Platform on Alternatives (REMA)	Spain	http://www.remanet.net/noticias/articulos/ncawesletter22032007.htm

Year	Organization	Location	URL
2000	European Consensus Platform for Alternatives (ECOPA)	Europe	http://www.ecopa.eu
2004	National Centre for the Replacement, Refinement and Reduction of Animals in Research (NC3Rs)	United Kingdom	http://www.nc3rs.org.uk
2004	International Centre for Alternatives in Research and Education (I-CARE)	India	http://www.icare-worldwide.org/india/doerenkamp-center.html
2005	European Partnership for Alternative Approaches to Animal Testing (EPAA)	Europe	http://ec.europa.eu/enterprise/epaa/index_en.htm
2005	Japanese Center for the Validation of Alternative Methods (JaCVAM)	Japan	http://jacvam.jp/en/
2007	Norwegian Consensus Platform for Replacement, Reduction and Refinement of Animal Experiments (Norecopa)	Norway	http://www.norecopa.no
2007	French Platform for the Development of Alternative Methods in Animal Testing (FRANCOPA)	France	http://www.francopa.fr
2007	Centre for Advanced Research & Development of Alternative Methods (CARDAM)	Belgium	http://www.cardam.eu/
2008	Finnish Centre for Alternative Methods (FICAM)	Finland	http://ficam.fi
2009	Center for Alternatives to Animal Testing (CAAT) – Europe	Germany	http://cms.uni-konstanz.de/leist/caat-europe/
2010	South Korean Centre for the Validation of Alternative Methods (KoCVAM)	South Korea	http://ihcp.jrc.ec.europa.eu/glossary/kocvam
2010	American Society for Cellular and Computational Toxicology (ASCCT)	United States	http://ascctox.org/index.cfm
2011	Brazilian Center for Validation of Alternative Methods (BraCVAM)	Brazil	http://www.altex.ch/en/index.html?id = 17&ncat = 1&nid = 192

Note: For additional listings, see http://www.frame.org.uk/page.php?pg_id = 263

Table 1.6 Representative developments related to pathway-based testing as exemplified by the National Research Council report on *Toxicity Testing in the 21st Century*.

Year	Event	Note	Source
2007	*Toxicity Testing in the 21st Century: A Vision and a Strategy* published	A US National Research Council report commissioned by the US EPA	http://www.nap.edu/openbook.php?record_id = 11970
2007	US government launches Tox21 program, including the EPA component, ToxCast	A partnership among several federal agencies	http://epa.gov/ncct/Tox21/ http://www.epa.gov/ncct/toxcast/
2010	Human Toxicology Project Consortium holds a conference on accelerating the transition to pathway-based testing	The Consortium is a multi-stakeholder effort	http://htpconsortium.files.wordpress.com/2012/09/stephenstoxscifeb2012.pdf
2010	AXLR8 established	Coordinates R&D to accelerate the transition to pathway-based testing	http://axlr8.eu
2011	SEURAT-1 multi-million Euro program established	"Safety Evaluation Ultimately Replacing Animal Testing"	http://www.seurat-1.eu
2011	Human Toxome Project established at CAAT	Mapping pathways of toxicity	http://www.altex.ch/en/index.html?id = 50&iid = 123&aid = 1 http://www.thehamner.org
2012	The Hamner Institutes for Health Sciences begins case study approaches to implementing the NRC vision	Designing pathway assays for use in risk assessment	
2012	OECD establishes Adverse Outcome Pathways (AOPs) as a basic principle of the Test Guidelines Programme	Issues Draft Guidance: Proposal for a Template and Guidance on Developing and Assessing the Completeness of [AOPs]	http://www.oecd.org/env/ehs/testing/49963554.pdf

taken too literally or to represent the only meaningful way to designate eras within this history.

1.4.1 Incubation (1959–1979)

As noted in Section 1.2, there is a clear time-lag between the publication of *The Principles* in 1959 and the emergence of publications in the toxicological literature that either cite this book (Figure 1.1) or mention the 3Rs concepts it described (Figure 1.2). This delayed uptake is consistent with the relative paucity of noteworthy events during this period, as revealed by the timelines (Section 1.3). Nonetheless, the main and enduring events during this phase were: (1) the establishment of several early funding sources for 3Rs research, the first of which was the (British) Humane Research Trust (Table 1.1), and (2) the founding of the alternatives center FRAME (Table 1.5) and its journal *ATLA* (Table 1.3). These early events occurred primarily in Great Britain, fittingly the home country of Russell and Burch and UFAW – the organization that launched their project.

These findings are consistent with the common understanding that, for toxicology and the life sciences in general, Russell and Burch figuratively wandered in the wilderness for decades before their book got the attention it deserved.[14,16,35] In hindsight, there are perhaps many reasons for this. For example, after the book was published, William Russell, Rex Burch, and UFAW each quickly moved on to other challenges, largely leaving it to others to take up the ideas in the book.

Another possible reason for the lag in attention to *The Principles* and its 3Rs framework in toxicology may have been the early appearance of the book relative to the emergence of the field of toxicology as a scientific discipline. *The Principles* was published in 1959 before toxicological societies had even been established in the United States (1961) and Britain (1979). It also appeared early in the history of laboratory animal science, the discipline that emerged to address issues related to the care, management, and use of laboratory animals. In the United States, the first Guide for the Care and Use of Laboratory Animals was not published until 1963.[b]

1.4.2 Increasing Acceptance and Spread (1980–early 1990s)

A second phase in the history of 3Rs activity in toxicology, roughly from 1980 through the early 1990s, is characterized by rising attention to the 3Rs, as evidenced both by increasing reference to *The Principles* (Figure 1.1), a greater number of 3Rs-related publications (Figure 1.2), and a clustering of notable developments.

Specifically, what we are characterizing as the increasing acceptance and spread of the 3Rs approach in Europe and North America are variously reflected in the establishment of alternatives centers such as CAAT and ZEBET

---

[b]http://www.aalas.org/association/history.aspx#Timeline

(Table 1.5), the incorporation of the 3Rs framework in legislation such as the European Union legislation governing animal experimentation (EC 86/609) (Table 1.2), the founding of journals such as *ALTEX* dedicated to the subject (Table 1.3), and the organization of conferences such as FRAME's "Animals and Alternatives in Toxicity Testing" conference and a series of conferences organized by CAAT, each of which resulted in an edited volume of proceedings (Table 1.4).[36] In the United States, another noteworthy milestone in the increasing acceptance of the 3Rs approach was the 1986 publication of the Congressional Office of Technology Assessment's lengthy report on *Alternatives to Animal Use in Research, Testing, and Education*, which included two chapters on toxicity testing.[37]

A key driver of the increasing acceptance of the alternatives approach, especially in toxicology, was the emergence of the animal rights movement and its criticism of animal experimentation, particularly procedures such as the Draize eye irritancy test that were used to assess cosmetics and other consumer products.[6] In the United States, such criticism led Revlon to fund an alternatives research program at Rockefeller University and the Cosmetics Toiletries and Fragrance Association to establish CAAT.[38] Animal advocacy also led to a general expansion of funding sources for alternative methods, and helped to create the political climate that led to the incorporation of alternatives provisions in federal legislation such as the Animal Welfare Act and the ICCVAM Authorization Act (Table 1.2).

1.4.3   Maturation (early 1990s–2007)

A third phase in the history of 3Rs activity in toxicology, from roughly the early 1990s until 2007, was one of maturation of the field, characterized by continuing growth in publication of 3Rs papers (Figure 1.2) and citations to *The Principles* (Figure 1.1), and the founding of over a dozen alternatives centers (Table 1.5). Also noteworthy was the considerable activity around validation principles and processes (see also Section 1.5.1), with the emergence of national or regional validation centers in the European Union, the United States, and Asia (Table 1.5). Not surprisingly, then, the topic of validation was also increasingly prominent in the 3Rs literature (Figure 1.4), and dozens of alternative tests were validated (see Section 1.5.1), thanks in part to the work of ZEBET, ECVAM, ICCVAM, and JaCVAM.

During this period, the aim of one-to-one replacement of an animal test with an alternative test began to slowly give way to an appreciation of the value of integrated testing strategies, especially for challenging endpoints such as eye irritation and chronic systemic toxicity. This transition is reflected in the increasing representation of (integrated) testing strategy papers in the 3Rs literature (Figure 1.4). Nonetheless, the historical animal tests were, by and large, still assumed to represent the benchmark against which the performance of alternative tests and strategies were to be judged.

Given that 3Rs work was becoming increasingly international, the organization of the first World Congress on Alternatives and Animal Use in the Life

Sciences was a notable early event of this period. While the first major meetings on alternative methods began in the 1970s (Table 1.4), it was the World Congresses, initiated in 1993 by CAAT and reconvened every 2–3 years to the present, that gave those committed to the 3Rs a sense of community. Those meetings, which are devoted in large measure to toxicology issues, have drawn several hundred to over one thousand participants – scientists, administrators, regulators, funders, or animal advocates – from dozens of countries, and have been held in Europe, North America, and Asia.

International efforts on alternative methods have also been furthered by the work of the Organisation for Economic Co-operation and Development (OECD), an economic alliance of over 30 developed countries. These countries participate in the OECD's influential test guidelines program, which issues harmonized test guidelines and guidance documents, develops and validates test methods, coordinates testing programs, and encourages mutual acceptance of data generated using its approved guidelines.[c]

1.4.4 Paradigm Shift (2007–present)

In many respects we are still in the maturation phase. However, a tipping point was reached in 2007 that leads us to designate a fourth (and current) era, one of a paradigm shift. The precipitating event was the publication of the NRC report on *Toxicity Testing in the 21st Century: A Vision and a Strategy* (Table 1.6; see also Sections 1.1 to 1.3). This report, commissioned by the U.S. EPA, gave us the phrase "21st Century Toxicology" with its emphasis on testing that is *in vitro*, focused largely on human biology, based on upstream biological pathways and perturbations to normal processes, often characterized using high-throughput methodology and supplemented with computational approaches.

In this framework, the focus is no longer on predicting apical endpoints in high-dose animal studies, such as tumors or death. Rather, the NRC report proposes developing assays that detect perturbations to fundamental biological pathways (e.g., DNA synthesis and repair) that would typically lead to adverse phenotypic outcomes. Precise predictions of those outcomes would be secondary to identifying upstream perturbations to be avoided. The NRC vision was proposed as a long-term transformation, and in the early years, any pathway-based testing would need to be heavily complemented by "targeted testing," which would be mostly *in vivo*. Even such supplemental testing could move towards *in vitro* systems – in this case systems of a more complex and integrated nature (e.g., "organs on a chip") - prior to the envisioned complete (or near complete) transition to pathway-based testing.

The proposed pathway-based framework was not the work product of the mainstream 3Rs community, although one of the charges to the NRC committee that developed the proposal was to consider ways to reduce animal testing, which itself is a reflection of the penetration of 3Rs/animal welfare ideas throughout the toxicology community. Not surprisingly, the proposal has been

---

[c]http://www.oecd.org/env/ehs/testing/

embraced by the 3Rs community and seen as consistent with earlier calls for *in vitro* approaches and criticisms of animal testing.[34,39]

Toxicity Testing in the 21st Century gained immediate recognition in the toxicological literature (Figure 1.5). Interestingly, citations to the report surpassed citations to *The Principles* in 2010 and 2011. The report perhaps also spurred the significant uptick in 3Rs-related toxicology papers since 2007.

A number of efforts have emerged seeking to promote "21st century," pathway-based testing (Table 1.6). These include research and development programs on a large-scale in the United States (ToxCast and Tox21) and in the EU (e.g., SEURAT) and smaller scale efforts led by CAAT and the Hamner Institutes for Health Sciences. On the policy level, the Human Toxicology Project Consortium in the United States and the AXLR8 project in the European Union have sought to promote 21st century toxicology by assessing gaps in current R&D efforts and spurring needed efforts.

1.5   Impact Assessment of 3Rs Activity

To this point in the analysis, we have looked at the history of 3Rs activity in toxicology as revealed by patterns in the toxicological literature and by various timelines. So what difference has all this activity made in regulatory toxicity testing? We will examine this from two perspectives.

First, we look at alternative tests that have successfully gone through the process of validation and regulatory acceptance. Second, we assess whether these successes have had any discernible impact on overall trends in the use of animals in toxicology.

1.5.1   Validation and Acceptance Status of Alternatives

In vitro and other alternative tests have a long history of use in corporate decision-making about chemical safety and product formulation.[40] However, for many years such testing was not necessarily considered definitive in the regulatory context. Corporations would often follow up on their alternative testing with the historical animal-based methods. Validation – the formal assessment of the relevance and reliability of a test method for a particular purpose[41] – came to be considered a prerequisite for regulatory use of alternatives.[42]

Moreover, validation needed to be followed by a declaration of regulatory acceptance by the relevant government agencies, as a way of encouraging industry to use the validated tests and submit data based on them. Indeed, the need for successful validation and regulatory acceptance are written into U.S. law through the ICCVAM Authorization Act of 2000.[d]

New and modified assays that have been validated and accepted for regulatory use are listed in Table 1.7. A number of patterns can be discerned. First, most of the assessments of validation status and regulatory acceptance have occurred since 2000, following the establishment of key alternatives centers

---

[d]http://iccvam.niehs.nih.gov/docs/about_docs/PL106545.pdf

Table 1.7 Alternative test methods and testing strategies: Their validation and regulatory acceptance status.

	Full (√) or Partial (√/*) Replacement[a]	Reduction	Refinement	Validation Status	Regulatory Acceptance[b]
Aquatic Toxicity					
Upper threshold concentration step-down approach		√		2006 (ESAC)	2010 (OECD)[b]
Acute Systemic Toxicity (Oral)					
Up-and-down procedure		√		2001 (ICCVAM)	2006 (OECD)
Normal human keratinocyte neutral red uptake (NHK NRU) assay	√/*			2006 (ICCVAM)	2010 (OECD)
Balb/c 3T3 NRU assay	√/*	√		2006 (ICCVAM)	2010 (OECD)
Acute toxic class method		√		2007 (ESAC)	2001 (OECD)
Fixed dose procedure			√	2007 (ESAC)	2001 (OECD)
Acute Systemic Toxicity (Inhalation)					
Acute toxic class method		√			2009 (OECD)
Fixed concentration procedure			√		draft (OECD)
Carcinogenicity (Non-genotoxicity)					
Cell transformation assays	√/*	√		2012 (ECVAM)	draft (OECD)
Chronic Toxicity					
Removal of 1 year dog study for pesticides		√		2006 (ESAC)	Revised US EPA Pesticide Data Requirements
Dermal Penetration					
In vitro skin absorption methods	√/*c			2002 (OECD)	2004 (OECD)
Endocrine Active Substances					
Androgen receptor binding assay (rat prostate cytosol)	Ex vivo*				2009 (EPA)
Aromatase inhibition assay (human recombinant)	√/*				2009 (EPA)
Estrogen receptor (ER)-alpha transcriptional activation assay for estrogen agonists (STTA)	√/*			OECD/EPA	2009 (OECD)

Table 1.7   (Continued)

	Full (√) or Partial (√*) Replacement[a]	Reduction	Refinement	Validation Status	Regulatory Acceptance
Estrogen receptor binding assay rat uterine cytosol (ER-RUC)	√*(Ex vivo)				2009 (EPA)
H295R steroidogenesis assay	√*			OECD/EPA	2009 (EPA), 2011 (OECD)
US EPA Tier 1 Screening Battery	√*	√			2009 (EPA)
BG1Luc ER TA test method for estrogen agonists and antagonists	√*			2012 (ICCVAM)	draft (OECD)
Eye Corrosion					
Bovine corneal opacity permeability (BCOP) test	√*			2007 (ICCVAM)	2009 (OECD)
Isolated chicken eye (ICE) test	√*			2007 (ICCVAM)	2009 (OECD)
Cytosensor Microphysiometer modified (cytotoxicity/cell-based assay)	√*			2009 (ESAC)	2010 (draft OECD)
Fluorescein Leakage (cytotoxicity/cell-based assay)	√*			2009 (ESAC)	2010 (draft OECD)
Hen's egg test – chorioallantoic membrane (HET-CAM)	√*				EU Competent Authorities for Dangerous Substances Directive
Isolated rabbit eye test	√*				EU Competent Authorities for Dangerous Substances Directive
Routine use of topical anesthetics, systemic analgesics, and humane endpoints			√	2009 (ICCVAM)	draft (OECD)
Eye Irritation					
Cytosensor Microphysiometer modified (cytotoxicity/cell-function based in vitro assay)	√*			2009 (ESAC)	2010 (OECD draft)
Rabbit low-volume eye test (LVET)			√	2009 (ESAC)	

Test / endpoint	3R	Validation	Acceptance
Routine use of topical anesthetics, systemic analgesics, and humane endpoints	√	2009 (ICCVAM)	2012 (Expected)
Genotoxicity			
In vitro sister chromatid exchange test	√*		1986 (OECD)
In vitro unscheduled DNA synthesis test	√*		1986 (OECD)
Saccharomyces cerevisiae gene mutation assay	√*		1986 (OECD)
Saccharomyces cerevisiae mitotic recombination assay	√*		1986 (OECD)
Bacterial reverse mutation (Ames) test	√*		1997 (OECD)
In vitro mammalian cell micronucleus test	√*	2006 (ESAC)	2012 (OECD)
Hematotoxicity: Acute Neutropenia			
Colony-forming unit granulocyte macrophage (CFU-GM) assay	√*	2006 (ESAC)	Submitted to EMA
Immunotoxicity/Skin Sensitization			
Local lymph node assay (LLNA)	√	1999 (ICCVAM)	2002 (OECD)
Reduced LLNA: rLLNA	√	2007 (ESAC)	2010 (OECD)
Nonradiolabelled LLNA: DA	√	2008 (JaCVAM)	2010 (OECD)
Nonradiolabelled LLNA: BrdU-ELISA	√	2009 (ICCVAM)	2010 (OECD)
LLNA for potency characterization	√	2011 (ICCVAM)	2009 (UN GHS)
Phototoxicity			
3T3 Neutral Red Uptake Phototoxicity Test	√	1997 (ESAC)	2004 (OECD)
3T3 NRU Phototoxicity Test: Application to UV filter chemicals	√	1998 (ESAC)	2004 (OECD)
Pyrogenicity			
Human whole blood IL-1	√*	2006 (ESAC)	EMA
Human whole blood IL-6	√*	2006 (ESAC)	EMA
Human cryopreserved whole blood IL-1	√*	2006 (ESAC)	EMA
PBMC IL-6	√*	2006 (ESAC)	EMA
MM6 IL-6	√*	2006 (ESAC)	EMA
Limulus amebocyte lysate (LAL) test	√*		European Pharmacopeia
Reproductive & Developmental Toxicity			
Embryonic stem cell test for embryotoxicity	√*	2002 (ESAC)	OECD draft

Table 1.7 (Continued)

	Full (√) or Partial (√*) Replacement[a]	Reduction	Refinement	Validation Status	Regulatory Acceptance
Micromass embryotoxicity assay	√*			2002 (ESAC)	
Whole rat embryotoxicity assay	√*			2002 (ESAC)	
Extended one-generation reproductive toxicity study		√			2011 (OECD)
Skin Corrosion					
EpiSkin® human skin model	√			1998 (ESAC)	2004 (OECD)
Rat skin transcutaneous electrical resistance (TER) assay	√			1998 (ESAC)	2004 (OECD)
Corrositex® noncellular membrane	√			1999 (ICCVAM)	2006 (OECD)
EpiDerm™ human skin model	√			2000 (ESAC)	2004 (OECD)
SkinEthic™ human skin model	√			2006 (ESAC)	2004 (OECD)
Vitrolife-Skin human reconstructed epidermis	√			2008 (JaCVAM)	2004 (OECD)
EST-1000 human reconstructed epidermis	√			2009 (ESAC)	2004 (OECD)
Skin Irritation					
EpiSkin® skin irritation test (with MTT reduction)	√			2007 (ESAC)	2010 (OECD)
EpiDerm™ skin irritation test (with MTT reduction)	√			2007 (ESAC)	EU test method B.46 in COM regulation 440/2008/EC
EpiDerm™ SIT model (EPI-200)	√			2008 (ESAC)	2010 (OECD)
SkinEthic RHE model	√			2008 (ESAC)	2010 (OECD)

[a]Whether or not a test method is a full or partial replacement is not always unambiguous.

[b]The upper threshold concentration step-down approach was issued as a guidance document because consensus could not be reached.

[c]The *in vitro* skin absorption methods do not apply to mixtures/formulations.

Abbreviations: EMA: European Medicines Agency; EPA: US Environmental Protection Agency; ESAC: ECVAM Scientific Advisory Committee; GHS: Globally Harmonized System of Classification and Labelling of Chemicals; ICCVAM: US Interagency Coordinating Committee on the Validation of Alternative Methods; JaCVAM: Japanese Center for the Validation of Alternative Methods; OECD: Organisation for Economic Co-operation and Development.

Adapted from http://alttox.org/ttrc/validation-ra/validated-ra-methods.html, last updated July 10, 2012.

(Table 1.5) and the development of the principles and procedures of validation and regulatory acceptance (see Section 1.4.3). Second, the bulk of this effort has been invested in replacement alternatives (full or partial), with acute systemic toxicity and skin sensitization being notable exceptions. This activity has been driven, in part, by the ban on animal testing for cosmetic ingredients, pursuant to the European Cosmetics Directive.

Much of what we might term "toxicological space" has been touched by alternative methods. This is especially true for acute toxicity endpoints, where replacement alternatives have become available for skin penetration, skin corrosion, skin irritation, and phototoxicity. However, the challenge of replacing animal use for chronic endpoints is much more formidable.[24] One prominent effort addressing the challenge of alternatives to chronic toxicity testing is the SEURAT program, a multi-million Euro partnership between the cosmetics industry and the European Union.[e] A roadmap for replacing animals in chronic and systemic toxicity testing has been published recently.[25]

Of course, validation and regulatory acceptance of new methods do not necessarily ensure full implementation of those methods in all cases, so the degree of implementation and any barriers to implementation would need to be addressed in a more definitive analysis of the impact of alternative methods on toxicology.[43–45]

What would Russell and Burch themselves have made of the record of achievement as reflected in Table 1.7? Rex Burch died in 1996 and William Russell in 2006, but their writings and statements towards the end of their lives suggest their pride in what their pioneering book set in motion.[3,46] Of course, much remains to be done, with formidable challenges ahead (see Section 1.6).

1.5.2 Historical Trends in Animal Use in Toxicology

Both reduction and replacement alternatives should result in fewer animals being used. However, animal use in any scientific field can be influenced by a host of other factors, including whether overall activity in that field is expanding or contracting over time. An expanding field could mask any gains from reduction and replacement alternatives, whereas a field in decline could exaggerate the perceived impact of such alternatives.

Given the trends we have seen in the history of the 3Rs in toxicology, we would expect the largest impact of the 3Rs on animal use in this field to be apparent since the beginning of the 2000s, following the establishment of the process of validation and of the validation centers themselves, as well as the actual validation and regulatory acceptance of individual tests (see Sections 1.4.3 and 1.5.1). Even prior to the validation and regulatory acceptance of alternatives, a general sensitivity to animal welfare and the 3Rs approach may have lead to greater scrutiny of animal use in toxicology and subsequent reductions in animal numbers, especially in pre-regulatory toxicology. On the

---

[e]http://www.seurat-1.eu

other hand, animal testing inevitably increases over the short term when new testing programs are launched (e.g., REACH), new endpoints are developed (e.g., endocrine disruption), and new types of chemicals are commercialized (e.g., nanoparticles).

Reports from several countries provide evidence that overall animal use (for any purpose, not just for toxicology) declined during the last quarter of the 20th century (Andrew Rowan, personal communication). To analyze international trends in animal use in toxicology, data need to be gathered in a way that allows comparison across countries. The best source for such data comes from the EU. Consistent statistics on animal use in toxicology began to be aggregated across EU member states (nations) in 1999, and since then have been compiled every three years, with 2008 the most recent compilation. These statistics do not allow us to assess the impact of early 3Rs activity, but they do cover the years when we would expect to see an impact from the development and validation of alternatives (see Sections 1.4.3 and 1.5.1).

We combined the statistics on numbers of animals used for toxicology with statistics on numbers of animals used for the production and quality control of products and devices for human medicine, dentistry, and veterinary medicine. We controlled for the fact that the number of EU member states has grown over the years in question by assessing data from the five EU countries that use the most animals overall: France, Germany, the United Kingdom (UK), Italy, and Belgium. These five countries, when taken together, account on average for 80% of the total animal use in toxicology by the EU, and each has used more than 200,000 animals for toxicity testing in at least one reporting year. Moreover, we normalized each country's numbers, setting the total animal use for each country during its first reporting year as a baseline (100%), in order to more easily compare trends across countries. Thus, we can see the relative change in animal use in each country since 1999 (2002 for Germany).

We found that there is little consistency in the levels of animal use in toxicology across countries over the decade in question (Figure 1.6). Italy's numbers consistently (and dramatically) dropped, Belgium's numbers steadily rose, Germany's numbers remained fairly steady, and France and the UK showed more complicated patterns. France's numbers increased initially and then ended with a substantial decline, whereas animal use in the UK declined for many years but ended with a substantial increase from 2005 to 2008.

Looking across these five countries as a whole, however, the overall trend is one of modest decline, with the number of animals used for toxicity testing in 2008 totaling 87% of those used in 2002. This recent decline could be evidence that, notwithstanding the influence of various factors tending to drive up animal numbers in toxicology (e.g., REACH), the 3Rs are actually bringing animal numbers down. This remains speculative in the absence of further evidence.

Yet even in the absence of such information, it is clear that much remains to be done in applying the 3Rs in toxicology. Although no precise estimates are available, animal use in toxicology worldwide is still counted in the millions of animals. In light of the unfulfilled potential of 3Rs activity within toxicology, we next address some of the major challenges ahead.

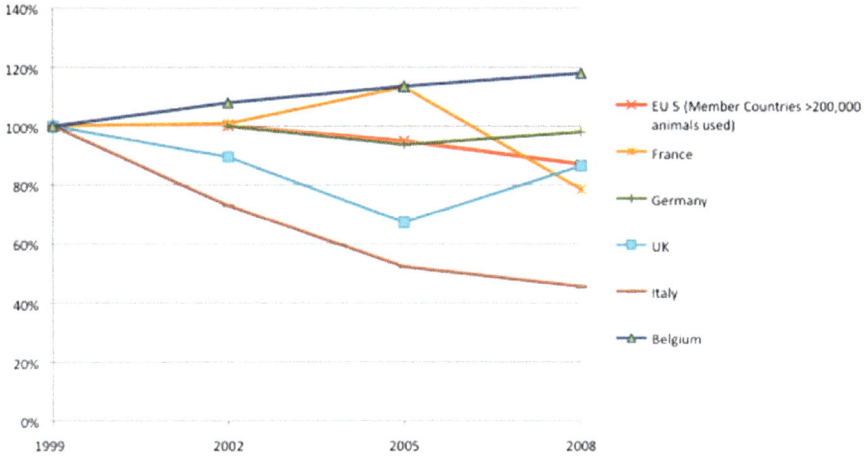

Figure 1.6 Animal use in toxicity testing in the European Union (EU) (1999–2008), based on European Commission reports to Parliament and working documents on the Statistics on the Number of Animals Used for Experimental and Other Scientific Purposes in the Member States of the European Union. Data shown only for those member countries using >200,000 animals in toxicity testing in a given year (France, Germany, United Kingdom, Italy, Belgium), which collectively account for approximately 80% of the total number of animals used in toxicity testing in the EU. No data are shown for Germany in 1999 because data were compiled differently that year, and therefore no data are shown for EU5 (the combined animal use from the 5 countries listed) for 1999. For ease of comparison, data are normalized such that animal use for the first reporting year is set to 100% for each country, and animal use in other years is expressed as a percentage of that year. Data tallied from boxes 2.4 (production and quality control of products and devices for human medicine and dentistry), 2.5 (production and quality control of products and devices for veterinary medicine), and 2.6 (toxicological and other safety evaluations, including safety evaluations of products and devices for human medicine and dentistry and for veterinary medicine) of the reports.

1.6 Remaining Challenges

In the decades following publication of Russell and Burch's *The Principles of Humane Experimental Technique* in 1959, the emerging 3Rs community devoted much of its energy to the field of toxicology. An infrastructure for making progress was slowly and steadily developed. This included alternatives-based organizations and centers, journals, websites, laws, conferences, and other activity. The 3Rs community also spearheaded the development of the scientific standards (e.g., validation) and worked with fellow toxicologists and others to improve the techniques (e.g., tissue culture) that facilitated progress. The result has been a progressive chipping away at traditional animal-based methods.

The publication of the NRC report on *Toxicity Testing in the 21st Century* in 2007 created a new point of reference for the 3Rs community, suggesting a new approach to replacing animal use in toxicology. Within the mainstream toxicology and public health communities, the NRC report engendered considerable enthusiasm for modernizing toxicology. Leading federal scientists began predicting a paradigm shift from *in vivo*- to *in vitro*-based methods (see Section 1.1). Many in the alternatives community have been seeking to facilitate this long-term effort.[39,47] As pathway-based approaches are further elaborated, they can be incorporated into ongoing 3Rs efforts.

However, implementing pathway-based approaches to predictive toxicology raises a host of formidable challenges.[48] These include developing fit-for-purpose assays to monitor pathway perturbations, distinguishing adverse responses in these assays from homeostatic responses,[49] accounting for metabolism of the parent compound, and shifting to a safety-based risk assessment paradigm, rather than one based on hazard.[10]

21st century toxicology efforts have the long-term goal of providing a new paradigm for safety assessment; however, for the time being, most of these efforts are being harnessed to predict the outcomes of traditional animal testing. Consequently, the first applications are supplementing rather than supplanting the current paradigm. We need to accelerate the transition to the new paradigm, based on human biology.[47]

20th century validation processes will need to be adapted to 21st century toxicology.[50] There are many reasons why the existing validation processes should not be deployed as is. First, pathway-based methods are not intended as one-to-one replacements of animal-based tests; multiple pathway-based assays, perhaps numbering in the hundreds, will be used to make predictions about individual chemicals. Second, pathway-based methods ultimately will be called on to mimic human biology, rather than to predict the results of animal testing – the standard assurance of relevance. Third, according to the NRC framework, pathway-based assays will be used to predict regions of safe exposures, rather than predict specific toxicities. And fourth, the science and technology of 21st century toxicology are changing too rapidly for an evaluation process that takes a year or more to complete.

Most of the existing validation *principles* may carry over and transcend a paradigm shift in testing, but the prevailing validation *procedures* will need to be translated to accommodate these new realities. Validation procedures must also somehow be speeded up to accommodate the new pace of change and be flexible enough to accommodate an ideal of continual improvement in testing methods. Some rethinking of validation has begun to occur in this context.[50–52]

And finally, perhaps the biggest real-world challenge to further progress on the 3Rs in toxicology is tackling systemic and chronic toxicity testing. Some large-scale programs in Europe have been seeking to address elements of this challenge, such as REPROTECT (reproductive toxicity)[53] and SEURAT-1 (repeat dose systemic toxicity testing),[54] with coordination and guidance provided by the AXLR8 program.[f] In the U.S., high-throughput testing is being used to identify biological signatures of chronic, systemic endpoints, such as in developmental and reproductive toxicity.[55,56] Clearly, integrated testing

---

[f]http://axlr8.eu

strategies involving multiple types of testing and approaches (e.g., pathway-based testing, high-throughput testing, toxicogenomics, organ-on-a-chip platforms, virtual organs) will play a role. Also critical will be a realistic assessment of the limitations (as well as the strengths) of the current animal-based assays.[25]

1.7 Conclusions

Progress on the 3Rs has been driven by a dual concern for animal welfare and scientific advancement. Much of this progress to date, especially in toxicology, can be attributed to the efforts of those who would identify themselves as part of the 3Rs or alternatives community and can be traced back in time to the pioneering efforts of William Russell and Rex Burch. What is especially exciting about the current era is that the 3Rs community is now working in parallel with a vanguard in the toxicology community seeking to usher in new approaches. Time will tell whether we are at the threshold of alternative approaches becoming the mainstream of the new toxicology.[34,39]

Appendix A: Citation Search Strategies

We searched Web of Science, BIOSIS, and SCOPUS databases for citations to Russell and Burch's *The Principles of Humane Experimental Technique*, whether in its original edition[4] or its 1992 reprint,[30] and the NRC's *Toxicity Testing in the 21st Century: A Vision and a Strategy*[9] from the year of their respective publication to 2011, the latest year for which complete records were available. In addition, we searched SciSearch and Google Scholar for citations to *The Principles* up to and including the year 1999 to help compensate for the likely under-representation of papers and book citations during these years in the main databases.

For Web of Science, BIOSIS, SCOPUS, and SciSearch, we limited our results to publications related to toxicity testing by using the appropriate database subject/research area limits. In some cases, databases grouped toxicology and pharmacology together as one subject area. In others, the subject areas were distinct and pharmacology was included only if it was defined to include topics of relevance to toxicity testing. For the Google Scholar search, no subject area limits were available, so we curated our results to remove any publications not clearly related to toxicology.

We adopted a search strategy that would be robust enough to capture variations in how the author names and titles were entered into the databases, including misspellings. Searches were conducted October/November 2012.

Databases

1   Web of Science

According to its website, "Web of Science® provides access to citation databases with multidisciplinary content including Open Access journals and over

150,000 conference proceedings. It includes current and retrospective coverage in the sciences, social sciences, arts, and humanities, with coverage to 1900."[g] The relevant subject filters were "Toxicology" and "Pharmacology & Pharmacy." "Toxicology" covers "resources that focus on the identification, biochemistry, and effects of harmful substances, including the side effects of drugs, in animals, humans, and the environment." "Pharmacology & Pharmacy" covers "resources on the discovery and testing of bioactive substances, including animal research, clinical experience, delivery systems, and dispensing of drugs. This category also includes resources on the biochemistry, metabolism, and toxic or adverse effects of drugs."[h]

2 BIOSIS

According to its website, BIOSIS Citation Index "covers all major areas in the life sciences, with broad coverage in molecular and cell biology, pharmacology, endocrinology, genetics, neurosciences, infectious diseases, ecology and organismal biology. It provides access to over 22 million records from journals, books, reports, meetings, and U.S. patents dating 1926 or later."[i] Subjects are defined in the same way as in Web of Science (above).

3 SCOPUS

According to SCOPUS, the database "is an abstract and citation database of peer-reviewed literature with tools that track, analyze and visualize research. Scopus includes over 20,500 titles from 5,000 publishers worldwide, 49 million records (78% with abstracts) and over 5.3 million conference papers."[j] The relevant subject filter in SCOPUS is "Pharmacology, Toxicology and Pharmaceutics" (no further description available).

4 SciSearch (Accessed via DialogWeb)

According to DialogWeb, SciSearch is "an international, multidisciplinary index to the literature of science, technology, biomedicine, and related disciplines produced by Thomson Scientific. SciSearch contains all of the records published in the Science Citation Index® (SCI®), plus additional records in engineering technology, physical sciences, agriculture, biology, environmental sciences, clinical medicine, and the life sciences. SciSearch indexes all significant items (articles, review papers, meeting abstracts, letters, editorials, book reviews, correction notices, etc.) from more than 6,100 international scientific and technical journals. Citation indexing allows for the searching of cited references."[k]

---

[g]http://thomsonreuters.com/products_services/science/science_products/a-z/web_of_science/#tab1
[h]http://ip-science.thomsonreuters.com/mjl/scope/scope_sci/
[i]http://wokinfo.com/products_tools/specialized/bci/
[j]http://www.info.sciverse.com/scopus/about
[k]http://library.dialog.com/bluesheets/html/bl0034.html

5   Google Scholar

According to Google, "Google Scholar includes scholarly articles from a wide variety of sources in all fields of research, all languages, all countries, and over all time periods. Google Scholar searches across many disciplines and sources: articles, theses, books, abstracts and court opinions, from academic publishers, professional societies, online repositories, universities and other web sites. To be considered for inclusion, website content needs to meet two basic criteria: 1. Must include scholarly articles – journal papers, conference papers, technical reports, or their drafts, dissertations, pre-prints, post-prints, or abstracts.... 2. Must show abstracts – websites must make either the full text of the articles or their complete author-written abstracts freely available and easy to see when users click on URLs in Google search results...."[1]

Search Terms

1   Web of Science

We selected the "Web of Science" tab and then the "Cited Reference" tab. To retrieve citations to *The Principles*, we entered:

"Russel*, W*" in Cited Author AND
"Burch, R*" in Cited Author AND
"1959–2011" in Cited Year(s)

This produced 13 entries, of which we selected the 11 that were relevant to our search and clicked "Finish Search." We refined the results by Research Area (on the left-hand side), selecting both Toxicology and Pharmacology/Pharmacy.

For citations to *Toxicity Testing in the 21st Century*, we located three entries, and the results were refined by Research Area to both Toxicology and Pharmacology/Pharmacy.

2   BIOSIS

We selected the "Select a Database" tab, then selected "BIOSIS Citation Index," and from there selected the "Cited Reference" tab. We entered the same search terms as for Web of Science (above).

This produced 15 entries, of which we selected the 13 that were relevant to our search and clicked "Finish Search." We refined the results by Research Area (on the left-hand side), selecting both Toxicology and Pharmacology/Pharmacy.

For citations to *Toxicity Testing in the 21st Century*, we located three entries, and the results were refined by Research Area to both Toxicology and Pharmacology/Pharmacy.

---

[1] http://scholar.google.com/intl/en-US/scholar/help.html#coverage

3   SCOPUS

To find citations, we first needed to enter a publication known to cite the work in question, then locate the reference in that publication's entry. Next to the reference, there is text stating "Cited X times," with a link to the citations. Doing this, we were able to retrieve citations to both editions of *The Principles* and to the *Toxicity Testing in the 21st Century* report. SCOPUS does not have the same complication with variations of author name(s) and title as do Web of Science and BIOSIS.

4   SciSearch

We used SciSearch only for *The Principles* and entered the following search terms:

S1   CR = RUSSELL WMS, 1959?

S2   S1 AND PY = 197?:1999

S3   S2 AND (TOX? OR TEST?)

And:

S1   CR = RUSSELL W??, 1992?

S2   S1 AND PY = 1992–1999

S3   S2 AND (TOXIC? OR TEST?)

5   Google Scholar: http://scholar.google.com

We searched for "Principles of Humane Experimental Technique," and then selected the top result, which was for "The principles of humane experimental technique; WMS Russell, RL Burch, CW Hume – 1959 - altweb.jhsph.edu." Several other entries were listed for the book, but these had relatively few citations (generally less than 15 each), compared to more than 1400 citations for the entry we selected. As the database does not allow results to be restricted to certain research or subject areas, citations through 1999 were hand-curated to remove any not related to toxicology or pharmacology.

Limitations

Not all journal articles, books, or other documents will be indexed by the databases we used, and coverage is typically weaker prior to the early- to mid-1990s. In addition, some publications with relevance to toxicology may not be indexed under the Toxicology or Pharmacology research areas. While database indexers do their best to apply all relevant terms, it may have been particularly challenging to apply the Toxicology/Pharmacology label during the nascent years of these fields, but also perhaps in more recent years as new specialties come to be identified as the primary focus.

To try to compensate for some of these factors, we included additional databases in our search for citations to *The Principles* prior to 2000. (There is likely to be greater overlap among the databases in more recent years.) While inclusion of these databases increased the absolute number of citations we were able to find, it did not significantly alter the general trend (data not shown).

Therefore, we expect that the total number of citations is likely higher than what we found, even with enhancement for the years prior to 2000, but the overall trend should remain valid. For *The Principles*, we see a low number of citations prior to 1980, a growth through the 1980s and 1990s, and a significant uptick starting in the mid-2000s. For *Toxicity Testing in the 21st Century*, we see an immediate uptake equaling, if not exceeding, reference to *The Principles*.

It is possible that the trend for *The Principles* largely mirrors the trajectory of toxicology publications in general. Advances in computer and internet technologies over the past one or two decades, for example, have had a tremendous impact on the number of and ease with which journals and papers can be published and indexed. As more papers get published, the number of papers citing *The Principles* has increased, but we do not know if the relative proportion of papers citing *The Principles* has decreased, increased, or remained steady. Notwithstanding this caveat, the trend in total number of citations could have looked very different, and our analysis shows a clear rise in influence of Russell and Burch's pioneering book during the last quarter of the 20th century and an enduring legacy into the 21st.

Appendix B: Literature Search Strategies

We conducted formal literature searches using Ovid Medline and Embase databases to identify a comprehensive, though not exhaustive, universe of 3Rs-related papers. Papers were considered to be related to the 3Rs if they used common 3Rs terminology or related synonyms, or if they were indexed as such by the databases. Several different search strings (queries) were devised to capture papers as comprehensively as possible while minimizing the number of irrelevant papers also captured. We did not, however, remove these "false positives" from our data set (of over 3000 records), which we estimated represented only 5–10% of our total.

We limited our results to those papers related to the field of toxicology, but did not include the field of pharmacology because these databases defined the field too broadly for our purposes. Instead, we included searches for vaccine safety and potency testing, which tended to fall in pharmacology rather than toxicology but are nonetheless related to toxicity testing.

We then categorized papers based on the occurrence of selected terms with 3Rs significance in the title, abstract, or key word fields of their database entries. Papers could be included in multiple categories if they contained more than one of the selected terms. While we selected a representative sample of important 3Rs-related approaches or concepts, such as humane endpoints and validation, no attempt was made to be exhaustive in the topics covered.

Searches were conducted during October 2012.

Databases

1   Ovid

According to Ovid, Ovid Medline is updated daily and "provides access to the latest bibliographic citations and author abstracts from more than 5,500 bio-medicine and life sciences journals in nearly 40 languages (60 languages for older journals). English abstracts are included in more than 80% of the records."[m] The "Toxicology Limit" was based on PubMed's Toxicology subset limit.[n]

2   Embase

According to Embase, the database "covers international biomedical literature from 1947 to the present day. The database contains over 25 million indexed records and more than 7,600 currently indexed peer-reviewed journals. All MEDLINE records produced by the National Library of Medicine (NLM) are included, as well over 5 million records not covered on MEDLINE."[o]

Embase uses "Areas of Focus" filters, of which the relevant one for our purposes was "Toxicology and Drug Dependence." According to Embase, this area "covers topics relating to toxic mechanisms and effects of both medicinal and non-medicinal substances. Included in this coverage are: Abuse of drugs, alcohol and organic solvents; Experimental pharmacology of addiction; Pre-dictive toxicology. Coverage: records from 1983 to present."[p] The "Pharma-cology and Pharmacy" area, as defined by Embase, dealt with topics largely outside toxicity testing.

Search Terms

Ovid

The following list of animal terms ("[list of animals]") appeared in most of our searches:

(animal$1 or rat or rats or mouse or mice or dog$1 or cat$1 or hamster$ or gerbil$ or "guinea pig$1" or monkey$1 or primate$1 or rodent$1 or rabbit$1 or bird$1 or fish$2 or zebrafish or chicken$1)

<u>Indexing terms:</u>
1) exp "Animal Use Alternatives"/ and (test or tests or testing or toxic$).ab,ti,jw.

<u>Alternatives to the use of animals:</u>
2) ((alternative$1 adj5 ("use" or "uses" or "using" or test or tests or testing)) and (("use" or "uses" or "using" or test or tests or testing) adj3 ([list of animals]))).mp.

---

[m]http://www.ovid.com/webapp/wcs/stores/servlet/
    ProductDisplay?storeId = 13051&catalogId = 13151&langId = -1&partNumber = Prod-901
[n]http://www.nlm.nih.gov/bsd/pubmed_subsets/tox_strategy.html
[o]http://www.embase.com/info/what-embase
[p]http://www.embase.com/info/helpfiles/search-forms/advanced-limits/areas-of-focus

3) ((alternative$1 adj5 ("use" or "uses" or "using" or test or tests or testing)) and (("use" or "uses" or "using" or test or tests or testing) adj3 ([list of animals]))).mp. AND toxic$.jw.

3Rs:

4) (3Rs or "Three Rs" or "Three R" or "3R principle" or "3R principles" or "3R approach" or "3R approaches" or "3R method" or "3R methods" or "3R concept" or "3R concepts" or "3R strategy" or "3R strategies" or (3R adj4 (alternative or alternatives))).mp.

Replace, reduce, and refine:

5) (Replac$ and reduc$ and refin$).mp.

Animal Refinement:

6) ((([list of animals]) adj5 refin$).ti,ab. and (alternative or alternatives or test or testing).mp.

Minimize or eliminate pain or distress:

7) ([list of animals]).ti,ab. and ((minim$ or eliminat$) adj4 (pain or distress)).ti,ab. and (alternative or alternatives or testing).mp.

Reduce, alleviate, or lessen pain or distress:

8) ([list of animals]).ab,ti. and ((reduc$ or alleviat$ or less$) adj4 (pain or distress)).ab,ti. and ((toxic$ and chemical$) or (toxic$ adj3 test$) or ((alternative or alternatives) and toxic$)).ab,ti

Reduce, alleviate, lessen, minimize, eliminate, or decrease suffering:

9) ([list of animals]).ab,ti. and ((reduc$ or alleviat$ or less$ or minim$ or eliminat$ or decreas$) adj4 suffer$).ab,ti. and (alternative or alternatives or test or testing).ab,ti.

Animal replacement:

10) (((([list of animals]) adj5 replac$) and (alternative or alternatives)).mp.

11) ([list of animals]).ab,ti. and (replace and (alternative or alternatives)).mp.

Animal reduction:

Reduce the number of animals/Number of animals reduced:

12) (((reduc$ or minim$ or fewer) adj4 number$) and (number$ adj3 ([list of animals])) and (alternative or alternatives or testing or toxicity)).ti,ab.

Reduce the use of animals/Use of animals reduced/Animal use reduced:

13) (((reduc$ or decreas$ or minim$ or eliminat$ or fewer) adj6 ("use" or "uses" or "used")) and (("use" or "uses" or "used") adj6 ([list of animals])) and (([list of animals]) adj6 (reduc$ or decreas$ or minim$ or eliminat$ or fewer)) and (alternative or alternatives or testing or toxicity)).ti,ab.

Reduce animal testing/Animal testing reduced:

14) (((reduc$ or decreas$ or minim$ or eliminat$) adj5 ([list of animals])) and (([list of animals]) adj1 (test or tests or testing)) and ((reduc$ or decreas$ or minim$ or eliminat$) adj4 (test or tests or testing)) and (alternative or alternatives or testing or toxicity)).ti,ab.

Refine and Reduce:

15) ((refin$ adj6 reduc$) and (alternative or alternatives or test$ or toxic$)).ti,ab.

Vaccine Safety and Potency Testing:
16) ((vaccine or vaccines) and (safety test$ or potency test$ or batch test$ or quality control) and (alternative or alternatives) and (in vitro or in vivo or ([list of animals]))).ti,ab.
17) Combine 1-2 and 4-16 with OR.
18) Limit 17 to toxicology.
19) Combine 3 and 18 with OR.

Embase

The following list of animal terms ("[list of animals]") appeared in most of our searches:

(animal OR animals OR rat OR rats OR mouse OR mice OR dog OR dogs OR cat OR cats OR hamster* OR gerbil* OR "guinea pig" OR "guinea pigs" OR monkey* OR primate* OR rodent OR rodents OR rabbit* OR bird* OR fish* OR zebrafish OR chicken*)

Indexing Terms:
1) "animal testing alternative"/exp AND (test OR tests OR testing OR toxic*):ab,ti

Alternatives to the use of animals:
2) ((Alternative or alternatives) NEAR/5 (use or uses or using or test or tests or testing)):ti,ab AND ((use or uses or using or test or tests or testing) NEAR/3 ([list of animals])):ti,ab AND [toxicology and drug dependence]/lim

3Rs:
3) (3Rs OR "Three Rs" OR "Three R" or "3R principle" or "3R principles" or "3R approach" or "3R approaches" or "3R method" or "3R methods" or "3R concept" or "3R concepts" or "3R strategy" or "3R strategies" or 3R NEAR/3 (alternative or alternatives)) AND [toxicology and drug dependence]/lim

Replace, reduce, and refine:
4) Replac* AND reduc* AND refin* AND [toxicology and drug dependence]/lim

Animal Refinement:
5) ((([List of animals]) NEAR/5 refin*):ti,ab AND (alternative or alternatives or test or testing) AND [toxicology and drug dependence]/lim

Minimize or eliminate pain or distress:
6) ([List of Animals]):ti,ab,de AND ((minim* OR eliminat*) NEAR/4 (pain or distress)):ab,ti AND (alternative OR alternatives OR test OR testing) AND [toxicology and drug dependence]/lim

Reduce, alleviate, or lessen pain or distress:
7) ([List of Animals]):ti,ab,de AND ((reduc* OR alleviat* OR less*) NEAR/4 (pain or distress)):ab,ti AND ((toxic* AND chemical*) OR (toxic* NEAR/3 test*) OR (alternative OR alternatives AND toxic*)):ab,ti AND [toxicology and drug dependence]/lim

Reduce, alleviate, lessen, minimize, eliminate, or decrease suffering:

  8) ([List of Animals]):ti,ab,de AND ((reduc* OR alleviat* OR less* OR minim* OR eliminat* OR decreas*) NEAR/4 suffer*):ab,ti AND (alternative OR alternatives OR test OR testing):ab,ti AND [toxicology and drug dependence]/lim

Animal Replacement:

  9) ((([List of animals]) NEAR/5 replac*) AND (alternative or alternatives) AND [toxicology and drug dependence]/lim

  10) ([List of animals]):ti,ab,de AND replace AND (alternative or alternatives) AND [toxicology and drug dependence]/lim

Animal Reduction:

Reduce the number of animals/Number of animals reduced:

  11) ((reduc* OR minim* OR fewer) NEAR/4 number* AND number* NEAR/3 ([list of animals])):ti,ab AND (alternative or alternatives or testing or toxicity):ti,ab AND [toxicology and drug dependence]/lim

Reduce the use of animals/Use of animals reduced/Animal use reduced:

  12) (((reduc* or decreas* or minim* or eliminat* or fewer) NEAR/6 (use or uses or used)) AND ((use or uses or used) NEAR/6 ([list of animals])) AND (([list of animals]) NEAR/6 (reduc* or decreas* or minim* or eliminat* or fewer)) AND (alternative or alternatives or testing or toxicity)):ti:ab AND [toxicology and drug dependence]/lim

Reduce animal testing/Animal testing reduced:

  13) ((reduc* or decreas* or minim* or eliminat*) NEAR/5 ([list of animals])): ti,ab AND (([list of animals]) NEAR/1 (test or tests or testing)):ti,ab AND ((reduc* or decreas* or minim* or eliminat*) NEAR/4 (test or tests or testing)):ti,ab AND (alternative or alternatives or testing or toxicity):ti:ab AND [toxicology and drug dependence]/lim

Refine and Reduce:

  14) (refin* NEAR/6 reduc*):ti,ab,de AND (alternative or alternatives or test* or toxic*) AND [toxicology and drug dependence]/lim

Vaccine Safety and Potency Testing:

  15) ((vaccine OR vaccines) AND ((safety NEXT/1 test*) OR (potency NEXT/1 test*) OR (batch NEXT/1 test*) OR "quality control") AND (alternative OR alternatives) AND ("in vitro" OR "in vivo" OR ([list of animals]))):ab,ti

  16) Combine #1-#15 with OR.

Categorization

We parsed our universe of 3Rs-related papers into various categories to assess the frequency of occurrence of select key topics. We assigned a paper to any categories for which the associated term(s) appeared in the abstract, title, or key word fields of the paper's database entry.

- Replacement: "replac"
- Reduction: "reduc"

- Refinement: "refin"
- In vitro/Cell culture/Tissue culture: "in vitro," "cell culture" and/or "tissue culture"
- Validation: "validat"
- In silico/(Q)SAR: "in silico," "SAR," "structure activity relation," and/or "structure-activity relation"
- (Integrated) testing strategy: "testing strateg"
- Humane endpoint: "humane endpoint"
- Enrichment: "enrichment"

Limitations

In addition to those limitations identified for the citation search for *The Principles* (Appendix A), which apply equally as well to the literature search for 3Rs papers, additional factors must be considered in interpreting these results.

Our search could only retrieve those papers employing our specific search terms and formulations. Some proportion of 3Rs-related papers, however, may use terminology that is less specific and therefore not easily identified by searches designed to minimize false positives. Even the main 3Rs terms "replace," "reduce," and "refine" are relatively non-specific, thereby increasing the difficulty of locating papers discussing these topics.

Relatedly, papers may address 3Rs concepts without using 3Rs-specific terminology or without being framed as such. The lack of uniform publication requirements to explicitly describe the 3Rs implications of a given work exacerbates this difficulty. The growing number of 3Rs-related papers may partly reflect the increasing penetration of 3Rs terminology, however, and this in itself would be an indicator of the growing influence and sophistication of 3Rs ideas.

Even if papers used the 3Rs terminology employed in our search, they may not have been identified if the search terms did not appear in the title, abstract, or key words. Also, many early papers are not indexed in the databases with their abstracts, so only their titles are amenable to searching. This contributes to our underestimate of 3Rs papers prior to the 1990s. Database indexing terms, such as Medline/Pubmed's "animal testing alternative," are also not entirely reliable, resulting in many alternatives papers not being indexed as such. In some cases, though, the term may be applied to papers that are not clearly discussing the 3Rs, contributing to our false positives.

With regard to categorizing papers based on the 3Rs concept they address (e.g., replacement, reduction, refinement, validation, etc.), our search does not retrieve papers addressing these concepts if they are not indexed as toxicology papers. Thus, work on topics like refinement, humane endpoints, and enrichment, which is more likely to be general to animal research rather than specific to toxicology, would be underrepresented in our search even though such work is of use to toxicologists.

Further, papers were categorized based simply on the occurrence of selected terms. It is possible that some papers may use a term but not be about that concept. For example, papers may mention "replacement, reduction, and refinement" to introduce the concept of the 3Rs, but may actually only address one of the Rs. Such papers, however, would get categorized under each of the Rs.

Despite these limitations, many of which are inherent to database searching, our analyses provide useful insight into the penetrance and explicit use of 3Rs terminology and concepts in toxicology over time.

Acknowledgements

The authors thank the following for their helpful comments on the project or manuscript: Michael Festing, Alan Goldberg, David Morton, Frauke Ohl, Andrew Rowan, Horst Spielmann, Bernard Unti, and Catherine Willet. We are especially grateful to Tim Allen, Daniel Butzke, and Lori Rosman for advice regarding literature searching, and to Sue Leary for her gracious support of our work.

References

1. J. W. Trevan, The error of determination of toxicity, *Proc. R. Soc. B: Biol. Sci.*, 1927, **101**, 483–514.
2. J. H. Draize, G. Woodard and H. O. Calvery, Methods for the study of irritation and toxicity of substances applied topically to the skin and mucous membranes, *J. Pharmacol. Exp. Ther.*, 1944, **82**, 377–390.
3. W. M. S. Russell, The progress of humane experimental technique, *Altern. Lab. Anim.*, 1999, **27**, 915–922.
4. W. M. S. Russell and R. L. Burch, *The Principles of Humane Experimental Technique*, Methuen, London, 1959.
5. D. M. Weary, A good life for laboratory animals – How far must refinement go?, *ALTEX Proc.*, 2012, **1/12**, 11–13.
6. A. N. Rowan, *Of Mice, Models, and Men: A Critical Evaluation of Animal Research*, State University of New York, Albany, NY, 1984.
7. W. H. Weihe, Use and misuse of an imprecise concept: alternative methods in animal experiments, *Lab. Anim.*, 1985, **19**, 19–26.
8. D. H. Smyth, *Alternatives to Animal Experiments*, Scolar Press Ltd, London, 1978.
9. National Research Council, *Toxicity Testing in the 21st Century: A Vision and A Strategy*, National Academies, Washington, DC, 2007.
10. M. E. Andersen and D. Krewski, Toxicity testing in the 21st century: Bringing the vision to life, *Toxicol. Sci.*, 2008, **107**, 324–330.
11. F. S. Collins, G. M. Gray and J. R. Bucher, Transforming environmental health protection, *Science*, 2008, **319**, 906–907.
12. E. Weise, Three U.S. agencies aim to end animal testing, *USA TODAY*, 2008.
13. M. A. Hamburg, Advancing regulatory science, *Science*, 2011, **331**, 987.
14. J. Zurlo, D. Rudacille and A. M. Goldberg, *Animals and Alternatives in Testing*, Mary Ann Liebert, New York, NY, 1994.

15. A. N. Rowan, Looking back 33 years to Russell and Burch: The development of the concept of the 3Rs (Alternatives), in *Alternatives to Animal Testing. New Ways in the Biomedical Sciences, Trends and Progress*, ed. C. A. Reinhardt, VCH, Weinheim, Germany, 1994, pp. 1–11.
16. M. Stephens, A. M. Goldberg and A. N. Rowan, The first forty years of the alternatives approach: Refining, reducing, and replacing the use of laboratory animals, in *The State of the Animals: 2001*, ed. D. J. Salem and A. N. Rowan, Humane Society, Washington, DC, 2001, pp. 122–135.
17. A. Goldberg, The science of alternatives: 25 years and tomorrow, *AATEX*, 2008, **14**, 29–36.
18. A. N. Rowan, The third R: Refinement, *Altern. Lab. Anim.*, 1995, **23**, 332–346.
19. M. H. Lloyd, B. W. Foden and S. E. Wolfensohn, Refinement: promoting the three Rs in practice, *Lab. Anim.*, 2008, **42**, 284–293.
20. L. H. Smaje, J. A. Smith, R. D. Combes, R. Ewbank, J. A. Gregory, M. Jennings, G. J. Moore and D. B. Morton, Advancing refinement of laboratory animal use, *Lab. Anim.*, 1998, **32**, 137–142.
21. M. F. W. Festing, V. Baumans, R. D. Combes, M. Halder, C. F. M. Hendriksen, B. R. Howard, D. P. Lovell, G. J. Moore, P. Overend, M. S. Wilson, Reducing the use of laboratory animals in biomedical research: Problems and possible solutions, *Altern. Lab. Anim.*, 1998, **26**, 283–301.
22. J. de Boo and C. Hendriksen, Reduction strategies in animal research: A review of scientific approaches at the intra-experimental, supra-experimental and extra-experimental levels, *Altern. Lab. Anim.*, 2005, **33**, 369–377.
23. M. Hudson and B. Howard, The FRAME reduction steering committee: Reflections on a decade devoted to reducing animal use in biomedical science, *Altern. Lab. Anim.*, 2009, **37**, 23–26.
24. S. Adler, D. Basketter, S. Creton, O. Pelkonen, J. van Benthem, V. Zuang, K. E. Andersen, A. Angers-Loustau, A. Aptula, A. Bal-Price, E. Benfenati, U. Bernauer, J. Bessems, F. Y. Bois, A. Boobis, E. Brandon, S. Bremer, T. Broschard, S. Casati, S. Coecke, R. Corvi, M. Cronin, G. Daston, W. Dekant, S. Felter, E. Grignard, U. Gundert-Remy, T. Heinonen, I. Kimber, J. Kleinjans, H. Komulainen, R. Kreiling, J. Kreysa, S. B. Leite, G. Loizou, G. Maxwell, P. Mazzatorta, S. Munn, S. Pfuhler, P. Phrakonkham, A. Piersma, A. Poth, P. Prieto, G. Repetto, V. Rogiers, G. Schoeters, M. Schwarz, R. Serafimova, H. Tähti, E. Testai, J. van Delft, H. van Loveren, V. Mathieu, A. Worth and J.-M. Zaldivar, Alternative (non-animal) methods for cosmetics testing: current status and future prospects-2010, *Arch. Toxicol.*, 2011, **85**, 367–485.
25. D. A. Basketter, H. Clewell, I. Kimber, A. Rossi, B. Blaauboer, R. Burrier, M. Daneshian, C. Eskes, A. Goldberg, N. Hasiwa, S. Hoffmann, J. Jaworska, T. B. Knudsen, R. Landsiedel, M. Leist, P. Locke, G. Maxwell, J. McKim, E. A. McVey, G. Ouédraogo, G. Patlewicz, O. Pelkonen, E. Roggen, C. Rovida, I. Ruhdel, M. Schwarz, A. Schepky, G. Schoeters, N. Skinner, K. Trentz, M. Turner, P. Vanparys, J. Yager,

J. Zurlo and T. Hartung, A roadmap for the development of alternative (non-animal) methods for systemic toxicity testing - t4 report, *ALTEX*, 2012, **29**, 3–91.

26. N. Berg, B. De Wever, H. W. Fuchs, M. Gaca, C. Krul and E. L. Roggen, Toxicology in the 21st century–working our way towards a visionary reality, *Toxicol. In Vitro*, 2011, **25**, 874–881.

27. M. Leist, N. Hasiwa, M. Daneshian and T. Hartung, Validation and quality control of replacement alternatives – Current status and future challenges, *Toxicol. Res.*, 2012, **1**, 8–22.

28. M. L. Stephens, K. Conlee, G. Alvino and A. N. Rowan, Possibilities for refinement and reduction: Future improvements within regulatory testing, *ILAR J.*, 2002, **43** Suppl., S74–9.

29. W. S. Stokes, K. Brown, J. Kulpa-Eddy, G. Srinivas, M. Halder, H. Draayer, J. Galvin, I. Claassen, G. Gifford, R. Woodland, V. Doelling and B. Jones, Improving animal welfare and reducing animal use for human vaccine potency testing: State of the science and future directions, *Proc. Vaccinol.*, 2011, **5**, 84–105.

30. W. M. Russell and R. L. Burch, *The Principles of Humane Experimental Technique*, Universities Federation for Animal Welfare (UFAW), Potters Bar, UK, 1992.

31. H. Spielmann, U. G. Sauer and O. Mekenyan, A critical evaluation of the 2011 ECHA reports on compliance with the REACH and CLP regulations and on the use of alternatives to testing on animals for compliance with the REACH regulation, *Altern. Lab. Anim.*, 2011, **39**, 481–493.

32. C. F. M. Hendriksen and D. B. Morton (eds.), *Humane Endpoints in Animal Experiments for Biomedical Research*, Royal Society of Medicine, London, UK, 1999.

33. C. S. de Azevedo, C. F. Cipreste and R. J. Young, Environmental enrichment: A GAP analysis, *Appl. Anim. Behav. Sci.*, 2007, **102**, 329–343.

34. M. L. Stephens, An animal protection perspective on 21st century toxicology, *J. Toxicol. Environ. Health B Crit. Rev.*, 2010, **13**, 291–298.

35. M. Balls, A. M. Goldberg, J. H. Fentem, C. L. Broadhead, R. L. Burch, M. F. W. Festing, J. M. Frazier, C. F. M. Hendriksen, M. Jennings, M. D. O. van der Kamp, D. B. Morton, A. N. Rowan, C. Russell, W. M. S. Russell, H. Spielmann, M. L. Stephens, W. S. Stokes, D. W. Straughan, J. D. Yager, J. Zurlo and B. F. M. van Zutphen, The three Rs: The way forward: The report and recommendations of ECVAM Workshop 11, *Altern. Lab. Anim.*, 1995, **23**, 838–866.

36. A. M. Goldberg (ed.), *In Vitro Toxicology*, Mary Ann Liebert, New York, NY, 1991.

37. U. S. Congress, Office of Technology Assessment, *Alternatives to Animal Use in Research, Testing, and Education*, U.S. Government Printing Office, Washington, DC, 1986.

38. A. N. Rowan, F. M. Loew and J. C. Weer, *The Animal Research Controversy*, Tufts University School of Medicine, North Grafton, MA, 1995.

39. T. Hartung, From alternative methods to a new toxicology, *Eur. J. Pharm. Biopharm.*, 2011, **77**, 338–349.

40. R. D. Curren and J. W. Harbell, Ocular safety: A silent (in vitro) success story, *Altern. Lab. Anim.*, 2002, **30** Suppl. 2, 69–74.

41. M. Balls and R. H. Clothier, Comments on the scientific validation and regulatory acceptance of in vitro toxicity tests, *Toxicol. In Vitro*, 1991, **5**, 535–538.

42. L. H. Bruner, G. J. Carr, R. D. Curren and M. Chamberlain, Validation of alternative methods for toxicity testing, *Environ. Health Perspect.*, 1998, **106** Suppl. 2, 477–484.

43. M.-J. W. A. Schiffelers, B. J. Blaauboer, J. M. Fentener, van Vlissingen, J. Kuil, R. Remie, J. W. G. M. Thuring, M. A. Vaal and C. F. M. Hendriksen, Factors stimulating or obstructing the implementation of the 3Rs in the regulatory process, *ALTEX*, 2007, **24**, 271–278.

44. R. A. Becker, C. J. Borgert, S. Webb, J. Ansell, S. Amundson, C. J. Portier, A. Goldberg, L. H. Bruner, A. Rowan, R. D. Curren and W. T. Stott, Report of an ISRTP workshop: Progress and barriers to incorporating alternative toxicological methods in the U.S., *Regul. Toxicol. Pharmacol.*, 2006, **46**, 18–22.

45. C. Eskes, K. Sullivan, M. Aardema, H. Spielmann, E. Hill, G. Schoeters, R. Curren, V. Mathieu and B. J. Blaauboer, Comparing the challenges in developing and implementing 3Rs alternative methods in Europe and the United States: Industrial and academic perspectives, *ALTEX Proc.*, 2012, **1/12**, 205–209.

46. R. L. Burch, The progress of humane experimental technique since 1959: A personal view, *Altern. Lab. Anim.*, 2009, **37**, 269–275.

47. M. L. Stephens, C. Barrow, M. E. Andersen, K. Boekelheide, P. L. Carmichael, M. P. Holsapple and M. Lafranconi, Accelerating the development of 21st-century Toxicology: Outcome of a Human Toxicology Project Consortium workshop, *Toxicol. Sci.*, 2012, **125**, 327–334.

48. M. E. Andersen and D. Krewski, The vision of Toxicity Testing in the 21st Century: Moving from discussion to action, *Toxicol. Sci.*, 2010, **117**, 17–24.

49. D. A. Keller, D. R. Juberg, N. Catlin, W. H. Farland, F. G. Hess, D. C. Wolf and N. G. Doerrer, Identification and characterization of adverse effects in 21st century toxicology, *Toxicol. Sci.*, 2012, **126**, 291–297.

50. T. Hartung, Evidence-based toxicology – The toolbox of validation for the 21st century?, *ALTEX*, 2010, **27**, 253–263.

51. R. Judson, R. Kavlock, M. Martin, D. Reif, K. Houck, T. Knudsen, A. Richard, R. R. Tice, M. Whelan, M. Xia, R. Huang, C. Austin, G. Daston, T. Hartung, J. R. Fowle, W. Wooge, W. Tong and D. Dix, Perspectives on validation of high-throughput assays supporting 21st century toxicity testing, *ALTEX*, 2013, **30**, 51–66.

52. T. Hartung, S. Hoffmann and M. Stephens, Mechanistic validation, *ALTEX*, 2013, **30**, 119–130.

53. L. Hareng, C. Pellizzer, S. Bremer, M. Schwarz and T. Hartung, The integrated project ReProTect: A novel approach in reproductive toxicity hazard assessment, *Reprod. Toxicol.*, 2005, **20**, 441–452.

54. M. Whelan, M. Schwarz and Scientific Expert Panel of the SERUAT-1 Research Initiative Panel, *SEURAT-1 Strategy Paper*, 2011, 1–10.

55. N. S. Sipes, M. T. Martin, D. M. Reif, N. C. Kleinstreuer, R. S. Judson, A. V. Singh, K. J. Chandler, D. J. Dix, R. J. Kavlock and T. B. Knudsen, Predictive models of prenatal developmental toxicity from ToxCast high-throughput screening data, *Toxicol. Sci.*, 2011, **124**, 109–127.

56. M. T. Martin, T. B. Knudsen, D. M. Reif, K. A. Houck, R. S. Judson, R. J. Kavlock and D. J. Dix, Predictive model of rat reproductive toxicity from ToxCast high throughput screening, *Biol. Reprod.*, 2011, **85**, 327–339.

57. National Research Council, *The Future of Animals, Cells, Models, and Systems in Research, Development, Education, and Testing*, National Academy of Sciences, Washington, DC, 1977.

58. A. M. Goldberg, L. F. M. van Zutphen and M. L. Principe (eds.), *The World Congress on Alternatives and Animal Use in the Life Sciences*, Mary Ann Liebert, New York, 1995.

59. Organisation for Economic Co-operation and Development, *Final Report of the OECD Workshop on Harmonization of Validation and Acceptance Criteria for Alternative Toxicological Test Methods*, OECD, Paris, France, 2009.

CHAPTER 2

Regulatory Testing To Inform Decisions: National and International Requirements

JOHN 'JACK' R. FOWLE III,*[a] ABIGAIL JACOBS[b] AND SUZANNE FITZPATRICK[c]

[a] Science to Inform, L.L.C. (United States Environmental Protection Agency, ret.), 155 Terrells Mtn, Pittsboro, NC 27312, USA; [b] United States Food and Drug Administration, Center for Drug Evaluation and Research, 10903 New Hampshire Ave, Silver Spring, MD 20993, USA; [c] United States Food and Drug Administration, Center for Food Safety & Applied Nutrition, 5100 Paint Branch Parkway, College Park MD 20740, USA
*Email: jackfowle@aol.com

2.1 Introduction

This chapter will identify some of the national and international government agencies and key non-governmental organizations that use and/or require toxicity testing. It will summarize major testing requirements, some of the notable efforts to move from conventional animal testing approaches to non-animal approaches, and highlight some of the key advances made to date. Most importantly, this chapter will convey the complexity and multi-faceted nature of regulatory safety testing such that anyone interested in improving regulatory testing approaches to inform decisions might wish to consider to help plan, conduct, and present his/her efforts so that they will be most acceptable and useful to regulatory authorities. The biggest challenges that must be overcome

Issues in Toxicology No. 19
Reducing, Refining and Replacing the Use of Animals in Toxicity Testing
Edited by David G. Allen and Michael D. Waters
© The Royal Society of Chemistry 2014
Published by the Royal Society of Chemistry, www.rsc.org

to advance the refinement, reduction, and replacement of animal tests with appropriate non-animal tests will also be discussed.

Much of this chapter will focus on US Federal agencies and their activities, especially those of the Food and Drug Administration (FDA) and the Environmental Protection Agency (EPA), since they have the most Congressionally delegated responsibilities for safety assessment and, thereby, have the greatest data requirements. While state governments also play a key role in environmental protection, they will not be considered in depth herein.

Clearly, international efforts are major drivers in shaping regulatory testing. Therefore, this chapter will also broadly summarize key international activities, including both multinational and selected country-specific efforts.

It is important to bear in mind that regulatory testing requirements change with time, so it is not possible to maintain a definitive, up-to-date description in a textbook chapter. Therefore, this chapter will describe the rationale associated with the development and application of regulatory tests to inform safety decisions, and use web hyperlinks throughout this text to provide readers with a means to obtain updated information.

Accordingly, a pivotal reference for this text is the AltTox Toxicity Testing Resource Center (http://www.alttox.org/), from which many of the descriptions of US and international organizations and agencies were obtained with permission.[1] It should not be construed that the authors of this chapter, the editors of this book, or the book publishers necessarily endorse the views and programs of AltTox. The internet resource provides an extensive summary of the various national and international agencies, their testing requirements, and progress towards implementing non-animal testing approaches on their website which is periodically updated.

The regulatory requirements in place today are the result of past efforts to meet society's desire to minimize the impact of the use of commercial products and services on human health and the environment through laws giving authority to government agencies to regulate and, through judicial findings from court cases, that affect regulatory approaches. Thus, the national and international sections of this chapter each begin by summarizing the legal context for regulatory testing to inform decisions.

2.1.1 Legal Context in the United States

The United States (US) Constitution gives Congress the power to pass legislation, power to the Executive Branch to implement the laws passed by Congress, and power to the judicial branch to arbitrate conflicts about how to interpret the laws and to punish those who break them. For those with an interest in regulatory testing to inform decisions in the US it is critical to bear this in mind as you read this chapter, because the current approach to inform regulatory safety decisions is the result of a long history about why Congress passed laws bearing on safety, how these laws were implemented by the various departments and agencies (hereinafter referred to as 'agencies') in the executive branch, and how the laws and their implementation have been refined based on judicial findings. Agencies can't do anything if there is no enabling legislation passed by Congress giving them the authority to do so. If the current regulatory testing practices are to be changed, the relevant agencies must have the Congressional authority to do so and any changes they make cannot conflict with judicial findings. It is important to note that given their limited expertise the members of Congress grant broad powers to the Executive Branch to apply their expertise to interpret the laws. Similarly, when courts find laws unconstitutional they send them back to the Congress for revision (see Additional Resources, ref. 1).

The key point to bear in mind is that while agencies work to ensure public safety through the best available information to inform their decisions, their decisions are not based on science alone. They must take into account Congressional mandates and legal precedent, as noted above, that reflect 'the will of the people.' Scientists, advocates, and other stakeholders can get frustrated that 'good science' doesn't drive decisions. A practical point to bear in mind, given the tripartite nature of the US government, and the requirement for public input, is that science is but one factor considered in decision making. Its role is to inform, not make decisions, and given the varied Congressional requirements in enabling legislation (e.g., some laws require benefit/cost considerations, some require the application of best available technology, some require consideration of environmental justice, etc.), legal precedent, and the inherent uncertainties in knowledge, science is not necessarily the most important factor considered for decision making. It does play a key role in informing policy, but so do a variety of factors such as economics, public values, political, technological, social, and legal (see Additional Resources, ref. 2).

Thus, anyone who wishes to move away from the status quo regarding regulatory testing will likely be most effective if he/she understands the regulatory testing requirements and why and how were they developed. It is also important to try to understand who are the key players in the various branches of government and other interested and engaged organizations involved as well as what motivates them, what do they view as threats to their vision, and where does the middle ground lie. Full treatment of these points is beyond the scope of this chapter and will only be noted in passing, but they are important to bear in mind.

In addition to the federal government, state governments play an important role in regulatory testing to inform decisions. The 'Federal' system (i.e. Federalism)[2] approach of US government provides for 'State's Rights,' meaning that states have the right to govern within their state's boundaries as long as they don't impede on the rights of other states such as limiting the flow of commerce across state boundaries. This is because individual states may have unique needs, for instance due to geography and climate. As a result federal laws tend to be less restrictive than state laws. However, each state must comply with federal law (e.g., California must comply with federal air pollution requirements, but their laws and regulations are more restrictive than are the federal air pollution laws and regulations). A discussion of state regulatory requirements is beyond the scope of this chapter but interested readers are encouraged to further look into state laws and regulations, as they do play an important role in regulatory testing. By way of example, links to state environmental laws can be found on the EPA's website.[3]

2.1.2 International Legal Context

Each internationally recognized country is a sovereign nation with the authority to perform the legislative, executive, and judicial functions, or their equivalent, that are necessary to govern their country with the full and exclusive right to do so within their boundaries. Each country is also recognized to have the authority to make binding treaties with other countries. The manner in which each country chooses to govern its nation is dependent on a variety of historical, social, geographical, and other factors. The important point is to note that 'one size does not fit all' for the purposes of governing, and each country has applied the approach to governance that works best for the general good of its citizens. As a result there is a diverse set of laws and regulations applied across the world, including those regarding regulatory test requirements, some of which conflict with those laws and regulations developed and applied by other nations.

This can create unintentional barriers to trade across the nations, because the global companies must ensure that their goods and services comply with the specific requirements of each and every nation in which they trade, and it can create disincentives to the development and use of new and better testing approaches. For instance, if one country requires conventional animal tests while another other does not, any company trading in both countries will have to do the conventional testing to satisfy the requirements of the country that requires such testing even though non-animal test results would have sufficed in the other country. To help address this issue, multinational governmental organizations have been developed via treaties, such as the Organisation for Economic Co-operation and Development,[4] to work across nations to remove non-tariff trade barriers by bringing them together to harmonize their approaches to the application of a wide range of regulations (e.g., trade, transportation, environment, food safety), including those dealing with regulatory test requirements to inform decisions.

2.1.3 Context: Use of Conventional Regulatory Testing to Inform Decisions

The historical background covering the development of alternatives to animal testing is described in Chapter 1. The history described in that chapter is important to bear in mind in terms of thinking about the use of alternatives to animal tests to inform decisions. It is largely because recent advances in high speed computing and in the understanding of the biological basis of disease over the past few decades have provided a basis for the development and use of a large number of high speed, high information content assays that mimic the key events that occur in disease processes, thus giving decision makers and others confidence that the approaches can inform decisions about appropriate levels of protection.

The ToxCast™ database[5] developed by the Environmental Protection Agency (EPA) and the Tox 21 Consortium[6] involving the EPA, the National Institute of Environmental Health Sciences (NIEHS), the National Institutes of Health (NIH) Chemical Genomics Center (NCGC) at the NIH National Center for Advancing Translational Sciences, and the Food and Drug Administration (FDA) represent some of the initial activities to achieve this goal. By mapping the events measured by these assays to the adverse outcome pathway (AOP) leading to the effect(s) of concern it is hoped that it will be possible soon to use a combination of computer models and non-animal assays to predict risk.[7] (It is important to note that some animal-based testing will likely continue to be needed for many years, for ecological risk assessment and for other purposes such as evaluating complex, multi-step disease processes like cancer and birth defects.)

To the extent that adoption of the 3Rs is successful, it will not occur overnight but will likely be implemented in an evolutionary fashion. The Organisation for Economic Co-operation and Development (OECD) has proposed an 'Integrated Approach to Testing and Assessment' (IATA) to accomplish this (see Additional Resources, ref. 3). In essence, IATA is an approach that allows regulatory authorities to add alternative approaches, stepwise and incrementally, to the testing toolbox to reduce, refine, and possibly replace the use of animal tests for that type of assessment when it has been demonstrated that a new approach is sufficiently good for its intended purpose such that it can replace the conventional test. This is because conventional assays have been used over many decades for safety assessment, and many thousands of decisions have been informed by the results. In order to make changes in the way testing is done strategic use of both conventional animal and human studies in tandem with alternative approaches will be required to test specific hypotheses about how well do the alternative to animal tests predict the outcome of interest in humans, or other species of interest, compared to the results of the conventional animal tests. Such information is needed to develop confidence in the minds of the executive, legislative, and judicial branches of government, and perhaps most importantly in the minds of the general public, including interest groups, that the alternatives are good enough for their intended use and, better yet that they are at least as good as, and preferably better than, conventional animal tests at informing decisions about risk.

Thus, the path to a future where conventional animal testing is replaced by alternative approaches will likely involve successive approximations employing a 'Yin and Yang' combination where computational toxicology and other alternative approaches are compared with traditional animal tests, as well as with outcomes in human and ecological species of interest, to define how well each approach works for each purpose. The use of AOPs through an IATA approach provides a systematic framework to make the case and a phased, evolutionary process to accomplish this.

2.2 US Regulatory Agencies, Research Agencies and Interagency Bodies

In the US, the EPA and the FDA account for most of the animal testing required under federal regulations. Their major data requirements will be briefly discussed, and the animal testing requirements of other US agencies will be summarized.

2.2.1 The Environmental Protection Agency (EPA)

The US Environmental Protection Agency (EPA) is an independent regulatory agency (meaning that it is not a cabinet-level department and it is not part of a cabinet-level department), established in 1970 with the mission 'to protect human health and the environment.'[8] The EPA is charged with administering all or part of 23 laws enacted by Congress and 3 executive orders issued by Presidents to protect human health and the environment.[9] Based on its expertise and public input, the EPA develops and enforces regulations and employs other means to implement the laws delegated to it by Congress, following the guidance in the Presidential executive orders, to prevent pollution and to protect human health and the environment from hazardous chemicals and other stressors.

The EPA has 13 headquarter offices and 10 regional offices.[10] Four of the headquarter offices (often referred to as program offices) are responsible for writing the regulations, and are basically organized around the major environmental recipients of pollutants (i.e., air, water, and land) as well as by the major source of pollution (i.e. commercial chemicals including pesticides). There is also an office to enforce compliance with these regulations although enforcement and compliance is largely handled by the regional offices. The other headquarter offices handle the administrative, legal, and research and development activities needed to support the program offices and regions.

The EPA has 24 laboratories. Seven are under the Office of Research and Development (ORD), another seven are under two program offices (three under the Office of Air and Radiation was well as four under the Office of Chemical Safety and Pollution Prevention), and there are 10 regional laboratories (one in each region).[11] The laboratories housed under the program offices and regions tend to be analytical support laboratories to assist monitoring compliance with regulations.

The ORD laboratories develop methods and models needed to improve the agency's capacity to evaluate risk to human health and the environment through research on better ways to assess exposure, to improve its ability to measure effects to humans and the environment, and to develop prevention and control technologies to prevent releases of pollutants to the environment as well as to control those that have already been released. The EPA supported the two National Academy of Sciences/National Research Council projects bearing on taking advantage of new advances in biology and computing to develop more informative, efficient, and inexpensive means to inform safety decisions. These resulted in 'Toxicity Testing in the 21st Century: A Vision and a Strategy' (see Additional Resources, ref. 4) and 'Exposure Science in the 21st Century: A Vision and a Strategy' (see Additional Resources, ref. 5).

The EPA has developed a strategy for using NAS' guidance to develop a more informative and efficient approach to testing that will rely less heavily on animals over time.[12] The EPA's National Center for Computational Toxicology has been the leader in moving towards predictive toxicology and away from the use of conventional tests to inform regulatory decisions.[13] Other EPA laboratories, such as the National Health and Environmental Effects Research Laboratory (NHEERL),[14] have collaborated with the Office of Pesticide Programs (OPP) and other program offices to develop AOPs for toxic endpoints of regulatory concern as a basis to link conventional toxicity tests to the rapidly evolving computational tests, providing a means to identify where they predict well and where improvements are needed as a basis to move in an orderly and informed fashion to replace less informative and efficient methods and models with more informative, predictive, and efficient ones that use few or no animals. While the ORD uses animals in its research, the major use of animals for regulatory testing to inform EPA decisions is conducted by industry to comply with program office requirements.

2.2.1.1 EPA's Office of Chemical Safety and Pollution Prevention

In the context of regulatory toxicology, the majority of EPA's animal use and/or testing requirements result from the Office of Chemical Safety and Pollution Prevention (OCSPP)[15] primarily under the Office of Pesticide Programs (OPP)[16] and to a certain extent under the Office of Pollution Prevention and Toxics (OPPT).[17]

2.2.1.1.1 Pesticides. All products that are intended to manage, destroy, attract, or repel 'pests' and that are used, sold, or imported into the US are regulated by the EPA's Office of Pesticide Programs (OPP), a subdivision of OCSPP. These products include synthetic and biochemicals, genetically engineered toxins, and even other organisms that are collectively referred to as 'pesticides.' The EPA's statutory authority for regulating pesticides is primarily derived from the Federal Insecticide, Fungicide and Rodenticide Act (FIFRA)[18] and the Food Quality Protection Act;[19] however, provisions in the Federal Food, Drug and Cosmetic Act (FFDCA)[20] and the Endangered Species Act (ESA)[21] also strongly impact the EPA's activities in the pesticides sector.

The goal of the EPA registration data requirements for pesticide chemicals is to mimic all the likely types of exposures that humans and species in the environment are expected to encounter during the manufacture, use, and consumption of pesticides. By their nature, pesticides are designed to kill, harm, or hinder organisms and so the assessment of pesticides is not about whether or not they are poisonous, but rather it is about characterizing the routes of exposures, the doses and combinations of pesticides that can be used safely. Thus, an extensive range of testing is required to characterize the chemical nature of pesticide products, their uses, their potential to cause a wide variety of adverse effects to humans, plants, and other animals, and their ability to cause effects from aggregate and, in the case of pesticides in the same chemical class that share the same mode of action, the cumulative exposure effects. These testing requirements are codified under Title 40, Part 158 of the Code of Federal Regulations (40 CFR § 158)[22] and are summarized in the EPA's 'Data Requirements for Pesticide Registration' webpage.[23]

The test requirements vary depending on the type of pesticide product being evaluated. The EPA generally accepts reduced data sets for naturally occurring biochemical and microbial pesticides (e.g., pheromones)[24] and for germ-killing antibacterial cleaning products[25] more often than it does for conventional pesticides.[26] The latter undergo more extensive testing including:

1 acute systemic toxicity in rodents and/or rabbits via oral, inhalation, and dermal routes;
2 eye and skin irritation in rabbits;
3 skin sensitization in mice or guinea pigs;
4 subacute and/or subchronic studies via the dermal route;
5 subchronic (3 month) feeding studies in rodents and dogs;
6 chronic (12–24 month) feeding studies in rodents;
7 lifetime (18–24 month) carcinogenicity studies in rats and mice;
8 mutagenicity and genotoxicity studies of at least two varieties;
9 reproductive toxicity in at least two generations of rodents;
10 prenatal developmental toxicity in rodents and rabbits;
11 metabolism and toxicokinetic studies in rodents;
12 acute aquatic toxicity to fish of two or more species;
13 partial or full life-cycle toxicity to fish of two or more species;
14 acute and/or dietary toxicity to birds of two or more species;
15 reproductive toxicity to birds;
16 and many more ...

In addition to mandatory testing of each active ingredient in a pesticide formulation, each finished product is also required to undergo separate acute toxicity testing via the oral, dermal, and inhalation routes, skin and eye irritation, and skin sensitization (known as the acute toxicity 'six-pack') for labeling purposes. Recognized testing methods to generate data to inform OPP's decisions include the OCSPP Harmonized Test Guidelines,[27] as well as the internationally harmonized OECD Guidelines for the Testing of Chemicals.[28] However, FIFRA grants the agency broad discretion to request testing data for any effect(s) of interest under the pesticide program, including by means of 'special,' non-guideline studies.

It is estimated that upwards of 12,000 animals and $10 m may be consumed during the extensive toxicological evaluation of each conventional chemical pesticide on the US market.[29] To take advantage of the recent advances in high speed computing and molecular biology, the EPA's pesticides program has adopted the NRC's Toxicity Testing in the 21st Century vision (see Additional Resources, ref. 4), and it has developed a strategic vision about how to apply this approach in conjunction with AOPs and IATA in its pesticide testing activities to make the process more informative and efficient.[30] In 2007, the EPA adopted the integrated testing strategy recommendations of a multi-stakeholder technical committee under the auspices of the International Life Sciences Institute (ILSI) Health and Environmental Sciences Institute (HESI) (see Additional Resources, ref. 6) to abolish the one-year general toxicity studies in dogs.[31] In 2009 it launched a non-animal testing approach to EPA labeling for eye irritation for antimicrobial cleaning products.[32] In June 2013 it concluded that the testing approach is acceptable for determining the appropriate eye hazard classification and labeling for AMCPs, and that the testing scheme satisfies the *in vivo* data requirement for eye irritation in 40CFR Part 158W for AMCPs.[33]

The agency solicits independent scientific input through the FIFRA Scientific Advisory Panel (SAP)[34] and advice from pesticide stakeholders through its Pesticide Program Dialogue Committee.[35] In May 2011, OPP, working in concert with its partners from ORD and members of the Pesticide Program Dialog Committee, presented its plans to move from its current animal testing approach to a more hypothesis based and predictive approach to evaluating safety at a SAP meeting entitled 'Integrated Approaches to Testing and Assessment Strategies: Use of New Computational and Molecular Tools' for guidance about how improve its more informative and efficient alternative testing program.[36] OPP used several case studies to illustrate how the new approach might proceed and its vision and plans for the future. The SAP concluded that the approach offers promise as a more informative, efficient, and economical approach to regulatory testing to inform decisions and that it should be pursued. The OPP has stated that it is committed to pursue this course of action into the future and those interested in keeping up-to-date are encouraged to periodically check the OPP's web page.

2.2.1.1.2   Endocrine Disruptors.   The Food Quality Protection Act of 1996 required the EPA to develop and implement 'a screening program, using appropriate validated test systems and other scientifically relevant information, to determine whether certain substances may have an effect in humans that is similar to an effect produced by a naturally occurring estrogen, or such other endocrine effect as [EPA] may designate.' Responsibility for developing a 'toolbox' of validated screens and tests for this 'Endocrine Disruptor Screening Program' (EDSP)[37] was delegated to the EPA's Office of Science Coordination and Policy (OSCP)[38] within the OCSPP, although decisions about how to use the results of testing was left with the regulatory units of the program offices both in the OCSPP and in other program offices as well. Using the Endocrine Disruptor Screening and Testing Advisory Committee (EDSTAC)[39] recommendations and working with the ORD, the OECD and others both in the United States and abroad, the OSCP developed a two-tiered testing approach to evaluating the potential of chemicals of interest to the EPA to cause endocrine disruption.

The Tier 1 tests are screening assays designed to determine if a chemical substance has endocrine disrupting properties and the tests being developed for Tier 2 are intended to characterize those substances identified by the Tier 1 tests as being possible endocrine disruptors with respect to whether they actually do cause such effects. When EDSTAC developed its recommendations, animal test approaches were the 'gold standard' for safety evaluation. Further, there are no generally accepted alternatives to animal tests for certain organism-wide functions that require interaction between organs, cognitive processing, hormonal action, and immunological effects or interactions between organisms such as neurobehavioral effects. Thus, the EDSP employs animal models in both the Tier 1 and proposed Tier 2 batteries.

In 2009, the EPA published an initial list of chemicals to be screened in Tier 1 tests for potential endocrine disrupting effects and the first test orders were issued late that year.[40] The initial focus is on pesticides and drinking water contaminants to which a large population may be exposed. Once the results are reviewed, the EPA will make them public and will identify those that must go through Tier 2 evaluation to characterize any endocrine disrupting effect that may be caused by a chemical at various doses. As part of that process the Agency described how it is working to integrate high throughput screening data with SAR, ER binding, and other approaches to develop AOPs for endocrine disruptors and incrementally move towards more predictive approaches through the use of IATA during a January 29–31, 2013 meeting of the Science Advisory Panel on the scientific issues associated with 'Prioritizing the Universe of Endocrine Disruptor Screening Program (EDSP) Chemicals Using Computational Toxicology Tools'.[41]

2.2.1.1.3 Commercial Chemicals. The OPPT administers the Toxic Substances Control Act (TSCA)[42] and the Pollution Prevention Act.[43] Before a new chemical is marketed in the US, the TSCA requires that a pre-manufacture notice be filed with the EPA providing information on the chemical's identity, intended uses, production volume, and anticipated exposure and emission levels, as well as any toxicity data in the company's possession.[44] The EPA uses the information in premanufacture notices, together with structure–activity relationship modeling, to determine whether there is a need to impose restrictions on the release and/or marketing of a chemical in order to ensure there is 'no unreasonable risk of injury to health or the environment.' No testing beyond this is required under TSCA for new chemical substances as defined under the act.

With respect to existing chemicals, the TSCA provides the EPA with the authority to require companies to submit 'all existing data concerning the environmental and health effects of [a chemical] or mixture.'[45] Under TSCA section 4, the EPA is further empowered to compel companies to generate test data when the agency finds that a chemical (1) may present an unreasonable risk of injury to health or the environment or (2) is or will be produced in substantial quantities and (a) there is or may be significant or substantial human exposure to the chemical or (b) it enters or may reasonably be antici- pated to enter the environment in substantial quantities. Many of the chemicals on the TSCA inventory, a list of chemical substances manufactured or im- ported into the United States, are polymers with chemical properties that make it unlikely that they will be absorbed into the body or cause adverse effects. Many others are made in quantities below 10,000 pounds (4500 kg) per year with such low exposure potential that risks are considered to be minimal. Thus, of the over 70,000 chemicals on the TSCA inventory only about 15 000 have the chemical characteristics and are made in such quantities that they may pose an environmental risk. Of these, about 540 chemicals have been issued test re- quirements and the EPA has made the decision not to test about 250 other chemical substances.[46] Most of 540 chemicals undergoing testing under Section 4 are high production volume or 'HPV' chemicals made or imported into the US in quantities of at least one million pounds (450,000 kg) per year.[47] There are about 2200 HPV chemicals on the TSCA inventory. While many HPV test rules are the result of EPA regulation, other testing has been done as a result Section 4 Enforceable Consent Agreements, and Voluntary Testing Agree- ments. Rather than 'one size fits all' testing which might have resulted in wasted resources, both in terms of dollars and animals, efforts have been made to evaluate what is known about the chemicals of interest as a basis to identify data gaps and to focus the testing on filling those gaps.

Testing to inform TSCA decisions is conducted using the tests identified in the OCSPP Test Guidelines, and the internationally harmonized Organisation for Economic Co-operation and Development (OECD) Test Guidelines which require a set of methods to evaluate the physical and chemical properties of chemical substance, its exposure characteristics including absorption, distri- bution, metabolism, and excretion in the body as well as its fate and transport in the environment, and its potential effects on humans and ecological species of interest as a basis to estimate risk. Depending on the type of tests required, such testing can use substantial number of animals.

As part of its efforts to protect human health and the environment, the EPA is working to ameliorate the need for testing through its Design for the En- vironment (DfE) Program.[48] Under DfE, the EPA works with chemical manufacturers to develop safer products by replacing toxic ingredients with less toxic substances and to inform substitution to safer alternatives and reduce the likelihood of unintended consequences that might result if poorly understood alternatives were chosen. The 'Design for the Environment Program Alter- natives Assessment Criteria for Hazard Evaluation' document contains guidelines for testing the alternatives and references to the OSCPP test guidelines.[49]

As is the case in the pesticides program, OPPT staff are developing a focused, hypothesis-based approach to testing using what is known about the chemical substance, including its physical and chemical properties, to identify data gaps and structure–activity relationship predictions coupled with high throughput testing to determine testing needs and priorities for testing. Without compelling reasons to do so, the requirements of TSCA do not allow for the EPA to require industry to conduct testing of chemical substances on the TSCA inventory for the wide range of human and environmental health endpoints that are evaluated for pesticides. As a result there are many chemicals on the inventory for which there is little or no exposure, hazard, or dose–response information that can be used to assess risk. As a means to obtain an approximation of what potential hazards might exist from exposure to chemical substances on the inventory, the OPPT is partnering with the EPA's National Center for Computational Toxicology (NCCT). The NCCT has developed a variety of tools to help advance the NRC's vision for Toxicity Testing in the 21st Century, including the ToxCast™ battery of high throughput assays covering the range of human health outcomes of concern that in concert with ToxRefDB,[50] a database that has systematically captured 30 years and over $2 billion worth of animal test data, ExpoCastDB,[51] a database of exposure data, and bioinformatics machine learning. Working in concert the OPPT, the NCCT, and others are applying these and other tools to screen previously untested chemicals in the TSCA inventory as a means to evaluate and prioritize them with respect testing needs. From the 3Rs perspective, an additional benefit of this effort is that even if authority existed to allow such testing, the use of conventional testing approaches to do so would require extensive use of animal studies. The NCCT approach uses no animals.

2.2.1.2 Other EPA Regulatory Programs

Regulatory testing requirements imposed by other EPA program offices are generally more limited than those of the OCSPP. However, a certain amount of acute and chronic testing of aquatic organisms is prescribed, for example, under water quality standards and whole effluent toxicity methods promulgated by the EPA's Office of Water.[52]

2.2.1.3 Adoption of Alternatives

The EPA is engaged in several activities to develop new and revised testing methods and guidelines that take full advantage of scientific and technical advances in an expeditious manner.[53] Its initial computational toxicology activities were launched in 2003 (see Additional Resources, ref. 7) and it formally established its computational toxicology research program in 2005. The Office of Pesticide programs was an early adopter and described its needs to the National Resource Council Panel that prepared the 2007 NAS document 'Toxicity Testing in the 21st Century: A Strategy and a Vision' (see Additional Resources, ref. 4) to inform their deliberations and development of the document. Subsequently the OPP has worked with the ORD, various nongovernmental organizations pursuing alternatives to animal tests and the regulated industry, as well as other federal agencies and international organizations to adopt more efficient and informative approaches to testing that reduce, refine, or replace the use of animals. In February 2008, the EPA took an additional step forward by signing a memorandum of understanding with the National Toxicology Program and National Institutes of Health Chemical Genomics Center outlining a strategy for interagency cooperation in 'the research, development, validation, and translation of new and innovative test methods that characterize key steps in toxicity pathways.'[54]

In May 2011 the OPP presented its 'Integrated Approaches to Testing and Assessment Strategies: Use of New Computational and Molecular Tools' to the SAP for guidance about how to improve its more informative and efficient alternative testing program and it has stated that it is committed to pursue this course of action into the future.

Because there is flexibility in the EPA regulations and guidance, the use of non-animal alternatives for evaluation of adverse effects to human health or the environment does not have to be formally validated by the OECD or validation organizations (e.g., ICCVAM) in order to be accepted for regulatory purposes. A possible exception is the endocrine disruptor screening program whose enabling legislation requires the use of validated methods. In general the EPA has the authority to review data on alternatives and decide that a particular alternative is acceptable for its context of use and is thus valid for that purpose. Nonetheless, the EPA and other national and international regulatory authorities prefer to use information from 'validated' methods and models to inform their decisions, not only because they help ensure scientific rigor and accuracy but also because they are generally more acceptable to the Congress, the courts, various stakeholders, and the public. Thus, extensive efforts have been conducted to 'validate' alternatives to animal tests and a number of others are currently underway and planned for the future as outlined in as outlined in Chapter 1.

2.2.2   The Food and Drug Administration (FDA)

The Food and Drug Administration (FDA) is a regulatory and research agency within the US Department of Health and Human Services and is responsible for 'protecting the public health by assuring the safety, efficacy, and security of human and veterinary drugs, biological products, medical devices, . . . , cosmetics, and products that emit radiation.'[55] The FDA's broad statutory authority[56] bearing on regulatory test requirements stems from two major laws the Federal Food, Drug, and Cosmetic Act of 1938 (FFDCA), as amended[57] and the Family Smoking Prevention and Tobacco Control Act.[58] The FDA also administers part or all of other legislation such as the Dietary Supplement Health and Education Act of 1994,[59] the Fair Packaging and Labeling Act,[60] the Food Safety Modernization Act,[61] and the Public Health Service Act.[62]

As with the EPA and all other regulatory agencies, Congress has charged the FDA to use its expertise to interpret the laws under its jurisdiction and to develop programs to implement them. The FDA's programs are organized into a number of centers and offices, each with its own regulatory, research, and/or enforcement mandate. In the context of regulatory toxicology, most regulatory testing requirements at the FDA can be linked to the following centers:

1 *Center for Biologics Evaluation and Research (CBER)*: regulates the licensing of human vaccines, human blood products, and animal-to-human xenotransplants as well as other human biologics;[63]
2 *Center for Devices and Radiological Health (CDRH)*: responsible for premarket approval of new medical devices and radiation-emitting electronic products;[64]
3 *Center for Drug Evaluation and Research (CDER)*: regulates the human testing, approval, licensing, and labeling of human drugs and therapeutic protein and monoclonal antibody biologics;[65]
4 *Center for Food Safety and Applied Nutrition (CFSAN)*: monitors the safety and labeling of cosmetics, food additives, processed foods, and shellfish;[66]
5 *Center for Veterinary Medicine (CVM)*: regulates veterinary drugs and devices, pet foods, and livestock feeds;[67] and
6 *National Center for Toxicological Research (NCTR)*: conducts intramural R&D and provides scientific support to other FDA centers and offices.[68]

Nonclinical studies for drugs and biologics are conducted in animals to support various activities, including:

1 selection of the first dose of new product given to humans;
2 selection of the route, dose, duration of human exposure to the product in all clinical trials and when the product is marketed;
3 to inform product labeling decisions for mutagenicity, carcinogenicity, fertility, and effects on the fetus if given to pregnant women or to women thinking of becoming pregnant.

The regulations covering investigational new drugs and new drug applications are in the US Code of Federal Regulations section 300 Title 21.[69] Section 312 covers investigational new drug applications, and section 314 covers approval to market a drug and labeling/package inserts. Biologic products are also covered by US Code of Federal Regulations Title 21.600. The regulations are somewhat vague regarding the details of the studies needed to inform decisions about biologic products. To help drug developers, the FDA has guidances on a variety of topics related to nonclinical drug development. How to think about issues is generally the goal of the guidances. They are not legally binding, and can be updated or replaced when the thinking on a topic changes. A list is given at the FDA website.[70]

To promote harmonization in drug development and to avoid the repetition of studies in various regions of the world, the FDA participates in the International Conference on Harmonization of the technical requirements for pharmaceuticals (ICH guidance) for nonclinical safety, product quality, and clinical aspects of drug development.[71] In Europe and Japan, these technical requirements are considered Guidelines, but in the United States, these are guidances, and not legally binding.

The FDA Redbook or OECD protocols are generally accepted but deviations from these protocols can be allowed, and where the FDA Redbook[72] or OECD[73] protocols differ from ICH guidances, ICH guidance advice is preferred. A few protocol details that are in ICH guidances include carcinogenicity study dose selection and the high concentration to be used in *in vitro* mammalian genotoxicity assays. ICH guidances encourage the 3Rs. Many non-animal methods are used by drug developers in the drug discovery phase prior to submission to the FDA. For human pharmaceuticals, there will eventually be human data at relatively large deliberate doses for products with a clinical benefit so the nonclinical studies are generally screens for potential human effects.

Studies with some validated alternatives for skin corrosion or irritation, ocular irritation or corrosion, phototoxicity, and skin sensitization have been accepted by the FDA. For pharmaceuticals, acute toxicity testing is not recommended, and separate assessment of local toxicity is not generally recommended, so phototoxicity, and skin sensitization are the areas of most interest for topical effects. The Bovine Corneal Opacity and Permeability (BCOP) assay for serious irreversible ocular toxicity is already accepted; assays for minimal (or no) ocular toxicity are still being worked on, but will find utility when available. Sometimes use of these validated alternatives (e.g., the 3T3 photocytotoxicity assay and the LLNA) with pharmaceutical actives and formulations reveals that pharmaceuticals give many more false positives than were seen in the validation studies. An *in vitro* battery for assessment of skin sensitization may soon be available. Although it is not problematic for screens of some product areas to have substantial false positive rates, too many false positives are not desirable for pharmaceuticals.

Similar to the EPA, there is flexibility in the guidances, and the use of non-animal alternatives for pharmaceutical evaluation does not have to be formally validated by the OECD or validation organizations (e.g., ICCVAM) in order to be accepted for regulatory purposes. ICH working groups or FDA working groups can review data on alternatives and decide that a particular alternative is acceptable for its context of use.

CDER/FDA also has a drug development tool program[74] and drug development tools, such as biomarkers, are a big part of the program.[75] The program:

- provides a framework for scientific development and regulatory acceptance of biomarkers for use in drug development;
- facilitates integration of qualified biomarkers in the regulatory review process;
- encourages the identification of new and emerging biomarkers for evaluation and utilization in regulatory decision-making;
- supports outreach to relevant external stakeholders to foster biomarker development.

Biomarkers used to inform the FDA decisions are 'qualified' rather than 'validated'. Specifically the biomarkers for a particular context of use are qualified, rather than tests for the biomarkers being qualified. There is an FDA 2011 draft guidance relating to biomarkers, 'Guidance for Industry Use of Histology in Biomarker Qualification Studies.'[76] There is also a 2011 FDA guidance on biomarkers, 'E16 Biomarkers Related to Drug or Biotechnology Product Development: Context, Structure, and Format of Qualification Submissions.'[77]

The major challenge to the reduction of animal use in tests to inform the FDA's decisions is the need to assess various systemic toxicities (e.g., chronic toxicity, carcinogenicity, fertility, embryo fetal development). These toxicities are multistep and often involve systemic interactions, and currently are not possible to assess in non-animal tests. Thus, a battery of non-animal tests, rather than a single test, will likely be needed for the near future to assess what cannot always be assessed in humans. Consequences to human health from poor predictivity for toxicities from pharmaceuticals can be serious. The criteria for acceptance of alternative methods for regulatory decisions regarding human pharmaceutical use differ markedly from criteria for the use of alternatives for labeling of material for which the probability of human exposure is very low.

Similar to the logic described for the EPA, the animal test requirements for food, drugs, and cosmetics under the FDA are designed to mimic exposures and effects anticipated for humans using the exposures that humans are likely to encounter when they consume or use these products. Thus, FDA's guidance calls for studies that provide information about the absorption, distribution, and metabolism of the substances of concern, their excretion, and their ability to cause a wide range of effects of concern over time periods that mimic the periods of time that humans are expected to encounter. In addition, medical devices are evaluated for their ability to function with living tissue without harming the tissue (biocompatibility). Stainless steel and ceramic devices are known to be biocompatible with human tissues and require no animal testing. Devices comprised of new materials do require biocompatibility testing in animals.[78]

2.2.2.1 Cosmetics

The marketing of cosmetics is regulated within the Office of Cosmetics and Colors in the FDA's Center for Food Safety and Applied Nutrition pursuant to the Federal Food, Drug, and Cosmetic Act and the Fair Packaging and Labeling Act. In contrast to most other products regulated by the FDA, cosmetics are not subject to specific testing requirements or pre-market approval by the agency. Detailed information about how the FDA regulates cosmetics can be found at the FDA's website.[79] The process includes a Voluntary Cosmetic Registration Program where cosmetic companies are encouraged but not required to report the ingredients in their products and safety data, and any animal study data on these ingredients to the FDA.[80] The International

Cooperation on Cosmetic Regulation (ICCR) facilitates this activity.[81] ICCR is a voluntary partnership among the health authorities of Canada (Health Canada), Europe (DG-Enterprise), Japan (Ministry of Health, Labor, & Welfare), and the US (FDA), with participation and technical support from the cosmetics industry associations of the four jurisdictions. It provides a forum for discussions on alignment of cosmetics regulations within the member jurisdictions and seeks to maintain the highest level of global consumer protection, while minimizing barriers to international trade.[82]

2.2.2.2 Food Additives

The FDA regulates the safety of all food products, except for meat and poultry products whose regulation Congress has delegated to the US Department of Agriculture. The Federal Food, Drug, and Cosmetic Act requires that manufacturers and packagers of processed foods demonstrate the safety (i.e., 'reasonable certainty of no harm') of all chemical additives and/or other materials that come into contact with food prior to marketing. The Office of Food Additive Safety within the FDA's Center for Food Safety and Applied Nutrition provides pre-market review of the safety of food/color additives and food contact materials.[83]

FDA guidance for the toxicological assessment of food ingredients (also referred to as the Redbook) prescribes often extensive toxicological testing based on 'concern levels' as determined by chemical structure and cumulative human exposure.[84]

Commonly required study types include the following:

1 acute oral toxicity;
2 short-term (28 day) toxicity in rodents and non-rodents;
3 subchronic (90 day) toxicity in rodents and non-rodents;
4 chronic (1–2 year) toxicity in rodents and non-rodents;
5 mutagenicity and genotoxicity;
6 carcinogenicity (including possible *in utero* exposure phase);
7 reproductive toxicity;
8 developmental toxicity;
9 neurotoxicity;
10 immunotoxicity;
11 metabolism and pharmacokinetics;
12 human clinical and/or epidemiology.

Exceptions to the above testing requirements are provided under a 1958 amendment to the FFDCA for two broad groups of substances:

1 substances that the FDA or the US Department of Agriculture had determined safe for use in food prior to 1958;
2 ingredients generally recognized as safe (GRAS) based on a long history of use and/or published scientific evidence.[85]

2.2.2.3   Human Biologics

A biological product, as defined under the Public Health Service (PHS) Act, can include a 'virus, therapeutic serum, toxin, antitoxin, vaccine, blood, blood component or derivative, allergenic product, or analogous product … applicable to the prevention, treatment, or cure of a disease or condition of human beings.' The PHS Act requires that all human biologics be licensed, appropriately labeled, and proven 'safe, pure, and potent.' Responsibility for ensuring the safety and efficacy of vaccines, cellular and gene therapy products, human blood and blood products, allergenics, tissues and tissue products, and all other biologics intended for human use in the US rests with the FDA's Center for Biologics Evaluation and Research.

A company wishing to begin clinical trials of a biological product must submit an Investigational New Drug application to the FDA describing the product, its method of manufacture, and quality control tests for release.[86] In the case of a vaccine, manufacturers must demonstrate the following characteristics:

1. *Purity*, meaning that it is not contaminated with viable bacteria, viruses, or fungi.
2. *Safety*, meaning that it is not dangerous or harmful – which is usually determined by means of 'abnormal toxicity' or similar studies, in which groups of animals are injected with a vaccine and monitored for clinical signs of toxicity.
3. *Potency*, meaning that it is effective in preventing infection – which is usually determined by means of a 'challenge study,' in which groups of animals are first inoculated with a vaccine and then exposed to a virulent strain(s) of the organism against which the vaccine is intended to protect; animals are then monitored for clinical signs of the infectious disease in question.[87,88] *In vitro* assays of potency of vaccine batches are being worked on to reduce animal use.[89]

A further pre-marketing requirement involves three phases of clinical trials in human volunteers to demonstrate safety and efficacy in the species of ultimate interest. If successful, a Biologics License Application is submitted for review by FDA regulators, as well as the agency's independent Vaccines and Related Biological Products Advisory Committee.[90]

Even after a vaccine is licensed, the FDA may require manufacturers to submit the results of their own tests for potency, safety, and purity for each vaccine lot and/or samples of each vaccine lot to the agency for testing.

2.2.2.4   Human Pharmaceuticals

The Federal Food, Drug, and Cosmetic Act provides for an extensive pre-market approval process for all pharmaceutical products – including generic, over-the-counter, and prescription drugs – to ensure their safety and effectiveness for human use.[91]

Responsibility for the regulation of human drugs, therapeutic proteins, and monoclonal antibodies in the US rests with the FDA's Center for Drug Evaluation and Research.

US regulations[92] and guidance for pharmaceutical safety assessments have been harmonized with those of other major pharmaceuticals markets (i.e., Europe and Japan) under the auspices of the International Conference on Harmonization Use (ICH) of Technical Requirements for Registration of Pharmaceuticals for Humans.[93] Harmonized ICH guidelines (considered to be guidances in the US) for the safety assessment of human medicines call for a wide array of pre-clinical (animal) studies before a drug candidate is deemed safe for human clinical trials or marketing. Commonly recommended animal tests include the following:

1 safety pharmacology studies (e.g., cardiac, CNS, and respiratory);
2 toxicokinetics and pharmacokinetics in rodents and/or other species in which toxicity was studied;
3 subacute (14–28 day), subchronic (90 day), and/or chronic (90+ day) studies in rodents, and in dogs – the duration of which is dependent on the duration of a proposed clinical trial or clinical use;
4 reproduction segment I – fertility studies in rodents usually before large-scale clinical trials;
5 reproduction segment II – prenatal developmental toxicity, usually in rodents and rabbits and generally before large-scale clinical trials;
6 reproduction segment III – postnatal development in rodents for marketing approval; for some products segment III may be combined with segment II studies.
7 mutagenicity and clastogenicity studies; clastogenicity studies can be combined with repeated-dose toxicity studies;
8 carcinogenicity studies in rats and mice which may be transgenic for chronically used drugs;
9 triggered specialized studies (e.g., immunotoxicity in rodents, phototoxicity and pyrogenicity for intravenous products).

Drug candidates that exhibit favorable toxicological profiles in pre-clinical testing may be candidates for human clinical trials. This step requires the submission of an Investigational New Drug Application (IND) to the FDA, detailing the results of *in vitro* and animal pharmacology and toxicology studies, manufacturing information, clinical protocols and investigator information.[94] Contingent upon no objection from the FDA and an institutional human research ethics committee, as many as three phases of human trials may be undertaken. For pharmaceutical products determined to be safe and efficacious in human trials, the final step prior to US commercialization is the submission of a New Drug Application, which must include sufficient information for the FDA to determine that (1) a new drug is safe and effective for its proposed use, (2) the benefits outweigh the risks, and (3) the proposed labeling requirements and the manufacturing methods and controls are adequate.

Federal law also requires that all prescription and over-the-counter drugs marketed in the US satisfy the quality standards established by the US Pharmacopeia. Once on the market, there are post-market surveillance controls with which a manufacturer must also comply.

Nonclinical testing of new pharmaceutical candidates consumes many animals and is required for both the active medicinal ingredient(s) as well as any variants in a formulation or delivery system.[95] Because of the increased availability of alternative methods such as *in vitro* and 'omic' techniques in preclinical safety testing, more and more drug developers have been using alternative methods in the drug discovery phase to reduce various types of toxicity seen for candidate drugs for human testing. In addition, various types of exploratory INDs are now available from the FDA, which allow testing in humans from microdoses and up into the pharmacologic range, with much less nonclinical data.[96]

2.2.2.5 Medical Devices

The Federal Food, Drug, and Cosmetic Act broadly defines medical devices as including any 'instrument, apparatus, implement, machine, contrivance, implant, *in vitro* reagent, or other similar or related article, including any component, part, or accessory, which is ... intended for use in the diagnosis of disease or other conditions, or in the cure, mitigation, treatment, or prevention of disease, in man or other animals ...'[97]

Responsibility for ensuring the safety and effectiveness of medical devices lies with the FDA's Center for Devices and Radiological Health.

The FDA classifies medical devices based on the level of control necessary to assure the safety and effectiveness of the device.

1 *Class I* devices present minimal risk to the user, such as tongue depressors, bedpans, elastic bandages, examination gloves, and hand-held surgical instruments and other similar types of common equipment. Devices in Class I are subject only to general controls, which cover issues such as manufacturer registration, good manufacturing techniques, proper labeling, pre-market notification, and general reporting requirements.
2 *Class II* devices require additional special controls to assure safety and effectiveness. Examples include surgical and acupuncture needles, suture material, dental implants, infusion pumps, and X-ray machines.
3 *Class III* devices require pre-market safety and efficacy review and approval by the FDA because they are 'purported or represented to be for a use in supporting or sustaining human life or for a use which is of substantial importance in preventing impairment of human health.' Examples include heart valves, implanted cardiac pacemakers and cerebral stimulators, and bone and breast implants.

FDA regulations for pre-market approval applications require the submission of summaries of all 'nonclinical laboratory studies' as well as 'clinical investigations involving human subjects' undertaken to demonstrate the safety and efficacy of a device for human use.[98] However, specific testing requirements are not specified. Federal law also requires that all medical devices marketed in the US satisfy the quality standards established by the US Pharmacopeia.[99] Once on the market, there are post-market surveillance and reporting requirements with which a manufacturer must also comply.

2.2.2.6   Nutritional Supplements

Vitamins, minerals, herbs, or other botanicals, amino acids, and substances such as enzymes, organ tissues, glandulars, and metabolites are regulated under the Dietary Supplement Health and Education Act of 1994 (DSHEA) by the FDA's Office of Nutritional Products, Labeling, and Dietary Supplements, a division of the Center for Food Safety and Applied Nutrition.[100] In contrast to most other products regulated by the FDA, nutritional supplements are not subject to specific safety/efficacy testing requirements or pre-market approval by the agency. Responsibility for ensuring that a nutritional supplement is safe rests with the manufacturer, who is only obligated to notify the FDA prior to marketing a supplement containing a new dietary ingredient. The FDA is responsible for: monitoring the market for potentially unsafe products and/or those that make false or misleading label claims; investigating reports of adverse effects; and where appropriate, taking action to remove unsafe products from the market.

2.2.2.7   Veterinary Pharmaceuticals

The marketing of veterinary pharmaceuticals and medicated livestock feeds in the US is regulated by the FDA's Center for Veterinary Medicine pursuant to the Federal Food, Drug, and Cosmetic Act (FFDCA). FFDCA requires that animal drugs and feeds be safe and effective and that edible animal products derived from treated animals be free of unsafe residues.

The FDA is responsible for the pre-market review and approval of all veterinary pharmaceuticals.[101] US requirements for veterinary pharmaceutical safety assessments have been harmonized with those of other major pharmaceuticals markets (i.e., Europe and Japan) under the auspices of the International Conference on Harmonization of Technical Requirements for Regulation of Veterinary Medicinal Products (VICH).[102] Some VICH guidelines call for a wide array of animal studies, including the following:

1. subchronic (90-day) and chronic (2-year) toxicity in rodents and/or dogs;
2. reproductive toxicity in two or more generations of rodents;
3. developmental toxicity in rodents and/or rabbits;
4. genotoxicity studies of at least three varieties
5. testing for effects on human intestinal flora;
6. pharmacological effects;
7. immunotoxicity;

   8. neurotoxicity;
   9. carcinogenicity studies in rats and mice;
   10. triggered 'special' studies.

Data from these studies are submitted to the FDA in the form of an Investigational New Animal Drug application, on the basis of which further testing may be required to demonstrate: safety and effectiveness of the candidate drug in the target species, safety for human consumption of drug residues in food derived from treated animals, and the effect of animal drugs on the environment.[103]

For veterinary pharmaceuticals determined to be safe and efficacious, the final step prior to US commercialization is the submission of a New Animal Drug Application, which must include sufficient information for the FDA to determine that a new drug is safe and effective for its proposed use, that the benefits outweigh the risks, and that the proposed labeling requirements and the manufacturing methods and controls are adequate.[104] Once on the market there are post-marketing surveillance controls with which a manufacturer must also comply.

2.2.2.8   National Center for Toxicological Research (NCTR)

Much of the FDA's intramural research activity is centralized within the National Center for Toxicological Research, which is mandated to conduct 'innovative, integrative research to support and anticipate the FDA's current and future regulatory needs.' The NCTR has established a number of Centers of Excellence to further the development and application of new (-omic and *in silico*) technologies into FDA programs.

2.2.2.9   Adoption of Alternatives

The FDA has undertaken a number of initiatives that are germane to the 3Rs. Perhaps the most far-reaching of these is the Critical Path Initiative – a 'national effort to modernize the scientific process through which a potential human drug, biological product, or medical device is transformed from a discovery or 'proof of concept' into a medical product.'[105] The Critical Path Initiative aims, among other things, to develop a 'better safety toolkit ... that can more reliably and more efficiently determine the safety of a new medical product' and was the impetus for the creation of Centers of Excellence for *in silico* and –omic technologies under NCTR. In addition, informatics and computational safety analysis staff in the FDA's Center for Drug Evaluation and Research have helped pioneer many of the *in silico* models now widely used within the agency. The FDA has signed a memorandum of understanding with the EPA, the NIEHS, and the NCGC on High Throughput Screening, Toxicity Pathway Profiling, and Biological Interpretation of Findings.[106]

In August 2011 the FDA developed a 'Strategic Plan for Regulatory Science.'[107] It focuses on eight key areas.

- Modernizing toxicology – the study of chemical, biological or physical agents that can be harmful – and improving the ability of tests, models, and measurements to predict product safety issues. This includes, where feasible, the development of new methods that could reduce or replace animal testing.
- Crafting new tools and approaches for the development of personalized medicine – getting the right medicine to the right person at the right time.
- Supporting new and improved manufacturing methods by researching how new technologies affect product safety, effectiveness, and quality.
- Ensuring that the FDA is ready to evaluate innovative and emerging technologies with the necessary expertise and infrastructure.
- Expanding and improving the FDA's information technology infrastructure and the application of those resources to support sophisticated analyses of data.
- Implementing the prevention-focused food-safety system mandated by the Food Safety Modernization Act.
- Speeding the development of safe and effective medical countermeasures to protect against threats to US health and security, such as chemical, biological, or nuclear threats or naturally occurring infectious disease outbreaks. Such countermeasures include drugs, vaccines, diagnostic tests, and personal protective equipment.
- Developing a communications strategy that will help the FDA adapt to rapidly evolving technologies that are changing how people receive and share information.[108]

A key component of the plan is to strengthen researchers' ability to detect chemical or biological substances that could be life threatening, or even fatal, before products containing them hit the market. Similar to the 'Toxicity Testing in the 21st Century' efforts described for the EPA, the FDA's goal is to use advances in biology and high speed computing to develop better ways to test drugs to more accurately predict important health risks. A key component of this is the FDA's collaboration with the Defense Advanced Research Projects Agency to create the 'human on a chip,'[109] a technology that could revolutionize drug testing by using human cells to create a miniature version of 10 different organ systems that mirror their function in a full-size human body. Its purpose is to allow researchers to test new drugs safely and more accurately without having to use animals as test subjects. Early success by the MIT Department of Biological Engineering[110] and the Wyss Institute for Biologically Inspired Engineering at Harvard University[111] is the development of a 'Lung-on-a-Chip' to mimic the physical and chemical behaviors of a living, breathing human lung.[112]

2.2.3 Other Agencies Involved in Regulatory Testing to Inform Decisions

2.2.3.1 *Department of Health and Human Services*

The Department of Health and Human Services (HHS) contains several institutes and centers that support regulatory efforts such as the Agency for Toxic

Substances and Disease Registry, the Centers for Disease Control and its National Institute for Occupational Safety and Health, the National Institute of Environmental Health Sciences, and the National Institute for Allergy and Infectious Diseases among others, including the FDA. Their efforts range from human surveillance, through vaccine safety evaluation and understanding the basis for disease to applying such knowledge for test development and characterization for use by decision makers. However, because their Congressional mandate is primarily to conduct research and knowledge-generation activities, and not to regulate, other than the FDA they will not be considered further in this chapter.

That said, it should be noted that some of these efforts, such as the conduct of chronic 2-year bioassays in rodents under the National Toxicology Program within NIEHS and vaccine safety testing program under NIAID, do inform regulatory decisions and are often conducted to meet a regulatory need. Nonetheless, given the focus of this chapter on regulatory testing to inform decisions it was decided to focus on the parts of government charged by Congress to serve a regulatory function and on the testing conducted by those agencies directly or as a result of the regulatory requirement that they promulgate. Bear in mind that HHS contains institutes and centers that conduct key research bearing on the use of testing used for regulatory decision-making and in the development of more predictive and efficient alternatives to conventional tests. Interested readers are encouraged to explore the activities of the efforts underway in HHS more fully.

2.2.4   Federal Departments and Independent Agencies

Congress has established several cabinet level departments, as well as several independent federal regulatory agencies, authorized to develop specific areas of expertise and to oversee aspects of commerce in their area of expertise such as banking, interstate commerce, consumer product safety, foods, environment, and the workplace, etc. Those established by Congress and charged with a mission involving regulatory testing are briefly described in this chapter.

2.2.4.1   Consumer Product Safety Commission (CPSC)

The US Consumer Product Safety Commission (CPSC) is an independent federal regulatory agency[113] created by Congress in 1972 pursuant to the Consumer Product Safety Act (CPSA). CPSC also administers several other laws including the Federal Hazardous Substances Act (FHSA) that can be found on the CPSC web page.[114]

CPSC is governed by a committee of Commissioners appointed by the President while its day to day operations are run by career civil servants. The CPSC mandate is to 'protect the public against unreasonable risks of injuries and deaths associated with consumer products.' Approximately 15,000 types of consumer products are regulated by the CPSC, including items such as children's toys and furniture, mattresses, household appliances, and some chemical components of common household products.[115]

The CPSC addresses its mandate to protect the public from hazardous consumer products by developing voluntary and mandatory standards, enforcing mandatory standards, banning and/or recalling unsafe consumer products, and educating consumers about product safety and responding to their inquiries. Another role of the CPSC involves 'conducting research on potential product hazards.'

The CPSC oversees two types of product safety standards: government-mandated and voluntary. The products covered under each type of standard are listed along with other pertinent information and documents on the CPSC website. The CPSC provides technical support for voluntary standards, but three external standards organizations coordinate and/or develop over 90% of the voluntary safety standards: the American National Standards Institute (ANSI), ASTM International, and Underwriters Laboratories (UL). 'These standards are not endorsed by the CPSC; however, failure to meet a voluntary safety standard could lead to a substantial product hazard determination by the Commission and result in a recall.'

In December 2012 CPSC issued a statement in the Federal Register that 'Overall, the Commission prefers test methods that reduce stress and suffering in test animals and that use none or fewer animals while maintaining scientific integrity.[116] The Commission's policy is to strongly support the use of validated alternatives to animal testing ...' The Commission's animal testing web page provides additional information about the alternative approaches acceptable to CPSC.[117]

In essence the Commission notes that the Federal Hazardous Substances Act (FHSA) does not require that animal testing be used for hazard evaluation although animal testing is an option. The December 2012 policy is to find alternatives to traditional animal testing and the Commission directed the CPSC staff to strongly encourage the use of scientifically validated alternatives to animal testing and the use of existing information, including expert opinion, prior human experience, and prior animal testing results, and particularly tests validated by the Interagency Committee for the Validation of Alternatives to Animal Testing (ICCVAM)[118] in the determination of hazard. Data from hazard tests not previously approved by the CPSC will be considered on a case-by-case basis. Data generated using Organisation for Economic Co-operation and Development (OECD) Guidelines were highlighted as appropriate for consideration by the CPSC in making a hazard determination as a result of the Mutual Acceptance of Data (MAD) agreement.[119] MAD allows OECD member countries to mutually accept data that comply with OECD Test Guidelines and Principles of Good Laboratory Practice.

2.2.4.2   US Department of Agriculture (USDA)

The US Department of Agriculture (USDA) is a cabinet-level department with the broad responsibility of 'protecting American agriculture.'[120] Within the context of regulatory testing to inform decisions, the USDA Center for Veterinary Biologics (CVB), a division of the Animal and Plant Health Inspection

Service (APHIS), 'regulates veterinary biologics (vaccines, bacterins, antisera, diagnostic kits, and other products of biological origin) to ensure that the veterinary biologics available for the diagnosis, prevention, and treatment of animal diseases are pure, safe, potent, and effective.'[121]

The Virus Serum Toxin Act (21 USC § 151–159 *et seq.*) proscribes the preparation or sale of 'any worthless, contaminated, dangerous, or harmful virus, serum, toxin, or analogous product intended for use in the treatment of domestic animals' and establishes a legal basis for the licensing and inspection of veterinary biologics and for the USDA's creation and enforcement of implementing regulations.[122]

USDA regulations concerning biological products are codified under Title 9 of the Code of Federal Regulations (9 CFR § 101–123).[123] Especially significant are the 'standard requirements' laid out in Part 113, which prescribe extensive *in vivo* toxicological testing for each batch product and which stipulate that '[n]o biological product shall be released prior to the completion of tests prescribed ... for the product to establish the product to be pure, safe, potent, and efficacious.' In addition, the regulations permit APHIS itself to conduct duplicative purity, safety, potency, and/or efficacy studies for confirmation purposes.[124]

The following are the principal categories of animal tests that may be required:

1 target species safety test (e.g., dog, cat, mouse, calf, swine, sheep);
2 live bacterial vaccines (e.g., *Brucella*, *Anthrax*, *Pasteurella*, *Chlamydia*);
3 inactivated bacterial products (e.g., *Leptospira*, *Clostridium*, *Tetanus*)
4 live virus vaccines (e.g., avian pox, blue tongue, measles, Newcastle disease);
5 killed virus vaccines (e.g., canine distemper, rabies, parvovirus);
6 antibody products (e.g., tetanus and clostridium antitoxins).

USDA has published an extensive series of Veterinary Service Memorandums and public notices that provide companies with additional detailed guidance concerning product- and agent-specific testing, licensing, and other regulatory requirements.[125]

2.2.4.3   Department of Transportation (DOT)

The US Department of Transportation (DOT) is a cabinet-level department whose mandate is to 'serve the United States by ensuring a fast, safe, efficient, accessible and convenient transportation system that meets our vital national interests and enhances the quality of life of the American people, today and in the future.'[126] The DOT is made up of numerous sub-organizations, each responsible for a distinct form of national transportation (e.g., aviation, highway, railway, maritime).

The DOT's Pipeline and Hazardous Materials Safety Administration oversees the safety of nearly one million daily shipments of hazardous material.

The Federal Hazardous Materials Transportation Law (49 USC § 5101 *et seq.*) states: 'The Secretary [of Transportation] shall designate a material or a group or class of materials as hazardous when the Secretary decides that transporting the material in commerce in a particular amount and form may pose an unreasonable risk to health and safety or property. The Secretary shall issue regulations for the safe transportation of hazardous materials.'[127]

US Hazardous Material Regulations are codified in Title 49, Parts 100–185 of the Code of Federal Regulations.[128] DOT requirements with a bearing on regulatory safety testing fall within Part 173 concerning the classification, labeling, and packaging of hazardous substances destined for intrastate, interstate, and global commerce.[129] DOT's hazard classification and labeling system closely mirrors that of the United Nations (UN) Globally Harmonized System (GHS).[130]

The GHS identifies nine classes of hazards that must be addressed on packing labels. Most deal with the inherent physicochemical properties of a substance and therefore are unlikely to involve any degree of *in vivo* testing; however, the following GHS hazard classes may involve animal testing to varying degrees.

1 *Poisonous Materials (Class 6, Division 6.1; 49 CFR § 173.132)*: 'Presumed to be toxic to humans because it falls within any one of the following categories when tested on laboratory animals ...' *In vivo* tests cited in this section are oral and dermal LD_{50} (median lethal dose) studies in rats and rabbits and inhalation LC_{50} (median lethal concentration) studies in rats.

2 *Infectious Substances (Class 6, Division 6.2; 49 CFR § 173.134)*: 'A material known or reasonably expected to contain a pathogen ... (i) *Category A*: An infectious substance in a form capable of causing permanent disability or life-threatening or fatal disease in otherwise healthy humans or animals when exposure to it occurs.'

3 *Corrosive Materials (Class 8; 49 CFR § 173.136)*: 'A liquid or solid that causes full thickness destruction of human skin at the site of contact within a specified period of time ... [T]he packing group must be determined using data obtained from tests conducted in accordance with the 1992 OECD Guideline for Testing of Chemicals, Number 404, 'Acute Dermal Irritation/Corrosion.'[131]

The DOT has authorized a regulatory exemption authorizing the use of alternative *in vitro* test methods in lieu of *in vivo* methods as a basis for classifying acute dermal irritation and corrosion for certain types of materials.[132]

2.2.4.4   Occupational Safety and Health Administration (OSHA)

The Occupational Safety and Health Administration (OSHA) is housed within the US Department of Labor. Congress created the OSHA to assure safe and healthy conditions for working men and women by setting and enforcing standards and providing training, outreach, education, and compliance assistance.[133] OSHA Regulatory Standards codified under Title 29 of the Code of Federal Regulations (29 CFR § 1910.1200) require 'that the hazards of all chemicals produced or imported are evaluated, and that information concerning their hazards is transmitted to employers and employees.'[134]

Health hazards identified as being of particular concern to the OSHA include 'chemicals that are carcinogens, toxic or highly toxic agents, reproductive toxins, irritants, corrosives, sensitizers, hepatotoxins, nephrotoxins, neurotoxins, agents which act on the hematopoietic system, and agents which damage the lungs, skin, eyes, or mucous membranes.'

The regulations further state that: 'If no relevant information is found for any given category on the material safety data sheet, the chemical manufacturer, importer or employer preparing the material safety data sheet shall mark it to indicate that no applicable information was found.' Thus, the general requirement that chemical hazards be 'evaluated' does not obligate manufacturers to conduct new testing of any kind to address the toxicological endpoints listed above (i.e., 'the *available results* of toxicological testing in animal populations shall be used to predict the health effects that may be experienced by exposed workers').

Hazardous properties are classified according to the United Nations' Globally Harmonized System (GHS) and are reported in the form of Material Safety Data Sheets (MSDS), International Chemical Safety Cards (ICSC), or other similar documents that among other things list health hazards, including signs and symptoms of exposure, and any medical conditions recognized as being aggravated by exposure and whether the chemical is listed in the National Toxicology Program's *Report on Carcinogens* or International Agency for Research on Cancer monographs.

OSHA regulations state: 'Chemical manufacturers, importers, and employers evaluating chemicals are not required to follow any specific methods for determining hazards, but they must be able to demonstrate that they have adequately ascertained the hazards of the chemicals produced or imported in accordance with the criteria set forth' in regulations.[135] However, the regulations go on to note that '[i]n vitro studies alone generally do not form the basis for a definitive finding of hazard under the HCS since they have a positive or negative result rather than a statistically significant finding.'[136]

2.2.5   Interagency Efforts

2.2.5.1   *Interagency Coordinating Committee for the Validation of Alternative Methods (ICCVAM)*

The ICCVAM Authorization Act (Public Law 106-545, 42 USC 285l-5) was passed by Congress in 2000.[137] NIEHS was designated to manage the committee and established the National Toxicology Program (NTP) Interagency Center for the Evaluation of Alternative Toxicological Methods (NICEATM)[138] to do so. Congress designated the 15 US Federal regulatory and research agencies that conduct or use animal test data to make up the ICCVAM, and a representative is appointed by the head of each agency head to serve on the committee. NICEATM provides administrative, technical, and scientific support to ICCVAM.

2.2.5.2   US Animal Welfare Oversight

The US system of regulation and oversight of animal care and use for testing and research consists of patchwork of legislative, regulatory, and guidance tools independently established and overseen by various federal agencies and independent scientific bodies.

2.2.5.2.1   Animal Welfare Act.   The Animal Welfare Act (AWA; 7 USC. § 2131–2156),[139] enacted in 1966 and amended periodically since then, serves as the primary federal law governing the care and use of at least some of the animals used in scientific experiments. The Act and its regulations set minimum standards for research facilities and animal dealers and transporters. The Act is enforced by the Animal and Plant Health Inspection Service of the US Department of Agriculture (USDA/APHIS).[140]

The USDA has promulgated a series of implementing regulations providing further details concerning institutional requirements the AWA.[141] USDA inspectors are required to perform at least one unannounced annual inspection at each facility that uses AWA-listed species. USDA personnel may also respond to public complaints regarding the care of regulated animals and are authorized to temporarily suspend a research facility's license in cases of clear non-compliance with the AWA or its regulations. The AWA regulations require that principal investigators consider alternatives to procedures that may cause more than momentary or slight pain or distress to animals, and to provide a written narrative description of methods and sources (e.g. the Animal Welfare Information Center, AWIC), used to determine that alternatives were not available.[142]

2.2.5.2.2   Health Research Extension Act.   The Health Research Extension Act of 1985 (Public Law 99-158)[143] requires that guidelines for the proper care and treatment of animals being used for research be developed. It also requires among other things that there be a justification for why animals are needed for the study, a requirement that an animal care committee be established '… at each entity which conducts biomedical and behavioral research with funds provided under this Act…', that each applicant for funding assure that they meet these requirements and that the personnel to be involved in these studies will be instructed '…in the humane practice of animal maintenance and experimentation…'

While federal inspections are not required, facilities are required to report deficiencies to the NIH Office of Laboratory Animal Welfare (OLAW) for review.[144]

2.2.5.2.3   Public Health Service Policy on Humane Care and Use of Laboratory Animals.   The Public Health Service (PHS) policy addresses the requirements of the Health Research Extension Act and applies to all scientific research involving vertebrate animals that is conducted or funded by any PHS agency. In addition, OLAW has direct oversight function for these requirements. OLAW also works to promote education and voluntary compliance by animal users through regional workshops and conferences.

PHS policy requires that federally funded laboratories follow the guidance contained in the *National Research Council Guide for the Care and Use of Laboratory Animals* (see Additional Resources, ref. 8). It is a 125-page publication prepared by the National Research Council (NRC) Institute for Laboratory Animal Resources to provide specific guidance and advice to research facilities in the following aspects of animal care and use:

1 animal housing; husbandry; social and behavioral enrichment;
2 animal monitoring; veterinary care; personnel qualifications and training; occupational health and safety;
3 animal procurement and transportation; preventive medicine; surgery; pain avoidance and management; euthanasia.

2.3   European Union: Programs & Policies

The European Union (EU) is a multinational organization focusing on the common interests of Europe. It was established as a means to govern the affected nation states for the good of all in that region. The resultant EU political system is unique in the world: a confederation of 27 countries operating in 23 official languages through a 'supranational' system of governance, under which member countries delegate some of their decision-making powers to the following EU institutions:[145]

1 Council of Ministers;
2 European Parliament;
3 European Commission;
4 Agencies.

Most EU legislation is a product of proposals made by the European Commission (EC), which are then subject to amendment and joint adoption by the Council of Ministers and the European Parliament, according to a 'co-decision' procedure. The acts of the council can take the form of regulations, directives, decisions, common actions or common positions, recommendations, or opinions. The council can also adopt conclusions, declarations, or resolutions.

The European Commission (EC) is an independent civil service with approximately 24,000 employees responsible for acting in the interests of Europe as a whole. Led by 27 commissioners, each appointed by one of the EU's member countries, the EC is responsible for drafting proposals for new

European laws, which are then presented to the European Parliament and the Council for approval and enactment.[146] While there are similarities between the US and the EU, a key difference is that while the laws passed by the US Congress apply to all states as soon as the law goes into effect, in the EU each country must also approve the laws adopted by the EU Council and Parliament.

The EC is divided into a number of Directorate Generals (DGs), each responsible for managing and/or providing technical support to different EU regulatory sectors. DGs with the greatest influence on regulatory testing to inform decisions are:

1 Enterprise Europe Network (pharmaceuticals, biologics, cosmetics, and consumer products);[147]
2 Environment (biocides and revision of the EU animal experimentation directive);[148]
3 Health & Consumer Protection (agrochemicals, food additives, GMOs, nanomaterials, etc.);[149]
4 Joint Research Centre (EURL, ECVAM, and European Chemicals Bureau);[150]
5 Research and Innovation (research policy and extramural funding).

A critical adjunct to the policymaking function of the DGs is the practical implementation of EU policy and legislation. This includes the development of testing guidelines or related guidance concerning acceptable approaches to satisfying information requirements prescribed under EU legislation. In certain regulatory sectors, this implementation function is increasingly being performed by several independent agencies, including;

1 European Chemicals Agency (ECHA);[151]
2 European Food Safety Authority (EFSA);[152]
3 European Medicines Agency (EMA).[153]

2.3.1   EU 3 R's Centers and Initiatives

A number of independent academic, government, and charitable centers also operate in the EU with more general 3Rs mandates that are summarized in Table 2.1 in alphabetical order.

2.4   Other International Regulatory/Research Agencies and Interagency Bodies

The US Department of State recognizes 195 independent nations in the world.[154] Each nation exerts sovereignty, controlling the geographic area within its borders and implementing a variety of laws pertaining to that geographic area. These laws vary from country to country, and it is not possible to

Table 2.1   EU 3Rs centers and initiatives.

Centers in the EU with 3Rs mandates	Tasks
3R Research Foundation Switzerland[162]	• Funds research • Publishes periodic 3Rs information bulletins online • Provides training
Doerenkamp-Zbinden Foundation for Animal Free Research[163]	• Promotes and rewards 'exceptional achievements in animal protection in biomedical research' • Funds research • Funds endowed chair in evidence-based toxicology
European Consensus Platform for Alternatives (ECOPA)[164]	• EU-wide umbrella group charged to facilitate exchange of scientific information, expertise and experience across nations • Organizes conferences and seminars • Supports scientific and educational initiatives • Conducts workshops and training
European Partnership for Alternative Approaches to Animal Testing (EPAA)[165]	• Voluntary collaboration between the European Commission, European trade associations, and companies from seven industry sectors. • Accelerate development, validation and acceptance of alternative approaches in regulatory testing
Fund for the Replacement of Animals in Medical Experiments (FRAME)[166]	• Registered charity to advance alternatives to animal experiments in the United Kingdom • Ultimate aim is elimination of the need to use laboratory animals in any kind of procedure • Conducts research • Publishes the peer-reviewed bimonthly journal *Alternatives to Laboratory Animals* (ATLA) • Campaigns and lobbies
German Center for the Documentation and Evaluation of Alternatives to Animal Experiments (ZEBET)[167]	• Government center mandated to 'bring about the replacement particularly of legally prescribed animal experiments with alternative test methods' • Documents and assesses 3Rs methods • Facilitates their recognition both nationally and internationally • Develops and validates *in vitro* and other alternative toxicological test methods
Dr Hadwen Trust for Humane Research[168]	• Supported by charitable organizations • Funds research • Promotes the 3Rs • Educates
International Society for *In Vitro* Methods (INVITROM)[169]	• Promotes development, application, and acceptance of *in vitro* models in the biomedical research • Informs regulatory bodies • Workshops and symposia • Working parties

Table 2.1   (*Continued*)

Centers in the EU with 3Rs mandates	Tasks
Netherlands Center for Alternatives to Animal Use (NCA)[170]	• Housed in the Utrecht University Faculty of Veterinary Medicine • Focal point to coordinate research and disseminate information on alternatives to animal experiments • Works closely with the Netherlands Organization for Health Research and Development • Educational outreach
Swedish Fund for Research Without Animal Experiments[171]	• Promotes the use of alternatives in both basic and applied research • Funds biological and social science research projects • Provides training
UK National Center for the Replacement, Refinement and Reduction of Animals in Research (NC3 Rs)[172]	• Funds research • Resource center • Inform culture shift • Work to change international regulations

present in this chapter an exhaustive summary of those dealing with regulatory testing to inform decisions. However, because countries share many of the same issues bearing on public health, food safety, transportation safety, worker safety, environmental protection, and the like they often cooperate to share expertise and foster harmonization of their respective laws to benefit from the work of other nations, to promote trade, and to provide other benefits to their citizens. Thus, the focus of this section will be on multinational intergovernmental organizations such as the Organisation for Economic Co-operation and Development (OECD), the International Conference on Harmonization of Technical Requirements for Registration of Pharmaceuticals for Human Use (ICH) and Veterinary Use (VICH), the United Nations (UN), and the World Health Organization (WHO), with emphasis on their guidelines and programs relevant to toxicity/safety testing (Table 2.2).

The major efforts of Japan will be highlighted as well, because, in addition to Europe and the US, it is a major user of information from animal regulatory testing. As with the previous section, the material in this section describing international activities is often derived from the AltTox Toxicity Testing Resource Center. The material is used with permission.

In addition to worldwide global harmonization activities, there are a number of regional organizations working to harmonize drug regulations.[155] These efforts include: ASEAN (Association of South-East Asian Nations),[156] CAN (Andean Community),[157] CADREAC (The Collaboration Agreement of Drug Regulatory Authorities in European Union Associated Countries),[158] the European Union, Gulf Cooperation Council (GCC),[159] MERCOSUR (Southern Common Market),[160] and the Southern African Development Community (SADC).[161]

Table 2.2 Multilateral organizations.

Multinational organizations with regulatory testing mandates	Tasks
Organisation for Economic Co-operation and Development (OECD)	• Intergovernmental organization with 34 member countries • Develops policy and harmonization on a broad range of topics of international economic importance • Evolved from the US Marshall Plan to reconstruct Europe after WWII • Mission to support economic growth, boost employment, raise living standards, maintain financial stability, assist other countries' economic development, and contribute to growth in world trade • Major work areas employment / education / social welfare, economy, environment / sustainable development, finance, governance, and innovation • Member countries are relatively affluent and democratic • Engages with civil society organizations, including the Business and Industry Advisory Committee to the OECD (BIAC) and the Trade Union Advisory Committee to the OECD (TUAC), and maintains official relationships with the International Council for Animal Protection in OECD Programmes (ICAPO), an NGO officially recognized by the OECD that represents animal protection interests within OECD's test guidelines program • OECD Guidelines for the testing of chemicals provide guidance on the key toxicity tests that are used worldwide
International Conference on Harmonization of Technical Requirements for Registration of Pharmaceuticals for Human Use (ICH)[173]	• Purpose to reduce or avoid duplicate testing for new medicine research and development • Harmonize technical requirements for human drug product registration • Members consist of one regulatory authority and one industry trade association each from the US, the EU, and Japan • World Health Organization, European Free Trade Association, and International Federation of Pharmaceutical Manufacturers & Associations serve as observers
The International Cooperation on Harmonization of Technical Requirements for Registration of Veterinary Products (VICH)[174]	• Goal is to harmonize technical requirements for veterinary product registration between the US, the EU, and Japan • Establish and monitor harmonized regulatory requirements for veterinary medicinal products

Table 2.2 (*Continued*)

Multinational organizations with regulatory testing mandates	*Tasks*
	• Provide a basis for wider international harmonization of registration requirements
	• Monitor and maintain VICH guidelines
	• Ensure efficient processes for maintaining and monitoring consistent interpretation of data requirements
	• Provide technical guidance enabling response to significant emerging global issues and science that impact on regulatory requirements within the VICH regions
	• 10 VICH Expert Working Groups (EWG): Quality, Safety, Ecotoxicity, Good Clinical Practice (GCP), Antihelmintics, Pharmacovigilance, Biologicals Quality Monitoring, Antimicrobial Resistance, Target Animal Safety, and Metabolism and Residue Kinetics
United Nations (UN)[175]	• Almost all nations belong to the UN, which currently has 193 member states
	• Purpose to maintain international peace and security; to develop friendly relations among nations; to cooperate in solving international problems and in promoting respect for human rights; to be a center for harmonizing the actions of nations
	• 16 specialized agencies, established as autonomous organizations, including the WHO, the Food and Agriculture Organization (FAO), and the UN Industrial Development Organization (UNIDO)
	• Plays an important role in the harmonization of standards and technical regulations among member countries
	• Certain UN committees provide guidance on acute health effects and hazards related to chemicals and the transport of goods between countries
	• Globally Harmonized System of Classification and Labeling of Chemicals (GHS) is a key product that addresses classification of chemicals by types of hazard and proposes harmonized hazard communication elements, including labels and safety data sheets
	• Developed to ensure that information on physical hazards and toxicity from chemicals be available to enhance the protection of human health and the environment during the handling, transport and use of these chemicals
	• GHS provides a basis for harmonization of rules and regulations on chemicals at national, regional, and worldwide level, an important factor also for trade facilitation
	• United Nations Environment Programme (UNEP) promotes sound management of chemicals globally including Persistent Organic Pollutants (POPs) and specific programmes and information for mercury, lead, and cadmium

- UN Model Regulations on the Transport of Dangerous Goods provide guidance to governments and international organizations on the safe transport of hazardous materials
- United Nations Environment Programme's (UNEP) 'Chemicals in Products' project established to help implement UN's Strategic Approach to International Chemicals Management (SAICM), an international policy framework to foster the sound management of chemicals

World Health Organization (WHO)[176]

- Specialized health agency of the United Nations
- Objective is 'the attainment by all peoples of the highest possible level of health'
- Mandated to 'develop, establish and promote international standards with respect to food, biological, pharmaceutical and similar products'
- Provides expertise and guidance on these issues through the development and implementation of international standards and guidelines for the safety, quality, and efficacy of medicines
- Key task to 'help countries consider the implications of the relevant harmonization agreements' such as those of the ICH (an organization that does not include all of the developing countries)

International Programme on Chemical Safety (IPCS)[177]

- Joint program between the WHO, the International Labour Organization (ILO), and the United Nations Environment Programme (UNEP), and is administered by the WHO
- Primary goals to establish the scientific basis for safe use of chemicals and strengthen national capabilities and capacities for chemical safety
- Key work areas are: chemicals assessment; methods for chemicals assessment; chemicals in food; poisons information, prevention, and management; chemicals incidents and emergencies; and capacity building
- Prepares risk assessments on specific chemicals – documents especially useful to developing countries
- Products include: concise international chemical risk assessment documents; environmental health criteria; health and safety guides; international chemical safety cards; pesticide data sheets; and WHO recommended classification of pesticides by hazard
- Plays an international role in developing and harmonizing hazard and risk assessment methods

Table 2.2 (*Continued*)

Multinational organizations with regulatory testing mandates	*Tasks*
Inter-Organization Programme for the Sound Management of Chemicals (IOMC)[178]	• Purpose to strengthen cooperation and increase coordination in the field of chemical safety • Composed of eight Participating Organizations, and one observer organizations (the United Nations Development Programme (UNDP); Food and Agriculture Organization (FAO); International Labour Organization (ILO); Organisation for Economic Co-operation and Development (OECD); United Nations Environment Programme (UNEP); United Nations Industrial Development Organization (UNIDO); United Nations Institute for Training and Research (UNITAR); World Health Organization (WHO); United Nations Development Program (UNDP); World Bank • Key work areas of the IOMC are risk reduction, knowledge and information, governance, capacity-building and technical cooperation, and illegal traffic • Developed Strategy for Strengthening National Chemicals Management Capacities to assist countries in strengthening their national chemicals management capacities throughout the lifecycle
Codex Alimentarius Commission[179]	• In WHO's Food and Agriculture Organization (FAO) • Develops harmonized standards, guidelines, and other recommendations to protect consumers regarding the quality and safety of food • Activities also ensure fair food trade practices and harmonization of international food standards
International Pharmacopoeia[180]	• The WHO Expert Advisory Panel on the International Pharmacopoeia and Pharmaceutical Preparations conducts the activities of the International Pharmacopoeia • Collaborates with industry and other organizations • *International Pharmacopoeia* provides a collection of monographs for pharmaceutical substances' active ingredients and excipients, dosage forms, radiopharmaceuticals, methods of analysis, and test reagents • Simple quality and analytical methods are adopted when possible, so the methods can be carried out by all countries • Can be used by WHO Member States, either as reference material or incorporated into national legislation by Member States as their pharmacopoeial requirements

	• Has no legal status unless incorporated into a nation's legislation • WHO also provides a list of national and regional pharmacopoeias, the *Index of Pharmacopoeias*
International Agency for Research on Cancer[181]	• Part of the World Health Organization • Mission to coordinate and conduct research on the causes of human cancer [and] the mechanisms of carcinogenesis, and to develop scientific strategies for cancer control • Research focused on the relationship of human cancer to man's environment • Not involved in policy or legislation regarding carcinogens
Quality, Safety and Efficacy of Medicines (QSM)[182]	• Part of WHO's Medicines Programme • Involved in quality assurance of medicines, regulation and legislation, and safety and efficacy of medicines • Activities are aimed at strengthening national drug regulatory agencies (DRAs) in developing countries • Provide relevant expertise and technical assistance
Interagency Pharmaceutical Coordination Group[183]	• Pharmaceutical advisors to leading world organizations such as the WHO, the World Bank, and the United Nations make up the Interagency Pharmaceutical Coordination (IPC) group • Purpose to coordinate pharmaceutical industry policy advice provided to countries as well as provide coordinated advice on the development of interagency technical documents
International Conference of Drug Regulatory Authorities[184]	• Fora for the drug regulatory authorities of Member States to work together to prioritize regulatory issues, exchange information, and discuss collaborative approaches • Important role in international harmonization efforts for drug regulations, and quality, safety, and efficacy standards
International Council for Animal Protection in OECD Programmes (ICAPO)[185]	• Non-governmental organization established in 2002 to provide formal representation of animal protection interests in certain activities within the test guidelines programme of the OECD • Officially recognized expert group • Participates in key OECD meetings

Table 2.2 (*Continued*)

Multinational organizations with regulatory testing mandates	Tasks
	• Goals include promoting greater use of alternative methods and reductions in animal use in OECD testing programs, and incorporation of more alternative methods in OECD Test Guidelines
	• Member organizations include: Animal Alliance of Canada; British Union for the Abolition of Vivisection; Deutscher Tierschutzbund e.V.Akademie für Tierschutz; Doris Day Animal League; Dr Hadwen Trust; Eurogroup for Animals; European Coalition to End Animal Experiments; The Humane Society of the United States; Japan Anti-Vivisection Association; People for the Ethical Treatment of Animals; and Physicians Committee for Responsible Medicine
	• Participates in key OECD activities such as: Working Group of National Coordinators of the Test Guidelines Programme, Task Force on Endocrine Disrupter Testing and Assessment, Task Force on Existing Chemicals, including (Q)SAR, Validation Management Group for Non-Animal testing, Validation Management Group for Mammalian testing
	• Comments on OECD draft test guidelines and other documents
	• Nominates outside experts to take part in OECD activities on ICAPO's behalf
	• Works to ensure that member countries abide by their OECD treaty obligations under the Mutual Acceptance of Data provision, which obligates member countries to accept data from OECD-approved test guidelines instead of requiring duplicative animal testing
International Council on Animal Protection in Pharmaceutical Programmes (ICAPPP)[186]	• Sister organization to ICAPO
	• Promotes animal protection in pharmaceutical testing guidelines developed internationally through tripartite discussions among Japan, Europe, and the United States under the banner of the International Conference on Harmonisation (ICH), the VICH and other similar harmonization programs
International Council of Chemical Associations (ICCA)[187]	• International alliance of chemical companies representing over 75% of the chemical manufacturing operations
	• Promotes and coordinates: (i) voluntary chemical industry initiatives such as Responsible Care and (ii) communications between the chemical industry and international organizations involved with health, environmental, and trade issues

The major goal for many of these organizations is to improve access within their regions to safe medicines by harmonization of drug quality, safety, and efficacy standards and guidelines. These harmonization efforts have the potential to reduce the numbers of animals used for drug registration in these countries by resulting in elimination of safety testing procedures that are outdated, duplicative, and/or of limited usefulness, or the introduction of new non-animal methods.

2.5 Japan

Japan has an active involvement in regulatory testing to inform decisions. Three governmental ministries, the Ministry of Health, Labour and Welfare (MHLW), the Ministry of Education, Culture, Sports, Science and Technology (MEXT), and the Ministry of Agriculture, Forestry and Fisheries (MAFF) are the regulatory agencies involved in science and technology in Japan. In addition, the Ministry of the Environment (MOE) provides environmental testing guidance and guidance for the humane handling of laboratory animals. These and other relevant Japanese ministries, bureaus, and programs including the Pharmaceuticals and Medical Devices Agency (PDMA) are summarized in Table 2.3.

Table 2.3 Japanese ministries, bureaus, and programs in regulatory testing to inform decisions.

Japanese Government institutions with 3Rs mandates	Tasks
Ministry of Health, Labour, and Welfare (MHLW)[188]	Many governmental branches including: • Pharmaceutical and Medical Safety Bureau • Industrial Safety and Health Department • The National Institute of Health Sciences is affiliated with MHLW
National Institute of Health Sciences (NIHS)[189]	• Ensures the 'quality, efficacy, and safety of chemical substances (including pharmaceuticals and food) that are closely related to people's lives' • Participates in international activities such as those of the ICH and the OECD to harmonize guidance • Includes the Japanese Center for the Validation of Alternative Methods (JaCVAM)
National Institute of Biomedical Innovation (NIBI)[190]	• Research branch of NIHS • Integrates organizations related to promoting the development of pharmaceuticals and medical devices distinct from the regulatory and promotion functions • Supports research supported by the NIBI including to develop standard and universal

Table 2.3   (*Continued*)

Japanese Government institutions with 3Rs mandates	Tasks
	technologies essential for drug discovery such as maintaining the Genome Medicine Database of Japan (GeMDBJ)

Japan Health Sciences Foundation (JHSF)[191]	• Nonprofit foundation backed by the MHLW
	• Funded by over 100 supporting member companies (pharmaceuticals, chemicals, food products, medical supplies, and textile goods)
	• Seven major activities
	  ○ Research Fund Project to: develop orphan drugs; regulatory science for drug development; develop of prevention, diagnosis and treatment for publicly important diseases; and use human tissues for drug development
	  ○ Conduct research to develop pharmaceutical products for the treatment of AIDS
	  ○ Health and Labour Science Research Promotion Project HLSRP to: conduct research on personalized Human Genome; regenerative Medicine; emerging and re-emerging infectious diseases; and research on intractable diseases
	  ○ Support programs such as scientist exchange and young researcher programs
	  ○ General project supports drug development, promotion, regulations and standards, and related activities
	  ○ Source of standardized cells and genes
	  ○ Center for Accreditation of Laboratory Animal Care and Use

Ministry of Education, Culture, Sports, Science and Technology (MEXT)[192]	• Science and Technology Policy Bureau plans and designs basic policies to promote science and technology, promotes the development of human resources in science and technology, international exchange, science and technology in local communities, and the safety of nuclear energy • Research Promotion Bureau promotes basic research, including life sciences and nanotechnology and promotes private sector, university, and government agency cooperation to promote the creation of new technologies and industries, and supports advanced large-scale research facilities • Regulates bioethics and safety precautions in the life sciences
Ministry of Agriculture, Forestry and Fisheries (MAFF)[193]	• Promotes Japanese agriculture • Develops policies for agricultural, forestry, and fisheries industries and products • Develops food labeling laws and standards

Table 2.3 (*Continued*)

Japanese Government institutions with 3Rs mandates	Tasks
Ministry of the Environment, Government of Japan (MOEJ)[194]	• Responsible for environmental policy, including air quality, water quality, and environmental chemicals and health • Administers Japan's environmental conservation policy • Administers Japanese environmental laws and programs including: Japan's Chemical Substances Control Law (CSCL); Japan HPV Challenge Program; and screening and testing approaches to chemicals suspected of having endocrine disrupting effects
Pharmaceuticals and Medical Devices Agency (PDMA)[195]	• Conducts drug and medical device reviews • Conducts post-marketing safety analyses • Provides relief services • Participates in international programs to facilitate regulatory harmonization
Ministry of Economy, Trade and Industry (METI)[196]	• Conducts many activities, most related to promoting Japan's industrial growth and trade
National Institute of Occupational Safety and Health, Japan (JNIOSH)[197]	• Conducts related to eliminating industrial accidents and diseases • Promote workers' health • Creating a safe and comfortable working environment • Develops and publishes guidelines and standards
Japan Science and Technology Agency (JST)[198]	• Independent administrative institution, subsidized primarily by the government • Promotes science and technology • Creates advanced technology • Promotes business using advanced technology • Promotes dissemination of scientific and technological information • Supports researcher exchange and provides research support • Promotes understanding of science and technology by the public

2.6 Conclusions

Regulatory testing has a long history of use to inform decision making in conjunction with economic issues, legal precedent, public values, and political considerations. Society wishes to benefit from a variety of products that improve living conditions, including agricultural, medical, and commercial products, but does not wish human health and the environment to be harmed by the use of such products nor the activities required for their manufacture. It has turned to science through regulatory testing to inform decisions to evaluate the risk from exposures to chemicals in foods, drugs, and the environment through a large and complex web of efforts by governments, industry, and non-governmental organizations.

Science plays an important role to inform decisions, but it does not stand alone. 'Science informs but policy decides' as a mix of economic, social, legal, and political considerations are considered in decision making. Further, knowledge is not complete and there are scientific uncertainties in the results from regulatory testing. Animal testing has been used extensively over the past half century and many thousands of decisions have been informed by such testing. It is generally recognized that conventional regulatory testing approaches have been helpful in advancing health and environmental health protection, but that it has many limitations too. Advances in high speed computing and biology offer the promise of more informative, efficient, and less expensive regulatory testing schemes in the future, as envisioned by the NAS (see Additional Resources, ref. 4), but a systematic linkage is needed between the new and conventional approaches to demonstrate that the new approaches are at least as, if not more, informative about health and environmental health risk as are the conventional approaches. Extensive dialog and heated debate is likely as the change proceeds given the large contributions of commercial products to the world economy and the many scientific uncertainties about what risks exposure to such products cause.

This chapter attempts to briefly describe the many players, government, industry, and non-governmental organizations, in the US and worldwide, that engage in some aspects of regulatory testing to inform decisions and to demonstrate the many regulatory approaches that exist, both in the US and internationally, to provide what is hoped to be a realistic, or at least not simplistic, context for regulatory testing to inform decisions. It is hoped that some understanding of this complex web might help facilitate the move to a more informative, efficient, and economical approach to regulatory testing to inform decisions through a recognition that science alone will not carry the day and that it must be considered in light of legislative and legal requirements and economic, political, and other social concerns. With time it is likely that society will move to a more informative and efficient approach to regulatory testing even if the focus of discussions is largely on science. However, it is also likely that the evolution to this new regulatory testing state will occur more quickly, perhaps in a punctuated evolutionary fashion if the discussion of science is placed in the context of the other key issues for decision making.

2.7   Additional Resources

1 S. Breyer, *Making Our Democracy Work: A Judge's View*, Vintage Books, NY, 1st edn, 2010, Chapters 8 and 9, pp. 88–120.
2 US EPA, *Risk Characterization Handbook*, EPA 100-B-00-002, Office of Science Policy, Office of Research and Development, Washington, DC, 2000, Chapter 5, pp. 51–54.

3 Organisation for Economic Co-operation and Development, *Workshop on Integrated Approaches to Testing and Assessment*, Economic Directorate, Organisation for Economic Co-Operation and Development, Paris, 2008.

4 National Research Council, *Toxicity Testing in the 21st Century: A Vision and a Strategy*, NAS Press, Washington, DC, 2007.

5 National Research Council, *Exposure Science in the 21st Century: A Vision and a Strategy*, NAS Press, Washington, DC, 2012.

6 N. G. Carmichael, H. A. Barton, A. R. Boobis, R. L. Cooper, V. L. Dellarco, N. G. Doerrer, P. A. Fenner-Crisp, J. E. Doe, J. C. Lamb, 4th and T. P. Pastoor, Agricultural chemical safety assessment: A multisector approach to the modernization of human safety requirements *Crit. Rev. Toxicol.* 2006, **36**(1), 1.

7 US EPA, *A Framework for a Computational Toxicology Research Program in ORD*, EPA/600/R-03/065, Office of Research and Development, Washington, DC, 2003.

8 National Research Council, *Guide for the Care and Use of Laboratory Animals*, NAS Press, Washington, DC, 1996.

References

1. AltTox Toxicity Testing Resource Center, 2013. Available at http://alttox.org/ttrc/.
2. Cornell Law School description of 'Federalism', 2013. Available at http://www.law.cornell.edu/wex/federalism.
3. EPA webpage links to state environmental laws, 2013. Available at http://www.epa.gov/epahome/state.htm.
4. Organisation for Economic Co-operation and Development, 2013. Available at http://www.occd.org.
5. ToxCast™ Data Base, 2013. Available at http://www.epa.gov/ncct/toxcast/.
6. The Tox 21 Consortium, 2013. Available at http://tox21.org/.
7. Adverse Outcome Pathway, 2013. Available at http://www.oecd.org/env/ehs/testing/49963554.pdf.
8. EPA Mission, 2013. Available at http://www2.epa.gov/aboutepa.
9. EPA Laws and Executive Orders, 2013. Available at http://www.epa.gov/lawsregs/laws/.
10. EPA Headquarter Offices and Regions, 2013. Available at http://www2.epa.gov/aboutepa/epa-organizational-structure.
11. EPA Labs and Research Centers, 2013. Available at http://www.epa.gov/about/#tabsmenu = 4.
12. EPA Strategic Plan for Evaluating the Toxicity of Chemicals, 2013. Available at http://www.epa.gov/spc/toxicitytesting/.
13. EPA National Center for Computational Toxicology, 2013. Available at http://www.epa.gov/ncct/.

14. EPA National Health and Environmental Effects Research Laboratory, 2013. Available at http://www2.epa.gov/aboutepa/about-national-health-and-environmental-effects-research-laboratory-nheerl.

15. EPA Office of Chemical Safety and Pollution Prevention, 2013. Available at http://www2.epa.gov/aboutepa/about-office-chemical-safety-and-pollution-prevention-ocspp.

16. EPA's Pesticide Programs, 2013. Available at http://www.epa.gov/pesticides/index.htm.

17. EPA's Office of Pollution Prevention and Toxics, 2013. Available at http://www.epa.gov/oppt/.

18. Federal Insecticide Fungicide and Rodenticide Act, 2013. Available at http://www.epa.gov/lawsregs/laws/fifra.html.

19. Food Quality Protection Act, 2013. Available at http://www.epa.gov/pesticides/regulating/laws/fqpa/.

20. Federal Food Drug and Cosmetic Act (FFDCA), 2013. Available at http://www.epa.gov/lawsregs/laws/ffdca.html.

21. Endangered Species Act, 2013. Available at http://www.epa.gov/lawsregs/laws/esa.html.

22. Title 40, Part 158 of the Code of Federal Regulations Covering Pesticide Registration, 2013. Available at http://www.ecfr.gov/cgi-bin/text-idx?c = ecfr&tpl = /ecfrbrowse /Title40/40cfr158_main_02.tpl.

23. Data Requirements for Pesticide Registration, 2013. Available at http://www.epa.gov/opp00001/regulating/data_requirements.htm.

24. Reduced Data Requirements for Biochemical and Microbial Pesticides, 2013. Available at http://www.epa.gov/fedrgstr/EPA-PEST/2007/October/Day-26/p20828.htm.

25. Reduced Data Set for Antimicrobial Pesticide Products, 2013. Available at http://www.epa.gov/oppfead1/cb/csb_page/updates/2012/testguidelines.html.

26. Pesticides data requirements for conventional chemicals, 2013. Available at http://www.epa.gov/fedrgstr/EPA-PEST/2007/October/Day-26/p20826.htm.

27. OCSPP Harmonized Test Guidelines, 2013. Available at http://www.epa.gov/ocspp/pubs/frs/home/guidelin.htm.

28. OECD guidelines for the testing of chemicals, 2013. Available at http://www.oecd.org/env/ehs/testing/oecdguidelinesforthetestingofchemicals.htm.

29. Estimating the Data Generation Costs for Registration of a New Conventional Pesticide Active Ingredient, 2013. Available at http://www.epa.gov/pesticides/ppdc/2011/october/new-conventional.pdf.

30. EPA OPP Strategic Directions for New Testing and Assessment Approaches, 2013. Available at http://www.epa.gov/pesticides/science/testing-assessment.html.

31. One-year Dog Study May Not Be Necessary for Pesticide Registration Testing, 2013. Available at http://www.gpo.gov/fdsys/pkg/FR-2007-10-26/pdf/E7-20826.pdf.

32. Non-animal Testing Approach to EPA Labeling for Eye Irritation, 2013. Available at http://www.epa.gov/oppad001/eye-irritation.pdf.
33. Use of an Alternate Testing Framework for Classification of Eye Irritation Potential of EPA Pesticide Products, 2013. Available at http://www.epa.gov/pesticides/science/eye-irritation.html.
34. EPA Scientific Advisory Panel (SAP), 2013. Available at http://www.epa.gov/scipoly/sap/.
35. EPA Pesticide Program Dialog Committee, 2013. Available at http://www.epa.gov/pesticides/ppdc/.
36. May 24–26, 2011 SAP review of EPA OPP's Integrated Approach to Testing and Assessment Strategies: Use of New Computational and Molecular Tools, 2013. Available at http://www.epa.gov/scipoly/sap/meetings/2011/052411meeting.html.
37. EPA endocrine disruptor screening program (EDSP), 2013. Available at http://www.epa.gov/endo/.
38. EPA OSCPP Office of Science Coordination and Policy (OSCP), 2013. Available at http://www.epa.gov/scipoly/.
39. Endocrine Disruptor Screening and Testing Advisory Committee (EDSTAC) Final Report, 2013. Available at http://www.epa.gov/endo/pubs/edspoverview/finalrpt.htm.
40. Endocrine Disruptor Screening Program; Tier 1 Screening Orders Issuing Announcement, 2013. Available at http://www.regulations.gov/#!documentDetail;D = EPA-HQ-OPP-2009-0634-0001.
41. January 29–31, 2013 meeting of the Science Advisory Panel on the scientific issues associated with 'Prioritizing the Universe of Endocrine Disruptor Screening Program (EDSP) Chemicals Using Computational Toxicology Tools, 2013. Available at http://www.epa.gov/scipoly/sap/meetings/2013/012913meeting.html#minutes.
42. Summary of the Toxic Substances Control Act (TSCA), 2013. Available at http://www.epa.gov/lawsregs/laws/tsca.html.
43. Summary of the Pollution Prevention Act, 2013. Available at http://www.epa.gov/lawsregs/laws/ppa.html.
44. EPA's New Chemicals Program and Premanufacture Notification Requirement, 2013. Available at http://www.epa.gov/oppt/newchems/.
45. EPA's Existing Chemicals Program, Including Testing Authority, 2013. Available at http://www.epa.gov/oppt/existingchemicals/pubs/basicinfo.html.
46. EPA Chemical Collection and Data Collection, 2013. Available at http://www.epa.gov/opptintr/chemtest/pubs/mtlintro.html.
47. EPA High Production Volume Chemical Challenge, 2013. Available at http://www.epa.gov/HPV/.
48. Description of EPA's Design for the Environment Program (DfE), 2013. Available at http://www.epa.gov/dfe/.
49. Design for the Environment Alternatives Assessment Criteria for Hazard Evaluation, 2013. Available at http://www.epa.gov/dfe/alternatives_assessment_criteria_for_hazard_eval.pdf.

50. EPA's ToxRef Database, 2013. Available at http://actor.epa.gov/toxrefdb/faces/Home.jsp.
51. EPA's ExpoCast™ Database, 2013. Available at http://actor.epa.gov/actor/faces/ExpoCastDB/Home.jsp.
52. EPA's Water Program Effluent Toxicity Guidelines, 2013. Available at http://cfpub.epa.gov/npdes/wqbasedpermitting/wet.cfm.
53. EPA's Strategic Plan for Evaluating the Toxicity of Chemicals, 2013. Available at http://www.epa.gov/spc/toxicitytesting/.
54. EPA and Other Federal Agencies Collaborate to Improve Chemical Screening, 2013. Available at http://www.epa.gov/agingepa/press/epanews/2010/2010_0719_1.htm.
55. Food and Drug Administration (FDA), 2013. Available at http://www.fda.gov/.
56. FDA Regulatory Programs, 2013. Available at http://www.fda.gov/regulatoryinformation/legislation/default.htm.
57. Federal Food Drug and Cosmetic Act, 2013. Available at http://www.fda.gov/RegulatoryInformation/Legislation/FederalFoodDrugandCosmeticActFDCAct/default.htm.
58. Family Smoking Prevention and Tobacco Control Act, 2013. Available at http://www.fda.gov/TobaccoProducts/GuidanceComplianceRegulatoryInformation/ucm246129.htm.
59. Dietary Supplement Health and Education Act of 1994, 2013. Available at http://ods.od.nih.gov/About/DSHEA_Wording.aspx.
60. Fair Packaging and Labeling Act, 2013. Available at http://www.fda.gov/RegulatoryInformation/Legislation/ucm148722.htm.
61. Food Safety and Modernization Act, 2013. Available at http://www.fda.gov/Food/GuidanceRegulation/FSMA/default.htm.
62. Public Health Service Act, 2013. Available at http://www.fda.gov/RegulatoryInformation/Legislation/ucm148717.htm.
63. FDA Center for Biologics Evaluation and Research (CBER), 2013. Available at http://www.fda.gov/AboutFDA/CentersOffices/OfficeofMedicalProductsandTobacco/CBER/default.htm.
64. FDA Center for Devices and Radiological Health, 2013. Available at http://www.fda.gov/AboutFDA/CentersOffices/OfficeofMedicalProductsandTobacco/CDRH/default.htm.
65. FDA Center for Drug Evaluation and Health (CDER), 2013. Available at http://www.fda.gov/AboutFDA/CentersOffices/OfficeofMedicalProductsandTobacco/CDER/default.htm.
66. FDA Center for Food Safety and Applied Nutrition, 2013. Available at http://www.fda.gov/AboutFDA/CentersOffices/OfficeofFoods/CFSAN/default.htm.
67. Center for Veterinary Medicine, 2013. Available at http://www.fda.gov/AboutFDA/CentersOffices/OfficeofFoods/CVM/default.htm.
68. FDA National Center for Toxicological Research (NCTR), 2013. Available at http://www.fda.gov/AboutFDA/CentersOffices/OC/OfficeofScientificandMedicalPrograms/NCTR/default.htm.

69. FDA Regulations US Code of Federal Regulations Section 300 Title 21, 2013. Available at http://www.accessdata.fda.gov/scripts/cdrh/cfdocs/cfcfr/cfrsearch.cfm.
70. FDA's Pharmacology and Toxicology Test Guidance, 2013. Available at http://www.fda.gov/Drugs/GuidanceComplianceRegulatoryInformation/Guidances/ucm065014.htm.
71. International Conference on Harmonization of the Technical Requirements for Pharmaceuticals (ICH), 2013. Available at http://www.ich.org/.
72. FDA Redbook: Guidance for Industry and Other Stakeholders, Toxicology Principles for the Safety Assessment of Food, 2013. Available at http://www.fda.gov/Food/GuidanceRegulation/GuidanceDocumentsRegulatoryInformation/IngredientsAdditivesGRASPackaging/ucm2006826.htm.
73. Organisation for Economic Co-operation and Development (OECD) home page, 2013. Available at http://www.OECD.org.
74. FDA Drug Development Tools (DDT) Qualification Programs, 2013. Available at http://www.fda.gov/Drugs/DevelopmentApprovalProcess/DrugDevelopmentToolsQualificationProgram/default.htm.
75. FDA's Biomarker Qualification Program, 2013. Available at http://www.fda.gov/Drugs/DevelopmentApprovalProcess/DrugDevelopmentToolsQualificationProgram/ucm284076.htm.
76. FDA Guidance for Industry Use of Histology in Biomarker Qualification Studies, 2013. Available at http://www.fda.gov/downloads/Drugs/GuidanceComplianceRegulatoryInformation/Guidances/UCM285297.pdf.
77. FDA Guidance On Biomarkers E 16, 2013. Available at http://www.fda.gov/downloads/Drugs/GuidanceComplianceRegulatoryInformation/Guidances/UCM267449.pdf.
78. FDA Description About Why Animals are Used for Testing Medical Products, 2013. Available at http://www.fda.gov/AboutFDA/Transparency/Basics/ucm194932.htm.
79. How FDA Evaluates Regulated Products: Cosmetics, 2013. Available at http://www.fda.gov/AboutFDA/Transparency/Basics/ucm262353.htm.
80. FDA's Cosmetics Animal Testing Program, 2013. Available at http://www.fda.gov/Cosmetics/ProductandIngredientSafety/ProductTesting/ucm072268.htm.
81. FDA's activities with the International Cooperation on Cosmetic Regulation, 2013. Available at http://www.fda.gov/Cosmetics/InternationalActivities/ConferencesMeetingsWorkshops/InternationalCooperationonCosmeticsRegulationsICCR/default.htm.
82. Terms of reference for ICCR, 2013. Available at http://www.fda.gov/InternationalPrograms/HarmonizationInitiatives/ucm114522.htm.
83. FDA Food/Color Additives and Food Contact Materials Guidance for Industry, 2013. Available at http://www.fda.gov/Food/GuidanceRegulation/GuidanceDocumentsRegulatoryInformation/IngredientsAdditivesGRASPackaging/default.htm.

84. Redbook, 2013. Available at http://www.fda.gov/downloads/Food/
    GuidanceRegulation/UCM222779.pdf.
85. FDA's Generally Recognized as Safe (GRAS) Chemicals, 2013. Available
    at   http://www.fda.gov/Food/IngredientsPackagingLabeling/GRAS/
    default.htm.
86. Links to FDA Investigational New Drug Application Form, 2013.
    Available   at   http://www.fda.gov/AboutFDA/ReportsManualsForms/
    Forms/default.htm.
87. FDA Investigational New Drug Application Form, 2013. Available at
    http://www.fda.gov/BiologicsBloodVaccines/GuidanceCompliance-
    RegulatoryInformation/default.htm.
88. FDA Investigational New Drug Application Form, 2013. Available at
    http://www.fda.gov/BiologicsBloodVaccines/GuidanceCompliance-
    RegulatoryInformation/default.htm.
89. Efforts to develop in vitro assays to evaluate the potency of vaccines,
    2013. Available at http://iccvam.niehs.nih.gov/meetings/BiologicsWksp-
    2010/VaccAgendaShort.pdf.
90. FDA's Vaccines and Related Biological Products Advisory Committee,
    2013. Available at http://www.fda.gov/AdvisoryCommittees/Commit-
    teesMeetingMaterials/BloodVaccinesandOtherBiologics/Vaccinesan-
    dRelatedBiologicalProductsAdvisoryCommittee/default.htm.
91. How FDA Evaluates Regulated Products: Drugs, 2013. Available at
    http://www.fda.gov/AboutFDA/Transparency/Basics/ucm269834.htm.
92. FDA's guidance for Drugs, 2013. Available at http://www.fda.gov/
    Drugs/GuidanceComplianceRegulatoryInformation/Guidances/
    default.htm.
93. Harmonized International Conference on Harmonization Use (ICH) of
    Technical Requirements for Registration of Pharmaceuticals for Humans.
    Available   at   http://www.ich.org/fileadmin/Public_Web_Site/ICH_Pro-
    ducts/Guidelines/Efficacy/E2F/Step4/E2F_Step_4.pdf.
94. FDA Investigational New Drug Applications Guidance, 2013. Available
    at http://www.fda.gov/Drugs/GuidanceComplianceRegulatoryInforma-
    tion/Guidances/ucm065008.htm.
95. FDA Nonclinical Testing of New Pharmaceutical Candidates, 2013.
    Available at http://www.fda.gov/downloads/Drugs/GuidanceCompliance-
    RegulatoryInformation/Guidances/ucm073246.pdf.
96. FDA Guidance for Industry M3(R2) Nonclinical Safety Studies for the
    Conduct of Human Clinical Trials and Marketing Authorization for
    Pharmaceuticals, 2013. Available at http://www.fda.gov/downloads/
    Drugs/GuidanceComplianceRegulatoryInformation/Guidances/
    UCM292340.pdf.
97. Medical Devices, 2013. Available at http://www.fda.gov/aboutfda/
    transparency/basics/ucm211822.htm.
98. Demonstration of Safety and Efficacy of Medical Devices, 2013. Available
    at   http://www.fda.gov/MedicalDevices/DeviceRegulationandGuidance/
    overview/default.htm.

99. US Pharmacopeial Convention (USP), 2013. Available at http://www.usp.org/about-usp.

100. Nutritional Supplements, 2013. Available at http://www.fda.gov/AboutFDA/Transparency/Basics/ucm195635.htm.

101. FDA's Animal and Veterinary Guidance, Compliance and Enforcement, 2013. Available at http://www.fda.gov/AnimalVeterinary/Guidance-ComplianceEnforcement/default.htm.

102. VICH International Cooperation on Harmonization of Technical Requirements for Registration of Veterinary Products, 2013. Available at http://www.fda.gov/InternationalPrograms/HarmonizationInitiatives/ucm114624.htm.

103. FDA Animal and Veterinary New Drug Applications, 2013. Available at http://www.fda.gov/AnimalVeterinary/DevelopmentApprovalProcess/NewAnimalDrugApplications/default.htm.

104. FDA Animal and Veterinary New Drug Applications Guidance, Compliance and Enforcement, 2013. Available at http://www.fda.gov/AnimalVeterinary/GuidanceComplianceEnforcement/GuidanceforIndustry/ucm123821.htm.

105. FDA's Critical Path Initiative, 2013. Available at http://www.fda.gov/ScienceResearch/SpecialTopics/CriticalPathInitiative/default.htm.

106. Memorandum of Understanding between EPA, NIEHS, NCGC and FDA on High Throughput Screening, Toxicity Pathway Profiling and Biological Interpretation of Findings, 2013. Available at http://www.fda.gov/AboutFDA/PartnershipsCollaborations/MemorandaofUnderstandingMOUs/DomesticMOUs/ucm219627.htm.

107. FDA Strategic Plan for Regulatory Science, 2013. Available at http://www.fda.gov/ScienceResearch/SpecialTopics/RegulatoryScience/ucm267719.htm.

108. FDA Modernizing Regulatory Science, 2013. Available at http://www.fda.gov/ForConsumers/ConsumerUpdates/ucm268201.htm.

109. DARPA Platform for More Effective Testing of Drugs and Vaccines, 2013. Available at http://www.darpa.mil/NewsEvents/Releases/2011/2011/09/16_DARPA_TO_DEVELOP_PLATFORM_FOR_MORE_EFFECTIVE_TESTING_OF_DRUGS_AND_VACCINES.aspx.

110. MIT Department of Biological Engineering's efforts to develop a human on a chip, 2013. Available at http://web.mit.edu/newsoffice/2012/human-body-on-a-chip-research-funding-0724.html.

111. Harvard's Wyss Institute's efforts to develop a human on a chip, 2013. Available at http://wyss.harvard.edu/viewpressrelease/91/.

112. DARPA, NIH and FDA collaboration to develop human on a chip, 2013. Available at http://www.fda.gov/ForConsumers/ConsumerUpdates/ucm317496.htm.

113. Consumer Products Safety Commission (CPSC), 2013. Available at http://www.cpsc.gov/.

114. CPSC Statutes, 2013. Available at http://www.cpsc.gov/en/Regulations-Laws--Standards/Statutes/.

115. Summary of CPSC's Programs, 2013. Available at http://www.cpsc.gov/ Safety-Education/Safety-Guides/General-Information/Who-We-Are--- What-We-Do-for-You/.

116. CPSC Animal Testing Policy, 2013. Available at http://www.cpsc.gov/en/ Regulations-Laws--Standards/Federal-Register-Notices/2013/ Codification-of-Animal-Testing-Policy/.

117. Recommended Procedures Regarding the CPSC's Policy on Animal Testing, 2013. Available at http://www.cpsc.gov/Business--Manu- facturing/Testing-Certification/Recommended-Procedures-Regarding- the-CPSCs-Policy-on-Animal- Testing/.

118. Interagency Coordinating Committee on the Validation of Alternative Methods (ICCVAM), 2013. Available at http://iccvam.niehs.nih.gov/.

119. OECD Mutual Acceptance of Data (MAD), 2013. Available at http:// www.oecd.org/env/ehs/mutualacceptanceofdatamad.htm.

120. US Department of Agriculture (USDA), 2013. Available at http:// www.usda.gov/wps/portal/usda/usdahome.

121. USDA Veterinary Biologics Program, 2013. Available at http://www. aphis.usda.gov/animal_health/vet_biologics/vb_mission_vision.shtml.

122. Virus Serum Toxin Act, 2013. Available at http://www.aphis.usda.gov/ animal_health/vet_biologics/vb_vsta_100.shtml.

123. USDA Veterinary Biologics regulations, 2013. Available at http:// www.aphis.usda.gov/animal_health/vet_biologics/vb_cfr.shtml.

124. USDA Veterinary Service biologics program, 2013. Available at http:// www.aphis.usda.gov/animal_health/vet_biologics/vb_vs_memos.shtml.

125. USDA Animal Health Guidance to New Biologics Applicants, 2013. Available at http://www.aphis.usda.gov/animal_health/vet_biologics/ vb_new_applicants.shtml.

126. Department of Transportation, 2013. Available at http://www.dot.gov/.

127. Federal Hazardous Materials Transportation Law, 2013. Available at http://www.phmsa.dot.gov/portal/site/PHMSA/menuitem.ebdc7a8a7e39f2e- 55cf2031050248a0c/?vgnextoid = ed62d0dfb2e87110VgnVCM1000009ed078- 98RCRD&vgnextchannel = 0e78a535eac17110VgnVCM1000009ed07898- RCRD& vgnextfmt = print.

128. US Department of Transportation (DOT) Pipeline and Hazardous Ma- terials Safety Administration Regulations, 2013. Available at http:// www.phmsa.dot.gov/regulations.

129. DOT Requirements for Tests to Determine Toxicity Classification, 2013. Available at http://www.ecfr.gov/cgi-bin/text-idx?c = ecfr&SID = e4ceed- 8f8d7eca4bafa4137fa7c3f9d6&rgn = div5&view = text&node = 49:2.1.1.3.9- &idno = 49.

130. United Nations Globally Harmonized System of Classification and La- beling of Chemicals (GHS), 2013. Available at http://www.unece.org/ trans/danger/publi/ghs/ghs_welcome_e.html.

131. Link to OECD Guideline for the Testing of Chemicals Number 404 in DOT Guidance Relevant to the Classification of Corrosive Materials, 2013. Available at http://www.phmsa.dot.gov/portal/site/PHMSA/

menuitem.ebdc7a8a7e39f2e55cf2031050248a0c/?vgnextoid = 0706fedc9-
c426310VgnVCM1000001ecb7898RCRD&vgnextchannel = aa8cd3c1-
af814110VgnVCM1000009ed07898RCRD& vgnextfmt = print.

132. DOT Pipeline and Hazardous Materials Safety Administration Test
Guidance regarding corrosive materials, 2013. Available at http://
www.phmsa.dot.gov/portal/site/PHMSA/menuitem.ebdc7a8a7e39f2e55-
cf2031050248a0c/?vgnextoid = 235802f4c5136310VgnVCM1000001ec-
b7898RCRD&   vgnextchannel = aa8cd3c1af814110VgnVCM1000009-
ed07898RCRD& vgnextfmt = print.

133. Occupational Safety and Health Administration (OSHA) web page, 2013.
Available at http://www.osha.gov/Publications/3302-06N-2006-English.
html.

134. OSHA regulations, 2013. Available at http://www.osha.gov/pls/oshaweb/
owasrch.search_form?p_doc_type = STANDARDS& p_toc_level = 1&p_
keyvalue = 1910.

135. OSHA Hazard Determination, 2013. Available at http://www.osha.gov/
dsg/hazcom/hazcom_appb_1994.html.

136. How OSHA Uses Information to Assess Hazard, 2013. Available
at http://www.osha.gov/pls/oshaweb/owadisp.show_document?p_table =
STANDARDS&p_id = 10372.

137. Description of ICCVAM, 2013. Available at http://iccvam.niehs.nih.gov/
about/about_ICCVAM.htm.

138. Description of NTP Interagency Center for the Evaluation of Alternative
Toxicological Methods (NICEATM), 2013. Available at http://www.
niehs.nih.gov/research/atniehs/dntp/niceatm/index.cfm.

139. Animal   Welfare   Act,   2013.   Available   at   http://awic.nal.usda.
gov/government-and-professional-resources/federal-laws/animal-welfare-
act.

140. USDA Animal and Plant Health Inspection Service (APHIS), 2013.
Available at http://www.aphis.usda.gov/.

141. Animal Welfare Act Regulations, 2013. Available at http://www.aphi-
s.usda.gov/animal_welfare/downloads/awr/awr.pdf.

142. The Animal Welfare Act Requirements to examine alternatives to animal
tests that minimize pain and distress, 2013. Available at http://library.
nymc.edu/informatics/bmsalternatives.cfm.

143. Health Research Extension Act of 1985, 2013. Available at http://grants.
nih.gov/grants/olaw/references/hrea1985.htm.

144. Office of Laboratory Animal Welfare (OLAW), 2013. Available at http://
grants.nih.gov/grants/olaw/olaw.htm.

145. European Union, 2013. Available at http://europa.eu/index_en.htm.

146. European Commission, 2013. Available at http://ec.europa.eu/index_
en.htm.

147. EU Enterprise Europe Network Directorate, 2013. Available at http://
een.ec.europa.eu/about/sector-groups.

148. EU Environment Directorate, 2013. Available at http://ec.europa.eu/
environment/index_en.htm.

149. EU Health and Consumer Protection Directorate, 2013. Available at http://ec.europa.eu/dgs/health_consumer/index_en.htm.
150. EU Joint Research Centre Directorate, 2013. Available at http://ec.europa.eu/dgs/jrc/index.cfm.
151. European Chemicals Agency, 2013. Available at http://echa.europa.eu/.
152. European Food Safety Authority, 2013. Available at http://www.efsa.europa.eu/.
153. European Medicines Agency, 2013. Available at http://www.emea.europa.eu/ema/.
154. 195 Sovereign Nations in the World, 2013. Available at http://www.state.gov/s/inr/rls/4250.htm.
155. Regional organizations working to harmonize drug regulations, 2013. Available at http://www.who.int/medicines/areas/quality_safety/challenges/en/index.html.
156. Association of South East Nations (ASEAN), 2013. Available at http://www.aseansec.org/.
157. CAN (Andean Community), 2013. Available at http://www.comunidadandina.org/ingles/who.htm.
158. CADREAC (The Collaboration Agreement of Drug Regulatory Authorities in European Union Associated Countries), 2013. Available at http://www.ema.europa.eu/docs/en_GB/document_library/Regulatory_and_procedural_guideline/2009/10/WC500004508.pdf.
159. European Union, Gulf Cooperation Council (GCC), 2013. Available at http://eeas.europa.eu/gulf_cooperation/index_en.htm.
160. MERCOSUR (Southern Common Market), 2013. Available at http://wits.worldbank.org/GPTAD/PDF/archive/MERCOSUR.pdf.
161. South African Development Community (SADC), 2013. Available at http://www.sadc.int/.
162. 3R Research Foundation Switzerland, 2013. Available at http://www.forschung3r.ch/.
163. Doerenkamp-Zbinden Foundation for Animal Free Research, 2013. Available at http://www.doerenkamp.ch/en/.
164. European Consensus Platform for Alternatives (ECOPA), 2013. Available at http://www.ecopa.eu/.
165. European Partnership for Alternative Approaches to Animal Testing (EPAA), 2013. Available at http://ec.europa.eu/enterprise/epaa/index_en.htm.
166. Fund for the Replacement of Animals in Medical Experiments (FRAME), 2013. Available at http://www.frame.org.uk/index.php.
167. German Center for the Documentation and Evaluation of Alternatives to Animal Experiments (ZEBET), 2013. Available at http://www.bfr.bund.de/en/zebet-58194.html.
168. Dr Hadwen Trust for Humane Research, 2013. Available at http://www.drhadwentrust.org/.
169. International Society for In Vitro Methods (INVITROM), 2013. Available at http://www.invitrom.org/.

170. Netherlands Center for Alternatives to animal use (NCA), 2013. Available at http://www.nkca.nl/.
171. Swedish Fund for Research Without Animal Experiments, 2013. Available at http://www.forskautandjurforsok.se/in-english/.
172. UK National Center for the Replacement, Refinement and Reduction of Animals in Research (NC3Rs), 2013. Available at http://www.nc3rs. org.uk/.
173. Mission and Terms of Reference for the International Conference on Harmonization of Technical Requirements for Registration of Pharmaceuticals for Human Use (ICH), 2013. Available at http://www.ich.org/about/vision.html.
174. Description of the International Cooperation on Harmonization of Technical Requirements for Registration of Veterinary Products (VICH), 2013. Available at http://www.vichsec.org/.
175. United Nations (UN) at a Glance, 2013. Available at http://www.un.org/en/aboutun/index.shtml.
176. World Health Organization (WHO), 2013. Available at http://www.who.int/about/en/.
177. International Programme on Chemical Safety (IPCS), 2013. Available at http://www.who.int/ipcs/en/.
178. Inter-Organization Programme for the Sound Management of Chemicals (IOMC), 2013. Available at http://www.who.int/iomc/en/.
179. Codex Alimentarius Commission, 2013. Available at http://www.codexalimentarius.org/.
180. International Pharmacopoeia, 2013 Available at http://www.who.int/medicines/publications/pharmacopoeia/overview/en/.
181. International Agency for Research on Cancer, 2013. Available at http://www.iarc.fr/.
182. Quality, Safety and Efficacy of Medicines (QSM), 2013. Available at http://www.who.int/medicines/areas/quality_safety/en/.
183. Interagency Pharmaceutical Coordination Group, 2013. Available at http://www.who.int/medicines/areas/policy/ipc/en/.
184. International Conference of Drug Regulatory Authorities, 2013. Available at http://www.who.int/medicines/areas/quality_safety/regulation_legislation/icdra/en/.
185. International Council for Animal Protection in OECD Programmes (ICAPO), 2013. Available at http://www.icapo.org/.
186. International Council on Animal Protection in Pharmaceutical Programmes (ICAPPP), 2013. Available at http://www.uia.be/s/or/en/1122275603.
187. International Council of Chemical Associations (ICCA), 2013. Available at http://www.icca-chem.org/.
188. Japanese Ministry of Health, Labour, and Welfare (MHLW), 2013. Available at http://www.mhlw.go.jp/english/org/detail/index.html.
189. Japanese National Institute of Health Sciences (NIHS), 2013. Available at http://www.nihs.go.jp/english/nihs/index.html.

190. Japanese National Institute of Biomedical Innovation (NIBI), 2013. Available at http://www.nibio.go.jp/english/laboratory/outline.html.
191. Japan Health Sciences Foundation (JHSF), 2013. Available at http://www.jhsf.or.jp/English/index_e.html.
192. Japanese Ministry of Education, Culture, Sports, Science and Technology (MEXT), 2013. Available at http://www.mext.go.jp/english/.
193. Japanese Ministry of Agriculture, Forestry and Fisheries (MAFF), 2013. Available at http://www.maff.go.jp/e/index.html.
194. Japanese Ministry of the Environment, Government of Japan (MOEJ), 2013. Available at http://www.env.go.jp/en/.
195. Japanese Pharmaceuticals and Medical Devices Agency (PDMA), 2013. Available at http://www.pmda.go.jp/english/index.html.
196. Japanese Ministry of Economy, Trade and Industry (METI), 2013. Available at http://www.meti.go.jp/english/.
197. Japanese National Institute of Occupational Safety and Health, Japan (JNIOSH), 2013. Available at http://www.jniosh.go.jp/en/index.html.
198. Japan Science and Technology Agency (JST), 2013. Available at http://www.jst.go.jp/EN/.

CHAPTER 3

Efforts Towards International Harmonization of Acceptable Alternatives to Animal Testing

CHANTRA ESKES

SeCAM Services & Consultation on Alternative Methods, Switzerland
Email: chantra.eskes@secam-ce.eu

3.1 Introduction

During the last 20 years extensive efforts have been carried out in order to develop, validate and implement reduction, refinement, and replacement, alternative methods to animal testing. Although progresses have been made, international harmonization and standardization is still key to leverage regulatory and trade barriers whilst ensuring the development and implementation of scientifically-based decision-making for the protection of human beings side-by-side with animal welfare considerations.

The present chapter provides with a description of the current practices around the world regarding the acceptance of alternative methods to animal testing, their similarities and differences and the current needs and challenges for the implementation of harmonized international standards in the acceptance of alternatives to animal testing.

Issues in Toxicology No. 19
Reducing, Refining and Replacing the Use of Animals in Toxicity Testing
Edited by David G. Allen and Michael D. Waters
© The Royal Society of Chemistry 2014
Published by the Royal Society of Chemistry, www.rsc.org

3.2 Acceptable Alternative Methods

Alternatives to animal testing include all procedures which can replace the need for animal experiments, reduce the number of animals required, or diminish the amount of distress or pain experienced by animals.[1] This definition embodies the '3Rs' concept proposed by Russell and Burch in *The Principles of Humane Experimental Technique*,[2] and the regulatory requirements of many countries related to the protection of animals used for scientific purposes (see Chapter 1).

An alternative method for the replacement (or partial replacement) of an animal test is generally defined as the combination of a 'test system', which provides a means of generating physicochemical or *in vitro* data for the chemicals of interest, and a 'prediction model (PM)' or 'data interpretation procedure'.[3] The prediction model or data interpretation procedure plays an important role in the acceptance process, as it allows converting the obtained data (e.g., *in vitro* or physicochemical) into predictions of toxicological endpoints in animals or humans.[4]

3.2.1 The Acceptance Process

The acceptability of a test method may depend on various factors such as national regulatory requirements, the test method purposes, uses, and applicability. In Europe, the original directive on the protection of laboratory animals for experimental and other scientific purposes stated that '*An* (animal) *experiment shall not be performed if another scientifically satisfactory method of obtaining the result sought, not entailing the use of an animal, is reasonably and practicably available*' (Directive 86/609; EC).[5] The same request is still present in the updated directive.[6] As such, for an alternative method to be regulatory accepted, it is critical to demonstrate that the method is scientifically satisfactory, i.e., valid, for the purpose sought. This is generally carried out through a validation process through which the scientific validity of a test method can be demonstrated.

If initially implemented in Europe, the concept of scientific validation for hazard assessment has since gained international acceptance, and represents nowadays a generally recognized requirement for the acceptance of a test method for safety assessment purposes. Test method validation is the process whereby the relevance and reliability of the method are characterized for a particular purpose.[4,7] In the context of a replacement test method, relevance refers to the scientific basis of the test system, and to the predictive capacity of an associated PM, whereas reliability refers to the reproducibility of test results, both within and between laboratories, and over time. The 'purpose' of an alternative method refers to its intended application, such as the regulatory testing of chemicals for a specific toxicological endpoint (e.g., eye irritation). Adequate validation (i.e., to establish scientific validity) of an alternative test requires demonstration for its stated purpose that:

- the test system has a sound scientific basis;
- the predictions made by the PM are sufficiently accurate; and

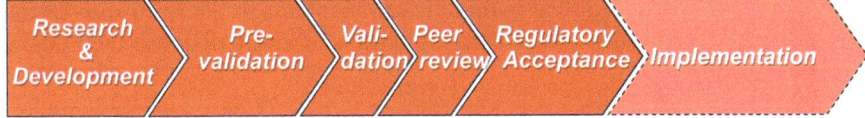

Figure 3.1 Process of validation going from basic research to the regulatory acceptance of alternative test methods for hazard assessment.[4,7,8]

- the results generated by the test system are sufficiently reproducible within and between laboratories, and over time.

The main steps in the process going from basic research to the validation and regulatory acceptance of alternative test methods[4,8,9] are summarized in Figure 3.1. Demonstration of the scientific validity of an *in vitro* method is usually required for its use within the regulatory framework for detecting both hazardous and non-hazardous effects as a replacement, reduction or refinement of the animal testing. Furthermore, depending upon the geographical regulatory framework and the stage of advancement, the data generated by non-validated alternative methods may also be acceptable. For example in the European Union, the safety assessment of chemicals (REACH regulation) foresees different uses for *in vitro* alternatives depending upon their validation status, as described below.[10,11]

1 Information from scientifically valid *in vitro* tests is accepted to determine whether a substance has hazardous properties or presents no hazard effects that warrants labeling and protection measures. In this case, the scientifically valid test methods may fully or partially replace an *in vivo* test. However, one important criterion for acceptance is the adequacy of the test method for the purpose of classification and labeling.
2 Information derived from suitable *in vitro* methods, i.e., methods fulfilling the criteria from the European Reference Laboratory – European Centre for the Validation of Alternative Methods (EURL ECVAM) to enter the pre-validation process (see Section 3.1.2.1), may only be used to indicate the *presence* of a certain dangerous property of a chemical or test article.
3 Finally, information from advanced methodologies (e.g., toxicogenomics) may also be used independently from their validation status to provide mechanistic insights.

Although having different format requests, the data requirements for establishing the scientific validity of an alternative test method are similar between Europe and the US[9,12,13] (Table 3.1). In contrast, experience has shown that there can be room for variation in the evaluation of the scientific validity of an alternative method depending, for example, on the regulatory framework as well as on the background and experience of the individuals carrying out the evaluations. For example, different rates of over- and under-predictions may be considered acceptable depending upon the context and foreseeable uses of the

Table 3.1   Comparison of the main data requirements in the European Union
and United States for the validation of alternative test methods.

United States[9]	European Union[12]
Rationale for the test method	Module 1: Test Definition
Test method protocol components	
Substances used in the validation	(within the respective modules)
In vivo reference data	(within Module 5 on predictive capacity)
Test method data and results	(within the respective modules)
Test method reliability	Module 2: Within-laboratory reproducibility
	Module 3: Transferability
	Module 4: Between-laboratory reproducibility
Test method accuracy	Module 5: Predictive capacity
	Module 6: Applicability domain
Test method data quality	Information on validation process
Other reports and reviews	(within respective modules)
Animal welfare considerations	(within Module 1 on test definition)
Practical considerations	(within Module 3 on transferability)

assays. Similarly, regarding the regulatory acceptance of alternative methods
different approaches may exist in different world regions. In the European
Union for example, perhaps due to the animal welfare legislation requirements[5,6] and the fact that its validation body (EURL ECVAM) is part of the
European Commission, responsible for proposing new and revised legislations,
past experiences have shown that scientifically valid alternative methods have
usually been included into the European legislation. In the United States,
perhaps due to the fact that the legislations are handled by specific regulatory
agencies independently, the regulatory acceptance may depend not only on
the scientific validity of the test method, but also on the acceptance of each
regulatory agency based on their specific needs and the specificities of their
regulated products.[13]

The next section of this chapter provides a detailed description of the
objectives and structures of the validation bodies in Europe, US, and
Japan, which play a pivotal role in the regulatory acceptance of alternative
methods.

3.2.2   Major Validation Principles

The criteria and processes for the validation of a test method were originally
developed and implemented in the European Union in 1991 by the European
Centre for the Validation of Alternative Methods (EURL ECVAM), followed
by the United States with the creation of the Interagency Coordinating Committee on the Validation of Alternative Methods (ICCVAM) in 1997, and
Japan in 2005 through the Japanese Center for the Validation of Alternative
Methods (JaCVAM). An overview of these three major validation bodies is
given below.

3.2.2.1 European Union

The European Centre for the Validation of Alternative Methods (see also http://ihcp.jrc.ec.europa.eu/our_labs/eurl-ecvam) was established in 1991 as part of the European Commission's Joint Research Centre General Directorate, to respond to the requirement from the original Directive on the protection of animals for scientific purposes that '*The Commission and Member States should encourage research into the development and validation of alternative techniques (...) and shall take such other steps as they consider appropriate to encourage research in this field*' (Directive 86/609/EEC).[5]

Directive 86/609/EEC was subsequently revised and replaced in 2010 by Directive 2010/63/EU where ECVAM is referred as the European Union Reference Laboratory (EURL) ECVAM and where its duties and tasks are further defined (Article 48/Annex VII of Directive 2010/63).[6] Briefly it encompasses the coordination and promotion of the development, validation and use of alternative methods; acting as a focal point for the exchange of information; setting up, maintaining and managing public databases and information systems on alternative methods; and promoting dialogue between legislators, regulators, and all relevant stakeholders with a view to the development, validation, regulatory acceptance, international recognition, and application of alternative approaches.

In 1995, based upon the experiences gained during earlier large-scale validation studies, and in consultation with various international experts, EURL ECVAM published recommendations on the practical and logistical aspects of validating alternative test methods.[14] This document represents one of the first basic principles for the validation of alternative methods including the management and design of a validation study that were later integrated at an international level.[4] Some of the key criteria encompassed by these principles include:

- an alternative method can only be judged valid if the method is reliable and relevant;
- the prediction model should be defined in advance by the test developer;
- performance criteria should be set in advance by the management team (for a prospective validation study);
- performance is assessed by using coded chemicals;
- there should be independence in:
  - the management of the study,
  - the selection, coding, and distribution of test chemicals,
  - the data collection and statistical analysis;
- laboratory procedures should comply with GLP criteria.

In the same year, EURL ECVAM published recommendations for a prevalidation scheme, which includes three main phases: protocol refinement, protocol transfer, and protocol performance.[15] The objective of the prevalidation process is to ensure that any method included in a formal validation study adequately fulfills the criteria defined for inclusion in such a study, so that financial and human resources are used more efficiently, and that there is a greater likelihood that the expectations will be met.

In 2004, a 'Modular Approach to the ECVAM Principles on Test Validity' was proposed with the objective to make the validation process more flexible by breaking down the various steps of validation into seven independent modules, and defining for each module the information needed for assessing the scientific validity of a test method.[12] These seven modules are (see also Table 3.1):

1 test definition;
2 within-laboratory reproducibility;
3 transferability;
4 between-laboratory reproducibility;
5 predictive capacity;
6 applicability domain; and
7 the definition of performance standards.

One of the main advantages of the modular approach to validation is that it gives the possibility to complete the different modules in any sequence, allowing the use of both prospectively and retrospectively generated data. This approach has the potential to increase the evidence gathered on a specific test method whilst decreasing the time necessary if only prospective data were to be considered.

Since its creation, the EURL ECVAM has actively contributed to the validation and regulatory acceptance of a number of alternative methods including, for example, full replacement tests for skin corrosion and irritation and phototoxicity, partial replacement tests for eye irritation, reduction tests for acute toxicity, and reduction and refinement tests for skin sensitization.

More recently, the EURL ECVAM established a formal procedure for the evaluation of the readiness of a test method to enter the (pre)validation process which follows two steps: a pre-submission and a complete submission step. The pre-submission step requires the compilation and submission of a Test Pre-submission Form to carry out a preliminary assessment of the status of development, optimization, and/or validation of the test method and its potential relevance with regard to the 3Rs (replacement, reduction, refinement of animal testing). Following successful conclusion of step 1 the test submitter is then invited to provide a complete submission based on a detailed Test Submission Template (TST), which is used for a comprehensive evaluation of the submitted method by EURL ECVAM.

3.2.2.2   United States

The National Institute of Environmental Health Sciences (NIEHS) created the ad hoc Interagency Coordinating Committee on the Validation of Alternative Methods in September 1994 to respond to requirements in the NIH Revitalization Act of 1993 (Public Law 103-43). This Act required NIEHS to establish criteria for the validation and regulatory acceptance of alternative toxicological testing methods, and that NIEHS recommend a process to achieve the regulatory acceptance of scientifically valid alternative test methods. The ICCVAM website includes a detailed account of the committee's history, which is included in part herein (see also http://www.iccvam.niehs.nih.gov/).

Subsequent to the report of the ad hoc ICCVAM in 1997, NIEHS established a standing ICCVAM committee to (1) implement a process by which new test methods of agency interest could be evaluated and (2) coordinate interactions among agencies related to the development, validation, acceptance, and national and international harmonization of toxicological test methods. ICCVAM was established as a permanent interagency committee of the NIEHS under the National Toxicology program (NTP) Interagency Center for the Evaluation of Alternative Toxicological Methods (NICEATM) in 2000 by the ICCVAM Authorization Act Public Law 106-545.[16]

The committee comprises representatives from 15 U.S. Federal regulatory and research agencies that use or generate toxicological information.[13] ICCVAM promotes the scientific validation and regulatory acceptance of toxicological test methods that more accurately assess the safety or hazards of chemicals and products and that refine (i.e., decrease or eliminate pain and distress), reduce, and replace animal use. NICEATM provides operational and scientific support for ICCVAM and ICCVAM-related activities. NICEATM and ICCVAM work collaboratively to evaluate new and improved test methods applicable to the needs of US Federal agencies.

Criteria for validation and regulatory acceptance have been published by the US Federal government.[17] The definition and principles of scientific validity are similar to those adopted in the European Union, although specific format of data compilation is required including for example (see also Table 3.1): test method protocol components, substances used for validation, *in vivo* reference data, test method data results, test method accuracy, test method reliability, test method data quality, additional reports and reviews, animal welfare considerations and practical considerations. For this purpose, ICCVAM has developed submission guidelines for communicating a proposed test method by a sponsor or nominator to ICCVAM for consideration and review.[9]

NICEATM, on behalf of ICCVAM, receives proposed test method nominations or submissions and communicates with the submitting organization or individual. Typically, the ICCVAM evaluation process involves an initial assessment by NICEATM of the adequacy and completeness of the proposed test method nomination or submission, and a determination by ICCVAM of the priority that the proposed test method will have for technical evaluation. Once a proposed test method has been accepted for evaluation, ICCVAM assembles an interagency working group of government scientists with scientific and regulatory expertise in the appropriate scientific disciplines to collaborate with NICEATM on the evaluation process. ICCVAM, in conjunction with NICEATM, develops recommendations and priorities for further efforts. Such efforts might include an expert workshop, an expert panel meeting, a peer review meeting, an expedited peer review process, or a validation study. Following this review process, ICCVAM develops and forwards recommendations on the usefulness and limitations of the proposed test method for regulatory purposes to Federal agencies. Based on their specific statutory mandates, each agency then makes a determination regarding the acceptability of the test method. Agencies are required to respond to ICCVAM within 180 days of receipt of an ICCVAM test method recommendation. If the test method is accepted, appropriate actions (e.g., revision of existing regulations, publication of guidelines, and/or guidance documents) are taken to inform the regulated community.[13]

Since its creation, ICCVAM and NICEATM have actively contributed to the validation and regulatory acceptance of a number of alternative methods including alternatives for skin corrosion testing, partial replacement tests for eye irritation, reduction tests for acute oral systemic toxicity, and reduction and refinement tests for skin sensitization.

3.2.2.3   Japan

The Japanese Center for the Validation of Alternative Methods (JaCVAM, see also http://jacvam.jp/en/) was established in 2005 as part of the Biological Safety Research Center (BSRC) of the National Institute of Health Sciences (NIHS). It aims at promoting the use of alternative methods to animal testing in regulatory studies, thereby replacing, reducing, or refining (the 3Rs) the use of animals wherever possible while meeting the responsibility of the BSRC to ensure the protection of the general public by assessing the safety of chemicals and other materials, as stipulated in the regulations of the NIHS. Its key objectives are to ensure that new or revised test methods are validated, peer reviewed, and officially accepted by regulatory agencies.[18] For this purpose, JaCVAM assesses the utility, limitations, and suitability for use of alternative test methods in regulatory studies for determining the safety of chemicals and other materials. JaCVAM also performs validation studies when necessary. Furthermore, JaCVAM establishes guidelines for new alternative experimental methods through international collaboration.

The evaluation of new and revised testing methods and establishment of guidelines for such methods are processed in the following order: open recruitment, receiving applications, selection for evaluation, preparation of Background Review Document (BRD) by an oversight committee, evaluation by a peer review panel, evaluation by a regulatory acceptance board, and proposal to regulatory bodies. If a validation result is found to be unsatisfactory in the evaluation process and a further validation is deemed necessary, JaCVAM will outsource the validation to an appropriate scientific society, such as the Japanese Society for Alternatives to Animal Experiments (JSAAE). An ad hoc management team for the validation study is formed for each method to perform a comprehensive validation process, including the preparation of reports. In addition, the oversight committee will collaborate with them to make the BRD available for the peer review panel when appropriate.

3.2.3 The Need for International Harmonization

Economic, legal, and scientific requirements call for a harmonized international acceptance of alternative methods. In the European Union, the Cosmetics Regulation prohibits animal testing (testing ban) on finished cosmetic products (since 2004) and cosmetic ingredients (since 2009), as well as the marketing (marketing ban) of finished cosmetic products tested on animals or containing ingredients which were tested on animals within or outside the European Union (complete marketing ban implemented in 2013).[19] In addition, the European chemicals regulation (REACH) calls for the use of alternative test methods and requires for example, that *in vitro* testing is carried out for eye and skin irritation for substances marketed in volumes between 1 to 10 tonnes per year. It also defines general rules for adaptation to the standard regimen, which comprise the use of alternative test methods.[10] Here again, such provisions apply not only to manufactured but also imported substance in quantities of 1 tonne or more per year.[10] As a consequence, the use of internationally accepted alternative methods to animal testing is important not only to comply with geographical regulatory requests, but also in order to favor and leverage international industrial commerce.

International harmonization on the acceptance of alternative methods is also critical to promote harmonized standards in different geographical locations and ensuring standardized criteria for the safety assessment of data derived from alternative methods. Such efforts are critical to favor harmonized definition of hazard properties and interpretation of potential risks to human health of test materials across countries in the world.

This might be especially important to accommodate the future outcome of current major US and EU initiatives to advance the use of non-animal methods for a more scientifically based regulatory decision making. In the US, the toxicology for the 21st century program is based on the consideration that recent advances in molecular biology (such as functional genomics, transcriptomics), the use of high throughput *in vitro* screening assays, systems biology, and predictive *in silico* approaches can significantly improve the current hazard evaluation of environmental chemicals. In particular, it is expected that such tools can lead to more efficient, informative, and less costly approaches for assessing the hazards posed by environmental chemicals and result in more informed regulations, reducing at the same time the need for animal testing as based on human cells and cell components.[20–22]

In the EU, a large research project co-funded by the European Commission and Cosmetics Europe (SEURAT) aims to identify, develop, and validate innovative non-animal safety testing methods for systemic toxicity assessment of chemicals and ingredients of cosmetic products.[23] It comprises six main research areas including the use of stem cells, hepatic microfluidic bioreactor, detection of biomarkers of repeated dose toxicity, integrated *in silico* models, computer models based on systems characterization of organotypic cultures, and supporting integrated data analysis, and servicing of alternative testing methods in toxicology. It represents one of the major coordinated EU research projects aiming at ultimately replacing animal tests for regulatory safety evaluation purposes.

Indeed, alternative methods may in some cases be mechanistically more relevant to predict a human health effect than the traditional animal testing. One well known example is the species differences related to the metabolic competence of liver cells which result in the generation of different metabolites derived from the parent compound due to differences in metabolism or so-called biotransformation.[24] Current international test guidelines acknowledge this difference and start to incorporate the information derived from *in vitro* models to allow a more scientifically based experimental design.[25] Such aspects are essential when dealing with the evaluation of systemic toxicities such as reproductive toxicity and chronic repeated-dose toxicity. Metabolism-mediated species differences account, for example, as one of the causes for failure of new pharmaceutical entities in the clinical phases.

As a consequence, international harmonization and standardized criteria for the acceptance of alternative methods is critical to: leverage potential trade barriers, help design and optimize newly developed methods for predicting adverse effects for regulatory purposes, and ensure having standardized criteria for the safety assessment and protection of consumers.

3.3   Efforts Toward International Harmonization

Regulatory acceptance of alternative methods can be viewed as the acceptance for safety assessment of a specific method in a given regulatory context. Indeed, as described earlier, the current mechanisms and procedures for regulatory acceptance across the world may differ depending on the geographical regions but also depending on the uses and purposes of the test methods (cosmetics, chemicals, pesticides, drugs), including:

- the recognition/tolerance by (control) authorities that manufacturers routinely use alternative approaches in their in-house safety assessments;
- the acceptance of scientifically valid safety alternative approaches as part of safety reviews by authoritative review bodies; and
- the formal recommendation/obligation to use certain validated alternative methods in the registration of chemicals.

Validation is an important step within the regulatory acceptance of alternative methods, so that several international efforts have been carried out to favor the harmonization of its processes and peer reviews with the ultimate aim of promoting harmonization of international acceptance and recognition of alternative methods. Some of the major efforts carried out are described in the following sections.

3.3.1   The Organisation for Economic Co-operation and Development

The Organisation for Economic Co-operation and Development (OECD) comprises 34 member countries and has engaged working relationships with most countries worldwide. For over 25 years, the OECD has recognized the need to protect animals in general and in particular those used in experimental work (www.oecd.org/env/ehs/testing/animalwelfare.htm). The Second High Level Meeting of the Chemicals Group addressed this ethical issue and adopted the following statement:

> *"The welfare of laboratory animals is important; it will continue to be an important factor influencing the work in the OECD Chemicals Programme. The progress in OECD on the harmonisation of chemicals control, in particular the agreement on Mutual Acceptance of Data, by reducing duplicative testing, will do much to reduce the number of animals used in testing. Such testing cannot be eliminated at present, but every effort should be made to discover, develop, and validate alternative testing systems."*

The Mutual Acceptance of Data states that data generated in one OECD member country in accordance with OECD Test Guidelines for chemicals and the OECD Principles of Good Laboratory Practice (GLP), shall be accepted in other member and adhering countries for purposes of assessment and other uses relating to the protection of human health and the environment. Accordingly, the MAD allows avoiding duplicative testing whilst at the same time promoting harmonized acceptance of test methods within the OECD member countries.

Since the adoption in 1981 of the first set of test guidelines, many of the short- and long-term toxicity tests, as well as the genetic toxicity tests, have been developed or revised to introduce aspects of the 3R principles. Some of the most noteworthy achievements include:

- The deletion in 2002 of Test Guideline 401 (Acute Oral Toxicity), and its replacement Test Guidelines 420 (Acute Oral Toxicity – Fixed Dose Procedure), 423 (Acute Oral Toxicity – Acute Toxic Class) and 425 (Acute Oral Toxicity – Up and Down Procedure) introducing reduction and/or refinement for this type of testing.[26–28]
- Test Guideline 428 (Skin Absorption: *In Vitro* Method) adopted in 2004,[29] offering a replacement alternative method to original Test Guideline 427 (Skin Absorption: *In Vivo* Method);
- Test Guideline 432 (*In Vitro* 3T3 NRU Phototoxicity Test) adopted in 2004,[30] offering a full replacement alternative test method;
- Test Guidelines 430 (*In Vitro* Skin Corrosion: Transcutaneous Electrical Resistance Test), 431 (*In Vitro* Skin Corrosion: Reconstructed Human *Epidermis* Test Method) and 435 (*In Vitro* Membrane Barrier Test Method for Skin Corrosion) originally adopted in 2004[31,32] and 2006[33] which offer a partial replacement of the Test Guideline 404 (Acute Dermal Irritation/Corrosion);
- Test Guideline 439 (*In Vitro* Skin Irritation: Reconstructed Human *Epidermis* Test Method) originally adopted in 2010,[34] which complements TG 430, 431 and 435 and offers a further partial replacement (or full replacement depending upon geographical regulatory requirements) of the Test Guideline 404;
- Test Guidelines 437 (Bovine Corneal Opacity and Permeability Test Method), 438 (Isolated Chicken Eye Test Method) and 460 (Fluorescein

Leakage Test Method) originally adopted in 2009[35,36] and 2012,[37] which offer partial replacements of the Test Guideline 405 (Acute Eye Irritation/Corrosion);

- Test Guideline 487 (*In Vitro* Mammalian Cell Micronucleus Test) adopted in 2010[38] to be used as a part of a test battery for genetic toxicology testing;
- Test Guideline 429 (Skin Sensitisation: Local Lymph Node Assay) adopted in 2002 and revised in 2010,[39] introducing refinement and reduction compared to Test Guideline 406 (Skin Sensitization);
- Test Guidelines 442A and 442B (Nonradioactive versions of the Local Lymph Node Assay) adopted in 2010,[40,41] which provide modifications to TG 429.

One of the major OECD achievements for the international acceptance of alternative methods was the adoption in 2005 of internationally agreed validation principles and criteria for the regulatory acceptance of alternative test methods as described in the OECD Guidance Document No. 34 on '*The Validation and International Acceptance of New or Updated Test Methods for Hazard Assessment.*'[4] This guidance document (GD) was the result of an OECD workshop organized in 1996 in Solna, Sweden, and the several ensuing expert meetings involving major stakeholders and the existing validation bodies at that time. The OECD GD 34 details internationally agreed principles and criteria for how validation studies of new or updated test methods should be performed. It represents a document of central importance for promoting harmonized approaches and procedures for the validation and regulatory acceptance of alternative methods within OECD member countries. Table 3.2

Table 3.2   Key factors for the validation and regulatory acceptance of new and revised toxicology test methods (extract from OECD GD34[4]).

Test method definition	– Select/ develop test – Provide detailed description – Collect background information supporting the test selection
Initial assessment of the relevance and reliability (Pre-validation)	– Intra-laboratory testing – Initial inter-laboratory testing – Test refinement/optimization
Broad assessment of the relevance and reliability	– Inter-laboratory follow-up testing – Accumulation of data on relevance and reliability – Finalization of appropriate test protocol
Overall evaluation and conclusion	– Independent peer-review – Conclusions of the validation study – Recommendations – Publication
Regulatory authority/review and conclusion	– Recommendations for or against proposed use

summarizes the main key elements for the validation and acceptance of new and revised toxicology test methods. Essential aspects include:

- Identify basis of test
- Define scientific purpose and relevance
- Define endpoints and measurement
- Define test limitations
- Define rationale and decision criteria in the interpretation of test results
- Define role of test
- Design validation study
  - Management structure
  - GLP procedures
  - Chemicals selection
  - Blind testing
  - Data collection and record keeping
- Identify laboratories
- Optimise test method and develop SOP
- Chemicals distribution
- Perform testing
- Data analysis
- Assess reliability
- Assess relevance
- Peer review
- Define role in testing strategies
- Regulatory acceptance and use

Finally, another commitment of the OECD to implement harmonized 3Rs principles into regulatory toxicity testing was the adoption in 2000 of the Guidance Document No.19 on '*the Recognition, Assessment and Use of Clinical Signs as Humane Endpoints for Experimental Animals Used in Safety Evaluations*'.[42] This document gives practical guidance on how to apply the 3Rs principles, with an emphasis on refinements when performing OECD test guidelines.

3.3.2 The International Cooperation on Alternative Test Methods

To strengthen collaboration and communication in the design, execution, and peer review of validation studies, the International Cooperation on Alternative Test Methods (ICATM) was created in 2009 through an official agreement between national validation organizations from Europe, USA, Canada, and Japan, following a recommendation from the International Cooperation on Cosmetics Regulation. This agreement formalized the existing collaborations between the validation bodies with the aim to enhance international cooperation and coordination regarding the scientific validation and evaluation of alternative non- and reduced-animal toxicity testing methods. Originally, four agencies took part of the ICATM's creation i.e., ECVAM, NICEATM-ICCVAM, JaCVAM, and The Canadian Environmental Health Science and Research Bureau within Health Canada. Other economies have later integrated ICATM such as the Republic of Korea's Center for the Validation of Alternative Methods (KoCVAM) in 2011.

ICATM aims at establishing international cooperation in the areas of validation studies, independent peer review, and development of harmonized recommendations for the global regulatory acceptance of alternative methods. The purpose is to promote consistent and enhanced international cooperation, collaboration, and communication among national validation organizations in order to:

- ensure the optimal design and conduct of validation studies that will support national and international regulatory decisions on the usefulness and limitations of alternative methods proposed for regulatory testing;
- ensure high quality independent scientific peer reviews of alternative test methods, and consistency in transparency and stakeholder involvement;
- enhance the likelihood of harmonized recommendations by validation organizations on the usefulness and limitations of alternative test methods for regulatory testing purposes;
- achieve greater efficiency and effectiveness by avoiding duplication of effort and leveraging limited resources; and
- support the timely international adoption of alternative methods.

ICATM's aim is also to ensure that new alternative test methods/strategies adopted for regulatory use provide equivalent or improved protection for people, animals, and the environment, while reducing, refining (causing less pain and distress) or replacing, animal use whenever scientifically feasible.

3.3.3 Further International Efforts

Other countries outside of the EU, Japan, and US are investing in creating validation centers for alternative test methods in order to ensure appropriate adoption and international dialogue on the acceptance of alternative animal testing.

South Korea established KoCVAM in 2010 as part of the National Institute of Food and Drug Safety (NIFDS) in the Korean Food and Drug Administration. KoCVAM was founded based on the Laboratory Animal Act of Korea to formulate and implement policies for promoting alternative test methods. KoCVAM's objectives are to institutionalize alternative methods through various activities, to build cooperative relationships with both domestic and foreign organizations, and to establish the scientific validity of proposed alternatives according to Korean regulatory requirements. KoCVAM also intends to keenly respond to the global trends by introducing globally accepted alternative test methods and promoting alternative methods developed by Korean organizations. Its main activities comprise:

- providing policy support regarding the development and acceptance of alternative test methods that replace animal testing;
- executing validation and peer review of new and revised alternative test methods and proposing related guidelines;
- building cooperation with both domestic and foreign organizations (e.g. ECVAM, ICCVAM and JaCVAM) and participating in international collaborative studies;
- providing education, training and information regarding alternative test methods.

Brazil is another country that has created a national validation center, with the establishment in 2011 of the Brazilian Center for Validation of Alternative Methods (BraCVAM). BraCVAM was founded through a cooperation between the Oswaldo Cruz Foundation (Fundação Oswaldo Cruz, FIOCRUZ) and the National Agency of Health Surveillance (Agência Nacional de Vigilância Sanitária, ANVISA), and is currently located at the National Institute of Quality Control in Health (Instituto Nacional de Controle de Qualidade em Saúde, INCQS). It is the first Latin American center dedicated to the development and validation of alternative methods. BraCVAM aims to promote national and international harmonization on the acceptance of alternative methods, and to be responsive to regional needs, such as the need to develop alternatives for the quality controls of biological products against specific Brazilian threats (e.g., antivenoms against venomous snakes, spiders, scorpions, and frogs). BraCVAM will also coordinate together with the National Institute of Metrology, Normalization and Industrial Quality (Instituto Nacional de Metrologia, Qualidade e Tecnologia, Inmetro), an Alternative Methods National Network (Rede Nacional de Métodos Alternativos, RENAMA) which aims at promoting collaboration of Brazilian research groups working with alternative methods both for testing and education purposes. Finally, BraCVAM plans to actively develop future collaborative efforts with similar centers around the world.

3.4 Conclusions

During the last 20 years extensive efforts have taken place in order to develop, validate, and implement reduction, refinement, and replacement alternative methods to animal testing. Although the successes may seem limited in numbers, they represent a huge step forward in the way scientists now face the use of alternative test methods for safety assessment purposes. The regulatory adoption of the 3Rs principles across the world and the adoption of harmonized principles of validation are critical to ensure that scientists from different countries approach the questions related to human safety assessment also in a more harmonized and standardized manner. These approaches are relevant to a wide range of stakeholders involved in the development, validation, and evaluation of data generated by alternative methods including scientists from academia, industry, validation bodies, and regulatory agencies. Implementation of internationally harmonized standards permits us to strive, in a joint effort, towards limiting animal testing while establishing scientifically driven decision-making approaches for safety assessments and the protection of consumers.

ICATM and OECD are key constituents in the international harmonization of the validation and regulatory acceptance of alternative methods. Although the criteria for establishing the scientific validity of alternative methods are based on internationally accepted scientific principles, there can be room for variations regarding their regulatory acceptance depending, for example, on the test method purposes, its status of validation, and the specific regulatory context.

ICATM has the potential to promote international harmonization regarding the criteria for establishing the scientific validity of alternative test methods, thereby promoting their international regulatory acceptance. For that purpose, a common ground of understanding including also new countries willing to implement alternative methods needs to be established possibly through extensive collaboration and discussions.

Through its health effects testing guidelines program, the OECD provides a forum for scientific discussions among its 34 member countries to share their knowledge and expertise to (ideally) achieve formal adoption of test guidelines. The last decade has demonstrated a successful track record for the adoption of alternative test method based on guidelines. Formal adoption indicates acceptance by all member countries and imparts the requirements under the Mutual Acceptance of Data agreement. The adoption of the most up-to-date test guidelines in the national regulations of OECD member countries further guarantees the standardized implementation of alternative test methods. As a consequence, the OECD represents a proven mechanism for achieving international harmonization in the regulatory acceptance of alternative methods for chemicals testing. However, other areas of safety assessment as well as OECD non-member countries may still benefit from engaging in similar international platforms in order to further enhance international harmonized standards.

Alternative approaches are gaining momentum as more and more test methods achieve worldwide acceptance thanks in part, besides the efforts of international organizations such as ICATM and OECD, the commitment of validation bodies around the world (ECVAM, ICCVAM, JaCVAM, KoC-VAM, BraCVAM). Such efforts are of great value to ensure a global understanding of the importance of progressing the protection of human beings side-by-side with the development and implementation of scientifically based decision making as well as the replacement, reduction, and refinement of animal testing whenever possible. The several initiatives that exist today allow for collaborative efforts in validating and adopting alternative methods, and it is hoped that new countries and other areas of safety assessment can join these initiatives so that international harmonization increases steadily and in parallel with the relevant global demands.

References

1. D. H. Smyth, *Alternatives to animal experiments*, Scolar Press, Royal Defence Society, London, 1978.
2. W. M. S. Russell and R. L. Burch, *The Principles of Humane Experimental Technique*, Methuen, London, UK, 1959.
3. G. Archer, M. Balls, L. H. Bruner, R. D. Curren, J. H. Fentem, H.-G. Holzhütter, M. Liebsch, D. P. Lovell and J. A. Southee, The validation of toxicological prediction models, *ATLA*, 1997, **25**, 505.
4. OECD Guidance Document No.34 on '*the Validation and International Acceptance of New or Updated Test Methods for Hazard Assessment*', OECD Series on Testing and Assessment, Organisation for Economic Co-operation and Development, Paris, France, 2005.
5. Council Directive 86/609/EEC of 24 November 1986 on the approximation of laws, regulations and administrative provisions of the Member States regarding the protection of animals used for experimental and other scientific purposes, *Official Journal*, 1986, **L358**, 1.
6. Directive 2010/63/EU of the European Parliament and of the Council of 22 September 2010 on the protection of animals used for scientific purposes, *Official J. Eur. Union*, 2010, **L276**, 33.
7. M. Balls, B. Blaauboer, D. Brusick, J. Frazier, D. Lamb, M. Pemberton, C. Reinhardt, M. Roberfroid, H. Rosenkranz, B. Schmid, H. Spielmann, A.-L. Stammati and E. Walum, Report and recommendations of the CAAT/ERGATT workshop on the validation of toxicity test procedures, *ATLA*, 1990, **18**, 313.
8. I. Manou, C. Eskes, O. de Silva, G. Renner and V. Zuang, Safety data requirements for the purpose of the Cosmetics Directive, *ATLA*, 2005, **33**(S1), 35.
9. ICCVAM guidelines for the nomination and submission of new and revised alternative test methods, NIH publication no. 03–4508, National Institute of Environmental Health Sciences, Research Triangle Park, North Carolina, USA, 2003.
10. Regulation (EC) No 1907/2006 of the European Parliament and of the Council of 18 December 2006 concerning the Registration, Evaluation, Authorisation and Restriction of Chemicals (REACH), establishing a European Chemicals Agency, amending Directive 1999/45/EC and repealing Council Regulation (EEC) No 793/93 and Commission Regulation (EC) No 1488/94 as well as Council Directive 76/769/EEC and Commission Directives 91/155/EEC, 93/67/EEC, 93/105/EC and 2000/21/EC, *Official J. Eur. Union*, 2006, **L396**,1.
11. ECHA, Chapter R.4, Evaluation of available information, in: *Guidance on Information Requirements and Chemical Safety Assessment*, 2011. Available

at: http://echa.europa.eu/documents/10162/13643/information_requirements_r4_en.pdf/ Accessed on 25.10.2012.

12. T. Hartung, S. Bremer, S. Casati, S. Coecke, R. Corvi, S. Fortaner, L. Gribaldo, M. Halder, S. Hoffmann, A. J. Roi, P. Prieto, E. Sabbioni, L. Scott, A. Worth and V. A. Zuang, Modular approach to the ECVAM principles on test validity, *ATLA*, 2004, **32**, 467.

13. W. S. Stokes, L. M. Schechtman and R. N. Hill, The Interagency Coordinating Committee on the Validation of Alternative Methods (ICCVAM): A Review of the ICCVAM Test Method Evaluation Process and Current International Collaborations with the European Centre for the Validation of Alternative Methods (ECVAM), *ATLA*, 2002, **30**(S2), 23.

14. M. Balls, B. J. Blaauboer, J. H. Fentem, L. Bruner, R. D. Combes, B. Ekwall, R. J. Fielder, A. Guillouzo, R. W. Lewis, D. P. Lovell, C. A. Reinhardt, G. Repetto, D. Sladowski, H. Spielmann and F. Zucco, Practical aspects of the validation of toxicity test procedures, The report and recommendations of ECVAM workshop 5, *ATLA*, 1995, **23**, 129.

15. R. D. Curren, J. A. Southee, H. Spielmann, M. Liebsch, J. H. Fentem and M. Balls, The role of prevalidation in the development, validation and acceptance of alternative methods, *ATLA*, 1995, **23**, 211.

16. ICCVAM Authorization Act of 2000. Public Law 106-545. Available at: http://iccvam.niehs.nih.gov/docs/about_docs/pl103_43.pdf/ Accessed on 25.10.2012.

17. Validation and Regulatory Acceptance of Toxicological Test Methods: A Report of the Ad Hoc Interagency Coordinating Committee on the Validation of Alternative Methods, NIH publication no. 97–3981, National Institute of Environmental Health Sciences, Research Triangle Park, North Carolina, USA, 1997.

18. H. Kojima, JaCVAM: An organization supporting the validation and peer review of new alternatives to animal testing, *AATEX*, 2007, **14**, special issue, 483.

19. Regulation (EC) No 1223/2009 of the European Parliament and of the Council of 30 November 2009 on cosmetic products, *Official J. Eur. Union*, 2009, **L342**, 59.

20. *Toxicity Testing in the 21st Century: A Vision and a Strategy*, National Research Council of the National Academies, Washington, DC, 2007.

21. D. J. Dix, K. A. Houck, M. T. Martin, A. M. Richard, R. W. Setzer and R. J. Kavlock, The ToxCast program for prioritizing toxicity testing of environmental chemicals, *Toxicol. Sci.*, 2007, **95**, 5.

22. R. J. Kavlock, C. P. Austin and R. R. Tice, Toxicity testing in the 21st century: implications for human health risk assessment, *Risk Anal.*, 2009, **29**, 485.

23. Towards the replacement of *in vivo* repeated dose systemic toxicity testing. Toxicology in the 21st century: mechanism-driven toxicology defines

the safe dose, ed. T. Gocht and M. Schwarz, Coach consortium, France, 2012.

24. S. Coecke, H. Ahr, B. J. Blaauboer, S. Bremer, S. Casati, J. Castell, R. Combes, R. Corvi, C. L. Crespi, M. L. Cunningham, G. Elaut, B. Eletti, A. Freidig, A. Gennari, J. F. Ghersi-Egea, A. Guillouzo, T. Hartung, P. Hoet, M. Ingelman-Sundberg, S. Munn, W. Janssens, B. Ladstetter, D. Leahy, A. Long, A. Meneguz, M. Monshouwer, S. Morath, F. Nagelkerke, O. Pelkonen, J. Ponti, P. Prieto, L. Richert, E. Sabbioni, B. Schaack, W. Steiling, E. Testai, J. A. Vericat and A. Worth, Metabolism: A Bottleneck in *In vitro* Toxicological Test Development, ECVAM Workshop 54, *ATLA*, 2006, **34**, 49.

25. OECD Guideline for the testing of chemicals 417, Toxicokinetics, Organisation for Economic Co-operation and Development, Paris, France, 2010.

26. OECD Guidelines for the Testing of Chemicals 420, Acute oral toxicity – fixed dose procedure, Organisation for Economic Co-operation and Development, Paris, France, 2001.

27. OECD Guidelines for the Testing of Chemicals 423, Acute oral toxicity – acute toxic class method, Organisation for Economic Co-operation and Development, Paris, France, 2001.

28. OECD Guidelines for the Testing of Chemicals 425, Acute oral toxicity: up-and-down procedure, Organisation for Economic Co-operation and Development, Paris, France, 2008.

29. OECD Guidelines for the Testing of Chemicals 428, Skin absorption: *in vitro* method, Organisation for Economic Co-operation and Development, Paris, France, 2004.

30. OECD Guidelines for the Testing of Chemicals 432, *In vitro* 3T3 NRU phototoxicity test, Organisation for Economic Co-operation and Development, Paris, France.

31. OECD Guidelines for the Testing of Chemicals 431, *In vitro* Skin Corrosion: Reconstructed Human Epidermis (RhE) Test Method, Organisation for Economic Co-operation and Development, Paris, France, 2013. Originally adopted in 2004.

32. OECD Guidelines for the Testing of Chemicals 430, *In vitro* Skin Corrosion: Transcutaneous Electrical Resistance Test (TER), Organisation for Economic Co-operation and Development, Paris, France, 2013. Originally adopted in 2004.

33. OECD Guidelines for the Testing of Chemicals 435, *In vitro* Membrane Barrier Test Method for Skin Corrosion, Organisation for Economic Co-operation and Development, Paris, France, 2006.

34. OECD Guidelines for the Testing of Chemicals 439, *In vitro* Skin Irritation: Reconstructed Human Epidermis Test Method, Organisation for Economic Co-operation and Development, Paris, France, 2013. Originally adopted in 2010.

35. OECD Guidelines for the Testing of Chemicals 437, Bovine Corneal Opacity and Permeability Test Method for Identifying i) Chemicals

Inducing Serious Eye Damage and ii) Chemicals Not Requiring Classification for Eye Irritation or Serious Eye Damage, Organisation for Economic Co-operation and Development, Paris, France, 2013. Originally adopted in 2009.

36. OECD Guidelines for the Testing of Chemicals 438, Isolated Chicken Eye Test Method for Identifying i) Chemicals Inducing Serious Eye Damage and ii) Chemicals Not Requiring Classification for Eye Irritation or Serious Eye Damage, Organisation for Economic Co-operation and Development, Paris, France, 2013. Originally adopted in 2009.

37. OECD Guidelines for Chemical Testing 460, Fluorescein leakage test method for identifying ocular corrosives and severe irritants, Organisation for Economic Co-operation and Development, Paris, France, 2012.

38. OECD Guidelines for the Testing of Chemicals 487, *In vitro* mammalian cell micronucleus test, Organisation for Economic Co-operation and Development, Paris, France, 2010.

39. OECD Guidelines for Chemical Testing 429, Skin sensitization: Local Lymph Node assay, Organisation for Economic Co-operation and Development, Paris, France, 2010.

40. OECD Guidelines for Chemical Testing 442A, Skin sensitization: Local Lymph Node assay: DA, Organisation for Economic Co-operation and Development, Paris, France, 2010.

41. OECD Guidelines for Chemical Testing 442B, Skin sensitization: Local Lymph Node assay: BrdU-ELISA, Organisation for Economic Co-operation and Development, Paris, France, 2010.

42. Guidance Document No.19 on '*the Recognition, Assessment and Use of Clinical Signs as Humane Endpoints for Experimental Animals Used in Safety Evaluations*', OECD Series on Testing and Assessment, Organization for Economic Co-operation and Development, Paris, France, 2000.

CHAPTER 4

Refinement Alternatives: Minimizing Pain and Distress in In Vivo Toxicity Testing

JON RICHMOND

Ethical Biomedical Research and Testing – Advice and Consultancy, Doubledykes, Carslogie Road, Cupar, Fife KY15 4HY, UK
Email: dr.jonrichmond@gmail.com

4.1 Introduction

Toxicology is the study, and prediction by inference and extrapolation, of the biological effects of chemicals and their metabolites on living organisms. Toxicity testing informs decisions to protect man and the environment, and comprises diverse objectives, study types and toxicological endpoints.

Traditional *in vivo* studies use small numbers of animals, high doses of test material, and relatively short exposure periods; eliciting overt adverse effects,[1-4] clinical signs and pathological changes as homeostatic mechanisms are stressed and overwhelmed. They give little insight into toxicity pathways, modes and mechanisms, making definitive inferences about relevant target population exposures difficult.

Some animals used for toxicology studies experience pain, suffering, distress and lasting harm: both 'direct' welfare costs due to the procedures and primary effects of the test materials, and 'indirect' welfare costs contingent upon the conditions under which the animals are bred, kept and used.

Issues in Toxicology No. 19
Reducing, Refining and Replacing the Use of Animals in Toxicity Testing
Edited by David G. Allen and Michael D. Waters
© The Royal Society of Chemistry 2014
Published by the Royal Society of Chemistry, www.rsc.org

It is essential that animal study data and conclusions only relate to direct effects of test materials uninfluenced by physiological or behavioural effects produced, promoted, modulated or prevented by husbandry and care systems, the procedures applied, and secondary and tertiary consequences of unnecessary, unrecognized or untreated pain, or distress. These non-test material related artefacts increase inter-animal variability and compromise the validity of study findings. High standards of animal welfare are a pre-requisite for high quality science[5,6] and the refinement of animal care and use is the means by which this is achieved.

Despite the on-going transformation of toxicity testing to more mechanistic approaches based primarily on non-animal studies providing insights into perturbations of defined toxicology pathways,[7] we cannot yet replace all animal use for toxicology research and testing without compromising scientific progress and human and environmental safety. However, our growing understanding of toxicological pathways and biomarkers sets the scene for reducing animal welfare costs and improving the resulting scientific insights by identifying and implementing earlier, more relevant, sensitive, specific, surrogate and humane endpoints.

4.2   The 3Rs: Replacement, Reduction and Refinement

The concept of a moral obligation to minimize animal welfare costs by replacing, reducing and refining the use of animals for experimental and other scientific purposes arose in the 19th century:

> ... We should never have recourse to experiment in cases which observation can afford us the information required; No experiment should be performed without a distinct and definite objective and without the persuasion that the object will be attained and produce a real and uncomplicated result; We should not needlessly repeat experiments and [only] cause the least possible suffering, using the lowest order of animals and avoiding the infliction of pain; We should try to secure due observation so as to avoid the necessity of repetition ....[8]

As detailed in Chapter 1, the conceptual framework of the 3Rs of replacement, reduction and refinement was subsequently developed by Russell and Burch,[9] who argued that the 3Rs both reduce animal welfare costs and enhance the quality of animal-based science.

Test methods based on an understanding of relevant toxicological mechanisms, early biomarkers and humane endpoints are elegant, valid science and replacement, reduction and refinement alternatives are better considered as 'advanced' rather than 'alternative' test methods. They extend the scope, and reduce the limitations, of traditional animal models and are more relevant, robust, valid, economical, faster and easily scalable.

The 3Rs are a pre-requisite for a successful and sustainable toxicology science base (Table 4.1).

Table 4.1 Public and political support for animal testing is conditional.

Political and public support for animal studies is conditional upon animal studies being conducted only out of scientific necessity and when;[10]

- the scientific objectives are timely, clearly defined and sufficiently important;
- there is no non-sentient replacement alternative;
- they are performed with care and compassion, with all relevant reduction and refinement options implemented in the design and conduct of the study;
- all reasonable efforts are made to minimize not just the number of animals used, but the total resulting animal distress;
- the scientific and societal benefits are maximized by yielding relevant and reliable results; and
- there is justification for, and transparency about, *what* animal testing is done, *why* it is done and *how* it is done.

Legislation, regulatory policy, testing requirements and preferred test methods take account of the 3Rs, requiring toxicologists to minimize the number of animals used and the suffering caused; with flexibility, providing expert judgement is demonstrably exercised, within policies, guidelines and requirements (for example with respect to dose levels, choice of species and endpoints) to use refined testing strategies, methods, and endpoints.

The 3Rs are supported by scientific societies such as the Society of Toxicology (SOT) and the British Toxicology Society (BTS); and OECD has published guidance on humane endpoints,[11] reflected in OECD Test Guidelines.

The 3Rs are consistent with:

- harmonization of international validation processes (See for example http://alttox.org/ttrc/validation-ra/international-harmonization/), regulatory testing requirements, test methods and decision making frameworks – eliminating the need to use different animal models and endpoints to inform multiple regulators about single toxicological endpoints; and
- protocols for the mutual recognition and acceptance of test data and regulatory assessments (see for example http://alttox.org/ttrc/validation-ra/international-harmonization/) – minimizing animal testing and ensuring regulatory decisions are taken in an ethical, consistent and timely manner.

Armed with increasingly powerful insights into toxicology pathways, mechanisms, biomarkers and endpoints, and a better understanding of animal welfare, the best animal welfare and best science must drive and shape future developments in *in vivo* toxicological policy and practice.[12,13]

4.3 Refinement

Refinement comprises measures to avoid, minimize, recognize and alleviate pain, distress and lasting harm – or to otherwise improve animal welfare and well-being – and should be viewed positively as 'improving welfare' rather than 'minimizing distress'.[10]

Table 4.2 Animal welfare costs.

Animal welfare costs are proportionate to the:

- number of animals used;
- degree of sentience and needs of the experimental subjects;
- nature, duration, intensity and frequency of the challenges;
- relevant biological systems and mechanisms;
- endpoints applied; and
- other factors aggravating or ameliorating the distress experienced.

In vivo toxicology studies impose animal welfare costs (Table 4.2), with husbandry practices and procedures causing pain, distress and lasting harm; and by withholding measures promoting positive welfare states.[12]

Refinement enhances the welfare of all animals bred, kept and used for experimental and other scientific purposes by reducing husbandry and procedure-related harms and promoting positive welfare experiences and emotional states; and thereby reducing inter-animal variability and improving the quality and reliability of experimental results.

Scientific progress leads inevitably to phasing out some forms of animal use whilst creating demand, or setting the scene, for new uses[14] and providing further scope for refinement by identifying earlier endpoints, enabling more relevant, targeted animal studies producing more information on the formation and activity of metabolites, relevant exposures and tissue concentrations, modes of action, toxicity pathways, mechanisms and target organ responses.

The practice of refinement requires a scientific, evidence-based approach built on an understanding of behaviours and findings in normal animals, the impact of care systems and scientific procedures, evaluation of welfare states, and the development and application of informed, practical solutions. Refinement in part incorporates progress in clinical practice: for example advanced analytical methods, imaging technologies, and the appropriate and effective use of anaesthetics, analgesics and antibiotics.

Refinement must be considered on a case-by-case basis to identify the options suitable for each toxicological endpoint, test material and protocol.

4.3.1 Refinement and Replacement

'Relative refinement' is the replacement of procedures causing animal pain or distress with non-painful procedures: for example the use of insentient or immature forms, and humane killing to provide material for *ex vivo* experiments.

4.3.2 Refinement and Reduction

'Reduction' minimizes the number of animals used to obtain suitably robust data: using more is wasteful, using too few produces inconclusive or misleading findings. Means of minimizing numbers include clearly defining objectives; structured literature searches; data sharing; tiered and hierarchical testing strategies; sound experimental design; and robust statistical analysis.

Table 4.3 A holistic approach to reduction and refinement.

Reduction and refinement must be considered in parallel when considering:[10]

- reducing animal numbers by applying aggressive, high-welfare cost protocols[15] which may increase cumulative welfare costs, produce welfare related artefacts, or impose unreasonable welfare costs on the animals that are used;
- using small numbers of higher species or larger number of a lower species;
- using models and endpoints resulting in small number of symptomatic animals or larger numbers of asymptomatic animals;
- accepting high incidences of low welfare costs (such as reduced weight gain) or preferring lower incidences of more substantial welfare costs; or
- re-using animals, thereby reducing animal numbers but increasing the welfare costs to the animals that are used.

Reduction and refinement are so closely linked that a holistic, rather than sequential, approach is required as balances are struck between the interests of individual animals and the total number of animals used (Table 4.3).

Experience and expert judgement are required to best balance competing reduction and refinement interests, with decisions on the best course of action taken on a case-by-case basis to minimize the total animal suffering likely to be caused.

4.4 The Challenges of Refinement

Russell and Burch[9] considered refinement the most challenging of the 3Rs: '... the central problem is that of determining what is and is not humane, and how humanity can be promoted without prejudice to scientific and medical aims ...'.

With respect to animal welfare, refinement requires an awareness of normality in the species, strain and individual animal, and objective measures of normal and abnormal states of wellbeing. Identifying more refined options requires objective and reliable assessment of which options are most welfare friendly;[16] however, identifying signs of pain and distress and assessing their severity can be difficult. Published welfare assessment schemes[17] typically rate poorer rather than higher welfare states, detect acute rather than chronic problems, and are not always sufficiently sensitive or specific to demonstrate subtle welfare gains.

Preference testing[18–21] and operant conditioning[22,23] do not identify welfare optima, rather they identify the least objectionable of the options evaluated.[24] In addition, the short-term preferences on which they are based may not correlate with long-term preferences and needs.[25,26]

There are difficulties validating the scientific benefits of refinement alternatives (with respect to their relevance and reliability for a defined purpose) against established, benchmark animal models. High quality human data is rarely available, current animal tests are not completely reliable and historical animal tests data are of variable quality. If refinement improves test-system performance in terms of sensitivity or specificity then the results, although more relevant and reliable, will be at variance with those of the less refined test systems.

Some resist refinement on the basis of there being an extensive, historical toxicological database – disregarding incremental refinements introduced as this data was complied, related to animal strains; microbiological and health status; housing, care, and diet; analytical methods; experimental design; and data analysis.

The fact that more refined methods and the higher welfare states they represent influence results is a reason to accept, rather than resist, progress and change. Concern that findings with refined practices might be different is not a criticism of the scientific validity of the more refined methods: it is an indictment of the reliability, relevance, limitations and inadequacies of the less refined methods.

Even when good faith measures are made to refine studies, other priorities may cause conflicts of interest. For example in chronic studies, in an attempt to ensure the required survival rates, there may be reluctance to humanely kill animals on welfare grounds even though findings in animals with serious welfare problems may not be relevant or reliable.

4.5   The Scientific Benefits of Refinement

Interpreting animal test data involves assumptions, inferences and extrapolations. To produce valid findings the animals used must be uniform, with physiological and behavioural parameters and responses within the normal range, and any departures from the normal state directly and intentionally attributable to the primary effects of the test materials. Inter-animal variation and atypical responses, whether genetic, physiological, procedural or incidental in origin, affect experimental data and its interpretation.

Stressors alter baseline behaviours, physical and clinical findings, and the nature and magnitude of responses to experimental and other interventions.[5,6] Refined procedures and optimized conditions of animal care and use minimize pain and distress, reduce inter-animal variation, and yield more reproducible and reliable data.

Variation between the responses of individual isogenic animals subjected to identical test material exposures do not reflect the intrinsic toxicological properties of the test material: it indicates the need to further refine animal care and use to minimize unwanted inter-animal variation and the artefacts this produces.

4.6   The Welfare Benefits of Refinement

Although stress, pain and distress are universal phenomena with physical and emotional components, their effects vary between species, strains, individuals and in the same individual at different times,[27] and it can be challenging to demonstrate that intuitively welfare friendly potential refinements actually reduce animal suffering or improve quality of life.

Table 4.4 Validating the scientific and animal welfare benefits of refinement.

When validating the scientific and animal welfare impact of refinement parameters assessed can include:

- variability of experimental results within a test group;
- ability to cope with changes to the social and physical environment;
- behavioural indices including the appropriateness of the behavioural repertoire, the balance of socio-positive and -negative behaviours, maladaptive or abnormal behaviours;
- signs of chronic fear and distress; and
- physiological parameters such as stress hormone levels, body temperature, heart rate, immune competence, body weight and food intake.

It is often presumed, in the absence of evidence to the contrary, that interventions, procedures, deprivations and pathologies representing or producing poor welfare states in humans also do so animals (Table 4.4). However, it cannot be assumed that they cause equal degrees of distress in humans and animals, or that procedures not distressing humans do not distress animals. Furthermore, animals with established pathologies may show no clinical signs during life; and animals may be distressed though not in pain, thus displaying signs of suffering not alleviated by analgesics.

Critical anthropomorphism is required:[16] empathy tempered with objective knowledge of normality in the animal, the test material and procedures, preceding events and the significance of observed findings.

Behavioural indices are often the earliest and most sensitive welfare indicators, and best suited to cage-side assessments.

4.7 Stress, Distress and Pain

Animal welfare is a measure of an animal's quality of life; its physical and mental wellbeing. It is an animal's emotional rather than physiological state that determines its sense of well being.[28]

Stress, distress and pain are all negative welfare states, but are different entities requiring different means of recognition and alleviation[29] by the appropriate use of environmental, non-pharmacological and pharmacological interventions, and humane endpoints. In toxicology research and testing distress is a more prevalent adverse effect rather than pain.

Toxicology research and testing should be planned and conducted to minimize exposure to stressors and ensure animals do not experience severe, ongoing or untreated pain or distress. Detecting the presence of pain and distress can be difficult, and judging severity is even more difficult, requiring training, experience and judgement, and taking account of behavioural changes, clinical appearance, and physiological responses.

4.7.1 Stress

Stress is a common event, caused by real or perceived perturbations to physiological homeostasis and psychological wellbeing, producing as coping mechanisms and adaptive changes a variety of non-specific, reversible, physiological, endocrine, neuroendocrine, immune, behavioural and psychological responses that vary between species, strains and individuals, and in the same individual at different times. Whilst these homeostatic changes may be beneficial to animals in the short-term, they can produce overt adverse effects if the stressors or the responses are prolonged.

Stress-induced changes compromise toxicology studies: inter-animal variation is increased, the uptake and clearance of test materials are altered,[30] and disease processes are modified.[31]

4.7.2 Distress

It is important to distinguish between stress and distress.[29]

Distress results when homeostatic mechanisms are overwhelmed, leading to subclinical or clinical decompensations and poor welfare states. Progression from stress to distress may be due to the nature, duration and intensity of the stressor; or the result of cumulative or serial stressors. The transition from stress to distress is more likely when animals have little control over their environment and when the stressors are unpredictable.[32] Distress may persist after the original stressors are removed.

To deal with distress it is important to identify its aetiology (non-painful causes include fear, nausea, hunger, dehydration, dizziness and weakness – and in toxicology studies these are more prevalent than pain) and whether it is primarily due to husbandry and care conditions, procedures performed or test materials.

Managing distress may require dealing with the primary cause, removing the animal from the study, or treating the signs with drugs (anxiolytics, antidepressants, sedatives or neuroleptics), or non-pharmacological measures (such as improving husbandry and care, socialization with competent and caring staff and compatible conspecifics, and procedural training).

4.7.3 Pain

Pain is an unpleasant sensory and emotional experience with psychological and somatic components; it is most commonly the result of tissue injury or inflammation. Species, strain, age, gender-specific and individual differences in responses to potentially painful events and their treatments are well documented.[33–37]

Pain in animals is often unrecognized and untreated because of lack of information about how it is best recognized and treated. Behavioural indices are often the best early indicators of pain.[38,39]

Anxiety, fear and stress accentuate experience of pain, and these should be managed by appropriate non-pharmacological measures such as environmental enrichment and acclimatizing the animal to its environment, handlers and test environment.

Untreated pain produces physiological, immunological, cognitive and behavioural effects detrimental to animal welfare and affecting experimental results.[40–42] Appropriate pain management decreases complications, lowers mortality, reduces variability and prevents scientific outputs being compromised by the side effects of the analgesics used.

When measures are in place to detect and manage pain, it is important that their effectiveness is confirmed,[43] even though the complete elimination of pain in conscious animals is not always achievable.

4.7.4 Severity Rating Systems

Animal suffering has emotional and physical dimensions and drawing timely, meaningful conclusions about animal welfare, refinements and the biological effects of test materials requires more than crude morbidity and mortality data (Table 4.5).

Behavioural findings and changes are often the earliest, most sensitive, specific and meaningful changes and indicators of welfare states, best suited to practical use in the laboratory setting.[38] It is generally assumed that the magnitude of clinical signs and behavioural changes is proportional to the degree of pain or distress. Behaviour-based pain assessment schemes have been developed for mice,[44,45] rats,[38,46] rabbits,[47] cats,[48] dogs,[49–51] sheep,[52] cattle[53] and horses.[54] However, these all rely on observers knowing what to look for, and when and where to look.[55]

Whilst some single parameters such as hypothermia[56] are predictive of significant morbidity or mortality, most assessment systems rely on combinations of observable discrete or continuous variables that can consistently be made by trained animal care and research staff.[57,58] Even with training and experience there may be some inter-observer variation; nevertheless, changes in successive observations by the same observer can be relied upon to indicate an animal's condition is stable, improving or deteriorating.

The better systems are consistent, easy to use, and specific and sensitive enough to detect poor welfare whether the impairment is direct or contingent, reflects immediate or delayed effects, is caused by local or systemic problems, or is the product of primary, secondary or tertiary changes. They provide a systematic approach to evaluating welfare; can assist to identify refined procedural options when planning studies; recognize, record, interpret and respond to findings during a study; encourage the use of standard documentation and plain, non-technical language with a limited range of standard keywords to

Table 4.5 The detection of adverse effects.

The detection of adverse and other test compound induced effects is based on consideration of behavioural, physiological, clinical and pathological findings, with reliance best placed upon multiple indicators which: • occur in appropriate contexts;   • progress with the severity of the insult;   • predict the ultimate welfare, clinical or pathological outcome; and   • can be controlled by appropriate specific, symptomatic or supportive treatments.

identify, describe and record findings; simplify staff training; and facilitate communication within and between research groups.

4.8 Refinement and Study Design

Refinement is relevant to planning and conducting animal-based toxicology studies.

4.8.1 Tiered, Step-wise, Hierarchical Approaches

The assessment and classification of test materials require serial assessments of information and findings to plan the sequence in which objectives are pursued and studies performed; reviewing study plans, protocols and decision-making as new datasets are generated.

The testing and evaluation of toxicants should proceed in step-wise fashion – considering the nature of the test material, reviewing existing information including findings with related materials with similar structures,[59] chemical characterization, *in vitro* followed by *in vivo* toxicity tests, and dose–response and extrapolation modelling.

Tiered, step-wise, hierarchical approaches to evaluating test materials constitute one important component of refinement (Table 4.6).

For regulatory testing when animal numbers specified for an individual test of an appropriate level of precision cannot be reduced below predetermined minima, there is scope to reduce the number of animals required for the larger programme of work by using tiered and step-wise approaches, with the right tests done at the right time and for the right reasons, and the careful use of control groups.

Tiered, step-wise, hierarchical approaches and decision trees have been developed for a range of test material classes and toxicological endpoints.[62–65]

4.8.2 Experimental Design

This chapter does not deal in detail with statistical methods: expert advice should be sought as required before studies are performed.

Table 4.6 Tiered, step-wise, hierarchical approaches and refinement.

Examples of refinement as part of tiered, step-wise, hierarchical approaches include:

- reviewing of physicochemical properties, considering *in vitro* test results, and the potential of materials to cause skin irritation may obviate the need to undertake animal studies to directly assess eye irritancy potential: and when animal testing is required a tiered approach, using one animal in the first instance pre-treated with a topical anaesthetic, systemic analgesics and applying humane endpoints, can reduce and minimize animal numbers and any suffering caused;[60]
- preliminary *in vitro* cytotoxicity test data informing dose-setting for *in vivo* toxicity tests;[61] and
- pilot studies and preliminary *in vitro* or animal test data, including dose-ranging studies, used to plan definitive studies.

There are three common experimental subtypes:

- pilot studies exploring logistics and gathering preliminary data to plan more substantive experiments;
- exploratory experiments producing data to generate and formulate hypotheses; and
- confirmatory experiments testing defined hypotheses.

Experimental design takes account of the degree of precision required (Table 4.7). Studies to identify hazardous properties, to assign materials to different toxic classes, and to inform robust risk assessments require different levels of precision, test strategies, methods and endpoints. When test materials are only to be assigned to general toxicological, hazard or risk categories requiring only an estimate of their biological properties the number of animals required may be smaller than that required to produce the more precise values required to place test materials in rank order.

Common study designs include: randomized between-subject studies which assume experimental subjects are uniform and use uniform group sizes to maximize power; randomized block designs, with randomization within blocks, to take account of suspected variables; and cross-over within-subject designs where the order of treatments is randomized (Table 4.8).

Variation and precision are important in determining the number of animals required and to properly interpret study findings.[67] The expected degree of variability within and between experimental subjects is an important determinant of the number of experimental subjects required; the more variable the animals' responsiveness the larger the number required for a statistically representative and relevant sample. Refinement minimizes inter-animal variation and reduces the numbers required.

Statistical elements of experimental design include the statistical method to be used, sample size (often determined by power analysis based on the expected effect size, the required significance level based on the probability of a false

Table 4.7 Principles of experimental design.

In all cases, as a pre-requisite for scientific validity and ethical animal use, the experimental design must meet five essential requirements:[66]

- the absence of bias; a pre-requisite for all inferential statistical techniques, usually achieved through randomization – applied not just in assigning individual animals to study groups and within groups and blocks, but also to the siting of cages and the order in which animals are dosed and sampled – and/or blinding;
- adequate statistical power; relevant to 'sensitivity', the ability to detect the effect of a test material;
- known range of applicability for extrapolation and generalization of findings; internal and external validity;
- simplicity; and
- amenability to statistical summarization and analysis to determine the extent to which differences within and between groups are dependent on chance.

Table 4.8   Principles of study design.

For each study or experiment it is necessary to:

- define the precise purpose of the study or experiment;
- select the appropriate model;
- define the dependent variables (the effects to be measured representing the animals' response to the treatment) and independent variables (the treatments or other conditions controlled and varied by the researcher), and the experimental unit (the smallest meaningful division of the experimental subjects relevant to statistical analysis: depending on the context this may the individual animal, cages mates or litter mates); and
- determine the correct statistical methods to use and how data will be collected, analysed and presented.

result, the sidedness of the test, and known or expected variability in the test subjects or test material), and power (probability of detecting the specified effect, showing it to be significant at the expected level). The power analysis depends critically on the estimated standard deviation, which can be difficult to judge before test data is available. The quality of statistics in the biomedical sciences is not universally high,[68–71] custom and practice are often poor indicators of future success when determining sample size, and expert advice should be sought at an early stage to select appropriate tests and statistical power, and ensure the data gathered is amenable to statistical summarization and analysis.

There may be some scope to reduce the number of animals required by taking account of the prevalence of the outcome of interest;[72] that is, it may require less data and fewer animals to identify extreme or common, rather than subtle or uncommon, toxicological properties.

There are equally important non-statistical refinements to consider at the planning stage such as the selection of the most suitable test method and experimental subjects, husbandry and care systems, procedural details, and other means of controlling and minimizing unwanted stressors and unnecessary variables.

4.8.3   Pilot Studies

Pilot studies are small-scale preliminary studies, sometimes using only single animals, undertaken to improve the design and conduct of definitive studies by:

- identifying potential interactions between test materials and study variables;
- determining the nature and variation of individual animal's responses allowing calculation of the number of animals required;
- demonstrating the relationship between dose/exposure and response;
- identifying the most appropriate dosing, sampling and monitoring regimens;
- providing insights into target organ toxicity and the nature, recognition, incidence, severity, timing and management of possible adverse effects and their potential reversibility;

- identifying early markers of the relevant toxicological endpoint; and
- defining humane endpoints minimizing animal suffering whilst maximizing useful data.

4.8.4 Control Groups

Control groups are used as standards for comparison – demonstrating that test systems are suitably responsive or indicating there are confounding variables within the test system. The use of control groups presupposes that the animals' initial baselines and responsiveness are as expected.

In some cases statistical power can be maintained, and resource costs and animal suffering reduced, by increasing the number of negative controls and reducing the number of animals in treatment groups. Other reduction and refinement options include using concurrent controls for multiple test materials, and using historical control data when test systems are in routine use within a laboratory and consistently produce appropriate positive and negative results.

4.8.5 Satellite Groups

Sometimes, with careful planning, several questions may be addressed in single study.[62,63] Satellite groups are one way of allowing more than one experimental objective to be pursued within a single experiment:[73] for example, toxicokinetic data can be gathered from animals in the course of single-dose toxicity studies.

4.8.6 Re-use

Re-use, the second or subsequent scientific use of an animal that has already been used when a naive, unused animal would also be scientifically satisfactory, can be considered when the first use has not compromised the animal's scientific suitability for the second or subsequent use, and the animal experienced only minimal pain, or distress, and no lasting harm.

When an animal is re-used it should be confirmed on a case-by-case basis by a competent person that it has been restored to a normal state of wellbeing, and that the cumulative suffering it may experience is not excessive.[74]

4.8.7 Replicating Studies

Duplicating animal studies due to a lack of awareness of what has been done before is not justified. Duplication, or rather replication, can be justified when introducing new model systems, evaluating procedural changes, changing laboratories or key personnel, and restarting testing after prolonged periods of inactivity.

4.9 Choice of Animal Models

Early science was curiosity driven, empirical and descriptive: experimentation to understand mechanisms and solve problems was a later development, and was followed by deductive science formulating and testing hypothesis.[75] This progression is recapitulated in toxicology where initial, developmental, basic research using exploratory models sets the scene for explanatory models; and explanatory models set the scene for predictive models which, if validated, can be considered for regulatory purposes.

Factors influencing model selection include its ability address the scientific question, the species and strain least likely to suffer, and the housing and care requirements required to obtain scientifically satisfactory results. In some cases account is also taken of accumulated historical data.

Animal models differ from the human condition with respect to 'fidelity' and 'discrimination' (Table 4.9).[9] 'Fidelity' relates to how closely a model approximates in nature or character to the original (face validity): 'discrimination' refers to how closely it reproduces the particular property of interest (predictive validity). High-discrimination models are required for toxicology studies if the results are to be extrapolated or generalized to humans, and the more we know of toxicological mechanisms, biomarkers and endpoints the easier it is to move from superficially attractive, but sometimes misleading, high-fidelity models to the required high-discrimination models.

Non-human primate models are high fidelity, but not always high-discrimination, models. The Limulus Amoebocyte Lysate test[76] used to detect Gram-negative endotoxins is an example of a low-fidelity/high-discrimination model: it does not use mammalian tissue or reproduce the pyrexia endpoint of the standard rabbit-based infusion-test; instead the presence of Gram-negative endotoxins consistently produces a specific change in horseshoe crab haemolymph.

Model selection requires that study objectives are carefully defined and that the properties of available animal models, including their scope and limitations and the nature of the test materials, have been taken into account. When dealing with complex questions it may be necessary to use more than one experimental approach and animal model.

Animal models are complex biological systems used to model a wide range of dynamic human biological systems and responses; and evaluating the data generated involves simplifications, generalizations and assumptions about genetic, developmental, metabolic, physiological and other characteristics, and how they impact on biotransformation, bio-activation and detoxification. Assessing hazards and risks associated with human exposure to toxicants has traditionally relied on extrapolating findings from animal models to humans,

Table 4.9 The properties of animal models.

The properties of animal models can be categorized on the basis of:

- face validity: the degree to which the model 'looks like' the condition or system of interest;
- predictive validity: the degree to which it responds like the system of interest; and
- construct validity: how well it accords with current knowledge and theories.

from one route of exposure to others, from high doses to likely real-life exposures, whilst making assumptions and trying to resolve uncertainties about biotransformation and other processes.[77]

Animal and non-animal models mimic only limited aspects of the human condition or system of interest.[9,78] This must be borne in mind when selecting relevant and reliable model systems and the data to be gathered, and in the interpretation, analysis, generalization and extrapolation of findings. For example, animal models of disease used for pharmaceutical research and development are seldom based on the same aetiologies as the disease processes in humans.[78]

Animal models of disease, animals expressing harmful genetic mutations and some lines of genetically modified animals[79] have specific problems and needs in addition to, or different from, normal animals. These must be addressed when such animals are bred, kept or used for scientific purposes.

4.10 Selection of Animals

High quality laboratory animal science requires high quality, high health status, purpose-bred animals free of clinical and sub-clinical disease, kept and maintained in ways which ensure they are uniformly and appropriately responsive to test materials. The use of purpose-bred animals from reputable laboratory animal breeders permits a high degree of control of genetic variability and microbiological status.

4.10.1 Species

Some species are afforded special legal protection, have a greater capacity to experience pain and distress (in which case the animals which will experience the least pain, suffering and distress should be selected), and have specific, complex husbandry requirements difficult to reproduce in the laboratory setting. Choosing the species whose needs can best be met may constitute refinement.

4.10.2 Genetic Status

Much historical toxicological data was generated using outbred animals[80] produced by random mating and large breeding pools.[81] Outbred animals are phenotypically similar but genetically different, and due to genetic drift similarly named strains from different sources are genetically distinct. This variability violates the fundamental assumption that in controlled experiments the experimental subjects are identical.

Inbred animals are produced by selective breeding to have a very high number of fixed homozygous alleles.[82,83] Although some strains can be considered isogenic, sub-strains can arise over time through genetic drift. Different strains have different properties and for some studies there may be a case for using more than one strain to take account of potential strain-based effects,[84] including strain-related spontaneous pathologies.

The availability of inbred and isogenic strains eliminates genetic variability and allows the use of smaller group sizes than with outbred or random-bred animals.[85] The case for using inbred or isogenic animals for toxicology studies has been repeatedly made.[80,86,87] However F1 inbred-strain hybrids are generally preferred for toxicology studies: they are genetically and phenotypically uniform with fewer strain-based limitations than pure inbred forms.

4.10.3   Microbiological Status

Microorganisms impact on physiological parameters, increase inter-individual variability, can influence tumour incidence and growth[88] and affect the properties and performance of tissues and cells used *ex vivo*. SPF (specific pathogen-free) status guarantees the absence of a defined list of microbiological agents.[89–91]

4.10.4   Age

Many toxicology studies use animals which are still growing and maturing, and care must be taken not to mistake age-related changes for the effects of test materials.

4.10.5   Sex

The need for single-sex studies should always be questioned, as this increases wastage rate of animals bred for scientific use and may miss gender-related effects.

4.11   Housing, Husbandry and Care

Animal housing, husbandry and care systems form integral parts of experimental protocols, and impact on animal welfare and scientific results by influencing physiological baselines, behaviours, gene expression and responsiveness.[92,93]

Minimizing inter-animal variation by ensuring uniform and appropriate behavioural and physiological baseline conditions and responsiveness depends not simply on the uniformity of housing and care conditions but on the conditions being well suited to the animals.[94–98] It is not sufficient merely that the animals used are in the same state; they must in a standard, stable and normal state physiologically and behaviourally. Inter-animal variation and sample size are minimized not by providing any standardized environment, but by providing a refined, high-welfare environment addressing physiological, social and behavioural needs, promoting high welfare standards and minimizing unwanted stressors. Standard texts provide detailed species-specific advice.[99]

Toxicological studies must have a high degree of external validity, producing results reproducible with minor variations in test conditions; reproducibility within highly standardized, non-enriched systems is not confirmation of external validity, as such systems promote standardization of inputs (what is provided for the animals) not standardized outputs (physiologically and psychologically normal experimental subjects) and outcomes (precise, and reliable data-streams).[100]

4.11.1 The Physical Environment

Laboratory animal housing must be sufficiently spacious and complex to permit a wide range of normal activities and behaviours. What is required varies with species, strain, age, physiological condition, stocking density, group size, and whether animals are breeding, are kept as stock, or are used for research and testing.[99] Basic physiological and ethological needs (such as freedom of movement, appropriate social contact, the ability to withdraw from social conflict, the opportunity to perform meaningful activities, and access to food and water) should never be restricted without good cause, and then only by the justified minima.

Minimum standards of care and accommodation have been published.[101–104] However, the evidence base for some of these is weak, and minimum standards are not optimum standards. The objective should not be to meet these minimum standards, but to devise and work to standards providing ever higher standards of animal welfare and higher quality science.

4.11.2 Environmental Enrichment

Environmental enrichment and refined housing and care systems are required wherever animals are bred, kept and used for experimental or other scientific purposes. Effective environmental enrichment is increasingly finding its place in toxicology research and testing (Table 4.10).[105,106]

The most important resources required to provide suitably enriched environments are caring, imaginative and competent animal-care staff. The best forms of environmental enrichment are not only species-, strain- and study-

Table 4.10 The purpose and benefits of environmental enrichment.

Environmental enrichment refers to modifications to the housing and care of captive animals:

- promoting physical and psychological wellbeing;[107]
- providing stimuli appropriate to species-specific needs, enabling the expression appropriate species specific behaviours,[108] or reducing or eliminating abnormal or undesirable behaviours;[109]
- allowing animals to control or manipulate their environment;[107,109–111] and
- facilitating the maintenance of homeostasis – failure of homeostasis being a potent stressor.[112]

Table 4.11 Environmental enrichment: measures of success.

Measures of success with environmental enrichment can include:

- purposeful use of the environment and improved control over the spatial, physical and social environment;
- normalization of the frequency, timing and duration of normal activities, and reduced frequency of abnormal behaviours;
- increased ability to adapt and cope with procedural and environmental changes and challenges; and
- more relevant and reliable scientific results.

---

specific, but depend on associated management systems and the measures in place to identify and remedy problems.

Species-appropriate social and physical environments enhance well being by allowing animals to cope with stress more effectively,[113–115] modulating their experience of pain and distress.[116,117]

The effectiveness of potential environmental enrichments should be critically evaluated in terms of the immediate and long-term impact on the animal well being and scientific outputs[6,92,111] as not all measures intended to achieve these ends are beneficial,[6] and selecting some options excludes the adoption of others (Table 4.11). Environmental enrichment is less effective once abnormal baselines, responsiveness and behaviours have become established.[118]

Environmental enrichment can be sub-divided into social enrichment and physical enrichment. Social enrichment is generally more effective than physical enrichment and a prerequisite for its effectiveness.

4.11.2.1 Social Enrichment

Early social experience has a major impact on phenotype and behavioural repertoire, and conditions at breeding and rearing facilities therefore play a large part in determining the suitability of animals as experimental subjects.

Animals, excluding species that are naturally solitary, and including dogs and non-human primates, should be socially housed in stable groups of compatible conspecifics; and single-housed only for veterinary, husbandry, other welfare or justified scientific needs, with appropriate additional resources targeted at their welfare.

Social enrichment includes the pair- or group-housing of socially compatible conspecifics, complemented by space of sufficient volume and complexity to permit appropriate species-specific interactions. This significantly reduces stress and inappropriate single-housing can compromise an animal's welfare and its suitability as a scientific subject:[5] singly-housed mice, for example, have abnormal immunocompetence and are more prone to mammary tumours.[119]

4.11.2.2 Physical Enrichment

To prevent or reduce stress-induced behaviours and adaptive changes animals should be given a degree of control over their environment by providing sufficient structured space, materials to manipulate (for example foraging or bedding material), appropriate sensory stimuli, and a varied diet, to facilitate species-appropriate exercise, foraging, and manipulative and cognitive activities.

4.12 Husbandry

4.12.1 Handling

Handling and restraint are potential stressors that have consequences both for the welfare of the animals and study results,[120,121] with more refined practices being associated with less stress to the experimental subjects and less impact on study results.

4.12.2 Transport and Acclimatization

Transporting animals, between or within establishments, can be stressful, producing effects which vary with the species, strain and individual animal.

Expert advice may be required to determine appropriate minimum recovery periods and, as the effects may be different for individual animals, animals should be shown to have returned to their normal, unstressed state before being used for scientific procedures.

4.12.3 Marking and Identification of Animals

Animals bred, kept and used for experimental and other scientific purposes need to be individually identifiable. This is generally achieved by marking individual animals rather than by recognition of individuals' physical characteristics. The preferred means of marking is that which causes the least pain or distress: mutilations are to be avoided.

4.12.4 Diet and Food Intake

Diet and food intake are powerful modulators of toxicity.

Ad lib feeding increases inter-animal variability and is associated with lowered life expectancy[122] and a range of pathologies[123–125] including obesity, renal and heart disease, and age of onset and progression of spontaneous tumours.

Adjusting and controlling body weight by restricting calorie intake (i.e. 80% *ad lib* calorie intake) can dramatically improve animal health, welfare and survival in long term studies[126–128] without compromising the detection of toxicological endpoints. Animals on appropriate calorie-restricted regimens show reduced inter-animal variability, better withstand husbandry and experimental stressors, and have less variable daily intake and body-weight range than *ad lib* fed animals making it easier to determine the effects of test materials on food intake.[122,125,129]

Care must be taken with food restriction regimens to ensure that food is available at appropriate times and, when animals are group-housed, to ensure that each animal receives its daily allowance to prevent abnormal behaviours including alterations in water intake and metabolic changes.

Short-term fasting affects the uptake and effects of orally administered test materials[130,131] by producing faster gastric emptying and uptake rather than by altering bio-activation or detoxification systems. This should be borne in mind if animal are fasted before being orally dosed.

4.13   Refining Procedures

4.13.1   Restraint

Animals have to be captured, handled, restrained and segregated from conspecifics to facilitate the performance of scientific procedures and for a variety of husbandry practices.

Capture, restraint, separation from cage-mates and removal from home enclosures are stressors, producing physiological and behavioural changes in the animals[132,133] and their cage-mates[134] of a character, magnitude and duration proportional to the nature, degree and duration of the interventions. Prolonged or inappropriate means of capture and restraint can cause overt welfare problems.[135]

The least stressful practices should be used to minimize the resulting stress-related physiological and behavioural changes, and consideration should be given to procedural training to encourage animals to allow the safe performance of routine procedures and practices without the need for capture and physical restraint.[136]

4.13.2   Procedural Training

Procedural training refers to reward-based training systems encouraging animals to cooperate with their handlers whilst procedures such as weighing, dosing and sampling are undertaken.[137] This reduces subsequent husbandry and procedural stressors, avoiding the need for capture or restraint.[138–140] Benefits include reductions in stress-related study artefacts[136] and reduced resource costs.[141]

4.13.3   Motivational Tools

Positive reinforcement and reward systems are the preferred means of motivating animals.

Food and water management systems are sometimes used to motivate animals, for example in behavioural and neuroscience research, which offer food or fluid as positive reinforcement during training and testing sessions with restricted access at other times. Although ideally such regimens would not be implemented without understanding how the resulting physical and psychological adaptations affect welfare and experimental results, it is not clear if food or water restriction have the same physiological and behavioural consequences and welfare costs.[128]

Consideration therefore needs to be given as to whether such protocols are necessary, the type and level of control to be used, the implications for animal health and welfare, effects on experimental results and opportunities for refinement. The regimen causing the least physical and psychological disturbance whilst sufficiently motivating the individual animals is to be preferred.

Details of good contemporary practice for food and water restriction have been published.[128]

4.13.4 Dosing and Sampling

The methods used for the administration of test substances and collection of blood samples affect animals' physiology and welfare and the results obtained,[142,143] particularly when unfamiliar or unpractised techniques introduce unnecessary stress and variability.[144]

Detailed advice on routes and limit volumes (which should always be considered to be the justifiable maxima rather than the norm) and other practical issues is available elsewhere.[144–147] Regularly updated information and detailed advice on contemporary good practice can be found at the NC3Rs website (http://www.nc3rs.org.uk/bloodsamplingmicrosite/page.asp?id = 426).

4.13.5 Dosing

Refinement is relevant to the route, frequency and duration of administration of test materials; the equipment used; and the dose, formulation, vehicle and volume administered. There are common pitfalls to avoided, and pilot studies may be required to refine and optimize definitive studies.

The frequency and duration of dosing are generally determined by the study objectives; the properties of the test material such as bioavailability and expected half-life and biological effects; and its interaction with the experimental subject, for example how it is metabolized, where it accumulates, and how it is detoxified and excreted.

Dose levels must be set with care to minimize unwanted toxicity whilst achieving the study objectives. If set too high or too low studies may have to be repeated and the identification of target organs and early indicators of toxicity may be missed.

4.13.5.1 Formulations

Vehicles should not affect the biophysiological properties of the test material, but should maximize exposure, lack pharmacological or toxicological effects, and be non-irritating and non-sensitizing.

Formulations and volumes of test materials are generally determined by the frequency of administration, the required accuracy of dosing, the nature and solubility of the test material, the required dose and the preferred concentration.

In addition to consideration of solubility, stability, purity and standard-ization, the closer the osmolarity, pH, buffer capacity, viscosity and tempera-ture of the test material to normal body fluids the greater the biocompatibility and less discomfort and stress will be caused.

4.13.5.2   Oral Dosing

With oral dosing, particularly by gavage, the timing of dosing and the volumes administered must not produce volume- or vehicle-related effects or com-promise normal food and fluid intake.

Gavage changes various physiological parameters[146,148–150] often in a vol-ume- and vehicle-dependent manner, the effects being more marked with lipophilic vehicles.[151]

The effects of orally administered test materials depend in part on whether the animal has been fasted before dosing, and decisions about whether animals are fasted before dosing and sampling should be informed by an under-standing of the physiological and behavioural consequences. Fasting acceler-ates gastric emptying, producing faster absorption and higher peak tissue concentrations;[130,152] and produces physiological, biochemical and metabolic changes, including lower metabolic rates.[153] In rodents, fasting reduces fluid intake producing hypovolaemia and cellular dehydration[154] affecting bio-chemical and haematological biomarkers and making blood sampling more difficult. Rodents fasted for 18 hours can lose more than 10% of their pre-fasting weights.[155] To minimize these effects water should always be available if food is withdrawn.

4.13.5.3   Parenteral Dosing

Test materials may be administered parenterally by injection, needleless in-jection systems or cannulae. For injection procedures the smallest bore needle capable of delivering the volume of test material in the required time should be used. Only sterile solutions or suspensions should be administered.

Other than administration directly into the circulation, the rate of test ma-terial uptake depends on the route of administration, the injection site and how well it is perfused, the general condition of the animal, and the dose and for-mulation used. For injection into closed spaces (for example intramuscular and intradermal injection) high pressure injection systems should be avoided, and the volumes and rates of administration should not produce adverse effects due to pressure effects or over-stretching of tissues.

For intravascular administration the volume and rate of administration should avoid unwanted volume-related effects, and minimize biological changes due to the nature and volume of vehicle used.

Intraperitoneal injection is a special case. The test material is taken up simultaneously by the hepatic portal circulation (where it may be metabolized by the liver before entering the systemic circulation) and the systemic circu-lation.[156] How test materials partition between the portal and systemic circulations depends on the nature and volume of the test material, the posture of the animal, varies from subject to subject, and in the same subject day to day.[157,158]

The need for multiple injections and the associated physical restraint procedures may be dispensed with by the placement and use of cannulae or osmotic minipumps to permit repeated (or continuous) administration; however, the potential refinement gain of fewer injections must be offset against the welfare costs of placing the devices, cannula-care procedures and potential device-related problems.

4.13.6 Blood Sampling

Blood sampling is commonly undertaken in toxicological research and testing. If multiple or frequent samples are required cannulation may be a means of minimizing the stress and simplifying the logistics of sampling.

Average total blood volumes and limit volumes are generally calculated on the basis of body weight or surface area, but safe limits are lower for animals with welfare problems. The volumes, rates of withdrawal and frequency of sampling must not cause hypovolaemia or anemia, and regularly updated guidance is available on good contemporary practice (see http://www.nc3rs.org.uk/bloodsamplingmicrosite/page.asp?id = 313). As it is not appropriate to pool samples for analysis, in animals with small blood volumes, to minimize stress artefacts and maximize successful samples, it may be necessary to limit the tests and analyses to those that are most relevant and to select some animals for hematological and others for clinical chemistry tests.

In planning blood sampling and comparing serial sample results remember that analytical findings vary according to the season and time of day the sample was taken, the part of circulation from which the sample was taken, and whether or not the animal was fasted, restrained or anesthetized. Seasonal and circadian changes[159] may be masked by the procedural-related stressors.

In many species venous blood can be obtained from superficial veins by venepuncture or venesection. Tail-snips in rodents are to be avoided as samples are contaminated with tissue fluid.

Arterial blood is obtained by direct arterial puncture or closed cardiac puncture (inserting a needle into the left ventricle of the heart of a terminally anaesthetized animal): however, this site of sampling and the administration of anaesthesia prevents direct comparison with earlier samples obtained by other means.

Samples obtained by retro-orbital puncture are not physiological fluids. Their haematological and biochemical parameters are not representative of blood anywhere in the circulation as samples comprise admixed capillary and venous blood contaminated with other tissue fluids in which a variety of clotting factors and other biologically active materials have been activated. The appropriateness of the use of such samples in the context of the scientific objectives must be considered when this route of sampling is proposed.

4.13.7   Non-invasive Sampling

A number of metabolites, hormones and other biomarkers can be measured in urine, faeces, saliva and hair[160] without the need to obtain blood samples. Such non-invasive methods, which can provide insights into both acute and chronic changes, produce fewer stress-induced artefacts and are increasingly used in toxicology and other studies as they are generally preferable to more invasive sampling methods both on welfare and scientific grounds.

4.13.8   Imaging Technologies

The rapid development of imaging technologies has provided important new tools to enhance data collection from animal studies. Advanced non-invasive imaging techniques have high degrees of spatial and temporal resolution; capable of providing anatomical and functional information, including gene expression, protein interaction, and the activity and fate of radio-labelled compounds; providing insights into toxicological mechanisms and potential pathological effects; and enabling serial data to be obtained from individual animals.

4.13.9   Analgesia and Anaesthesia

Analgesic and anaesthetic regimens should be suited to the nature and duration of the procedures and their possible consequences – but tailored to the findings in, and needs of, individual animals.

A detailed review of current best practice in the use of analgesics and anesthesia is beyond the scope of this chapter. Authoritative information can be found elsewhere.[161,162]

4.13.9.1   Analgesia

Although at the planning stage decisions about the potential need to administer analgesics can be made on anthropomorphic assumptions, once studies are underway decisions about their administration should be based on an assessment of the needs of individual animals mindful that animals may be distressed by adverse effects other than pain (for example nausea, fear, dehydration, hunger) that will not respond to analgesics. Some behavioural indices have been validated for the detection and quantification of post-procedural pain.[38,39,163] However, with chronic pain the behavioural signs can be subtle and non-specific and may include decreased appetite, weight loss, reduced activity, altered sleeping patterns and irritability.[163]

Effective pain management requires appropriate observation schedules, knowledge of the signs of pain in the animals being used, and treatment based on the findings in, and responses of, individual animals. The responsible use of analgesics reduces the depressant effect of pain on food and water consumption and body weight.[164]

Not all animals experience the same degree of pain and distress after the same procedure. Different species, strain, sexes and individuals respond differently to the same dose of test material and will have different analgesic and anaesthetic requirements.[165,166] Not all animals express pain and distress in the same way and some analgesics are effective in some species but not in others.[167] Thus the pre-scheduled provision of the same arbitrary standard doses of analgesics to all members of a study group risks some animals being under-dosed and having residual treatable pain, whilst others receive a relative excess risking potential analgesic-related adverse effects and other pharmacological effects on behaviours and clinical findings including sedative effects reducing food and water intake.[168,169]

Pre-emptive analgesia before painful procedures reduces subsequent analgesic needs[170–172] and this should be the norm with surgical procedures.

The use of topical anaesthetics and systemic analgesics improve animal welfare but do not alter the outcome of eye irritancy tests.[173]

4.13.9.2   General Anaesthesia

Care must be taken to ensure that anaesthetic regimens produce the required combination of hypnosis, narcosis, analgesia and areflexia for the nature and duration of the procedure without compromising experimental data or animal welfare.

Appropriate steps must be taken to monitor and maintain the circulation, respiratory function and body temperature of the anaesthetized subject within normal physiological limits until the anaesthetic effects have worn off.

Recovery from general anaesthesia can be hazardous, and animals should not be left unattended until the effects of the anaesthetic agents have worn off, and any necessary specific, supportive or symptomatic treatments have been given and their effectiveness determined.

4.13.10   Surgery, Surgically Prepared Animals and Post-operative Care

To minimize morbidity and mortality, all surgical procedures should be carried out by trained and competent persons following aseptic techniques; in facilities designed and dedicated to this purpose; using the best available surgical and animal care techniques; with the anaesthetic and analgesic practices best suited to the species, the nature and duration of the procedure and the scientific objective.

Surgical procedures in the context of toxicology research and testing commonly take the form of surgical preparations to facilitate dosing, sampling and data capture. These surgical preparations reduce the number of animals required, minimize the stress to which the animals are subsequently subjected, and improve the relevance and reliability of the experimental findings as data is then obtained from unstressed animals with normal baseline values and responsiveness.

Implantable telemetry devices[174] and biosensors have been validated to continuously monitor a range of physiological, biochemical and pathophysiological parameters in free-living animals within their home enclosures; minimizing stress-related artefacts[175,176] caused by serial handling, restraint, social disruption and sampling procedures; and resulting in baseline values for parameters such as blood pressure and heart-rate lower and more physiological than those seen with more conventional data capture methods.[177] Radiotelemeters are commonly increasingly used for detecting test material changes in cardiovascular parameters.[176–179]

Animals with implanted telemeters and biosensors can be used for combined general toxicology and safety pharmacology studies, enable blocked study designs to be used to filter-out inter-animal variability, reduce the number of animals required for statistical power, and can be re-used for different or sequential studies. It is estimated these technologies have the potential to reduce animal numbers in some programs of work by between 60 and 90%.[180,181]

However, there are welfare costs associated with implanting the devices, it may be several weeks before normal baseline values and responsiveness are re-established, and test materials with potent short-term pharmacological effects may produce changes outside the devices' dynamic ranges.

More complex surgical procedures should be performed as early in the working day and working week as possible, and the availability of competent staff to oversee post-operative care confirmed before procedures are performed.

4.13.11   Observation Schedules

Arrangements must be made for competent persons to check study animals at appropriate times, with all animals checked at least daily. All staff concerned with animal care and use should be competent in recognizing animal pain and distress, aware of the action to the taken when problems arise and empowered to take appropriate action.

Observations should be scheduled for when adverse effects are anticipated, and schedules adjusted as necessary to ensure the prompt identification and management of adverse effects. The frequency of observations should be intensified when welfare problems are likely, when serial observations show an animal's condition to be deteriorating, and when problems come to light.

Monitoring and observation are on-going processes, with new findings integrated with those from preceding observations. Good communication and teamwork are essential: records should be maintained of positive and negative findings, any remedial action taken and its effectiveness.

Instances where animals are killed *in extremis* or are found dead should be investigated, and study protocols revised as necessary, as opportunities for refinement have been missed and useful data lost.

4.14   Humane Killing

The majority of animals used for toxicological research and testing are humanely killed as part, or at the conclusion, of their scientific use. Humane killing methods ensure death with a minimum of pain and distress, and should be aesthetically acceptable.

Humane methods of killing[182–185] incorporate careful and compassionate animal handling regimens and typically ensure rapid loss of consciousness without producing signs of pain or distress, resulting in death with the minimum of physical and mental suffering, with death being confirmed before removing tissues or disposing of cadavers. Their performance requires expertise developed by appropriate training and the provision and maintenance of appropriate equipment.

Decisions to humanely kill animals whose death is anticipated depends on the experience of the responsible staff and their ability to recognize the onset, nature, duration and severity of clinical signs in individual animals.

4.15 Humane Endpoints

Humane endpoints, incorporating all reasonable, relevant, practical steps to minimize or place limits on justifiable suffering, by avoiding or promptly remedying adverse effects, are important parts humane research and testing (Table 4.12).

Humane endpoints may be defined and recognized by molecular markers, non-clinical physiological changes, the onset of clinical signs or established clinical signs.[10] The intention should be to develop and work to surrogate pre-clinical endpoints, shown to be reliably predictive of outcomes, which can be detected before overt clinical signs or other welfare problems result. Such

Table 4.12 Humane endpoints: common misconceptions.

There are three misconceptions with respect to humane endpoints that must be dispelled.[10]

1 Some mistakenly consider humane endpoints to represent '... *the earliest indicator in an animal experiment of severe pain, severe distress, suffering, or impending death* ...'.[11]
   - However, humane endpoints must not be set only to limit the severest adverse effects as their primary purpose is to prevent all unnecessary suffering.
   - They must be invoked as soon as the scientific objective has been achieved (and thus different endpoints may be appropriate when the same animal model is used with different test materials or for different purposes or objectives); when it is realized they cannot be achieved (for example when intercurrent problems have compromised the suitability of experimental subjects); when the harms exceed the potential benefits; and when the nature, duration or intensity of the welfare problems are impermissible.
   - In many cases, and in particular when the scientific objectives have been achieved or it is recognized they cannot be achieved, humane endpoints are invoked when the level of animal welfare costs is not high.
2 Although some judge or classify the animal welfare costs of procedures solely in terms of the nature of the interventions applied, it is the humane endpoints, that is the outcomes rather than the inputs, which limit and define the distress caused.
3 Properly considered surrogate endpoints do not limit the detection of relevant adverse effects caused by the test material.

endpoints are particularly well suited to studies where the relevant toxicity pathways and mechanism are understood and molecular and cellular markers can be developed.[17]

The murine local lymph node assay for skin sensitization[186] is a case in point. It relies on subclinical changes in local lymph nodes during the induction phase of skin sensitization, rather than on the clinical changes and pathologies produced when sensitized animals are re-challenged with the test material.

Humane endpoints must be contextualized for individual projects, experiments and experimental groups, but are best thought of as being applied to individual animals (Table 4.13).

The planning process is informed by prior knowledge of the properties of the test material, including its physiochemical properties, reactivity, structure–activity relationships, and other available *in vitro*, *in vivo* and human data; the properties of related materials; relevant toxicological mechanisms; tissues and organ systems likely to be affected; expected adverse effects, their timing, and how they will be recognized and managed; how unexpected adverse effects will be recognized and remedied; and what findings will confirm that the scientific objectives have been attained. Particular care must be taken to identify potential pharmacological effects not be indicative of poor welfare despite producing clinical signs.

Humane endpoints must be described in terms that are meaningful to those who will observe the animals, and observation schedules must facilitate their prompt detection and implementation. The action taken when humane endpoints are reached need not be humane killing or the animal ceasing to be an experimental subject – in some cases symptomatic or supportive treatment or procedural changes, such as reducing the dosing schedule, may resolve the welfare problem and enable the scientific objective to be achieved.

The primary changes produced by test materials are often subtle in nature and overshadowed by unwanted and unnecessary less specific and less subtle secondary changes (for example pain or anorexia) and tertiary changes (for example weight loss or dehydration). Untreated these secondary and tertiary effects compromise both science and welfare; however, symptomatic and supportive treatments remedy the welfare problems and permit the processes of interest to continue. The rationale for withholding specific, symptomatic and supportive treatments should be justified and documented.

Table 4.13 Planning humane endpoints.

Humane endpoints should be developed at the planning stage and must:

- be objective and evidence based to avoid culling animals whose welfare is less compromised than is believed or before the scientific objectives have been achieved;
- prevent significant welfare problems being missed;
- minimize welfare costs;
- inform judgements about the relative severity of different procedures and models; and
- play a part in the validation of potential refinements.

Death is seldom if ever required as an endpoint for toxicological studies and in practice is seldom due to the primary action of the test material. Study-related deaths are generally produced by secondary and tertiary adverse effects, most of which can be managed with appropriate symptomatic or supportive treatment without compromising the primary process.

The death of study animals, rather than their being humanely killed, is not consistent with good science or good welfare: valuable data may be lost, and opportunities for refinement missed.

4.15.1 Surrogate Endpoints: Physiological Data and Biomarkers

Once the toxicological mechanisms and endpoints of interest have been defined a number of validated toxicological biomarkers in a range of body fluids can be detected and quantified before the onset of clinical signs or established pathological changes, giving insights into whether changes are adaptive, reversible or irreversible. Early specific biomarkers minimize animal welfare costs and safeguard the data streams and findings from artefacts introduced by unrecognized or unwanted secondary and tertiary effects as homeostatic mechanisms are stressed and overwhelmed.

In some cases surrogate endpoints can be set on the basis of general, non-specific changes (e.g. appearance, behaviour, weight loss, food and water intake, or body temperature). Examples include the use of the HID_{50} (hypothermia-induced dose 50) as an indicator of impending, overwhelming infection or toxicity as an alternative to the morbidity and mortality that follow.[56] Body weight on its own is not always a good index of poor welfare: animals with tumours, ascites, renal failure or heart failure can gain weight whilst their health is failing; and inappropriate administration of analgesics may reduce food and water intake and body weight.

4.16 Personnel, Teamwork, Training and Competence

Refinement requires commitment, competence, collaboration and communication.

The most important factors in implementing refinement are the expertise and 'culture of care' of those responsible for animal production, care and use. Those involved must be open-minded; actively seek information on best current practice; anticipate, recognize and deal promptly with adverse effects; and disseminate good practice.

Toxicology research and testing are undertaken by multidisciplinary teams, and require a range of knowledge and skills not found in any individual. Those undertaking animal-based toxicology studies commonly require access to expertise in regulatory affairs, study design, statistics, toxicology and metabolism, laboratory animal science and medicine, animal care and use. Many organizations provide institutional-level expert support on animal care and use, experimental design, statistics, laboratory animal science, and veterinary medicine – and even the largest organizations seek outside expert advice and input on these and other issues as required.

Those responsible for assessing animal welfare must be consulted and involved at the planning stage to make best provision for animal housing, care and use. They must be aware of contingency plans to deal with expected and unexpected adverse effects and empowered to take prompt action when scientific or welfare endpoints are approached or reached: animal welfare and good science are both compromised if decision making and remedial action are delayed.

4.16.1   Individual Responsibility

Individuals involved with animal studies must:

- obtain appropriate training and continued professional development (including periodic revalidation of skills);
- keep abreast of technical and regulatory progress and good practice through involvement with professional societies and visiting and benchmarking against others;
- take responsibility for their personal effectiveness in making timely contributions to the planning, performance and review of studies; and
- show an awareness and acceptance of the need for expert input from others when required.

4.17   Review and Continuous Improvement

The practice of refinement does not end until lessons learned are incorporated into on-going and future practice.

- Interim review of on-going studies may show that the number of animals might be reduced, earlier endpoints applied, or other measures applied to reduce animal welfare costs without loss of precision or otherwise compromising the scientific objectives.
- Review of completed studies, including linking *in vivo* to *post mortem* findings, can determine whether the likely clinical and sub-clinical manifestations of the toxicities and pathologies produced were detected during life, and whether the scientific objectives could have been achieved at a lesser welfare cost if earlier endpoints had been applied. This information should be used when further studies are planned and performed.

4.18   References and Resources

Access to information has never been easier. However, in practice, this makes it difficult to find and quality assure the best and most up-to-date information on refinement, and internet searches for 'alternatives' and 'refinement' may not find the most relevant information. Guidance on efficient and effective internet searching strategies has been developed by the European Commission.[187]

Although part of mainstream science, the 3Rs has its own specialist literature, networks and websites which may not be known to or regularly referenced by toxicologists. Insufficient prominence is given to progress with the 3Rs in some specialist toxicology resources, with many papers lacking husbandry, procedural and welfare details.

A number of national and international centers have been established to promote the 3Rs: a number of these maintain websites providing up-to-date information on good practice relevant to the refinement of the use of live animals for toxicological research and testing. See for example the UK National Centre for the 3Rs (http://www.nc3rs.org.uk/), The Johns Hopkins University Center for Alternatives to Animal Testing (http://caat.jhsph.edu/), the NTP Interagency Center for the Evaluation of Alternative Toxicological Methods (NICEATM) (http://www.niehs.nih.gov/research/atniehs/dntp/assoc/niceatm/) and the Interagency Coordinating Committee on the Validation of Alternative Methods (ICCVAM) (http://iccvam.niehs.nih.gov/) the Universities Federation for Animal Welfare (http://www.ufaw.org.uk/), the Canadian Council on Animal Care (http://www.ccac.ca/en), and the Institute for Laboratory Animal Research (http://dels.nas.edu/ilar/).

4.19   Concluding Remarks

Although the development and use of non-animal technologies for toxicology research and testing are seen as the future, the immediate priority is to continuously refine *in vivo* animal studies to comply with legal imperatives and ethical norms, reduce and minimize animal welfare costs, and improve the quality of toxicity data generated to protect man and the environment.

References

1. IRIS (Integrated Risk Information System), *Glossary of IRIS Terms. Integrated Risk Information System*, U.S. Environmental Protection Agency [online], 2007. Available at: http://www.epa.gov/iris/gloss8.htm.
2. A. G. Renwick, S. M. Barlow, I. Hertz-Picciotto, A. R. Boobis, E. Dybing, L. Edler, G. Eisenbrand, J. B. Grieg, J. Kleiner, J. Lambe, D. J. D. Müller, M. R. Smith, A. Tritscher, S. Tuijtelaars, P. A. Van den Brandt, R. Walker and R. Kroes, Risk characterisation of chemicals in food and diet, *Food Chem. Toxicol.*, 2003, **41**, 1211–1271.
3. A. Sergeant, in *Human and Ecological Risk Assessment: Theory and Practice*, ed. D. J. Paustenbach, John Wiley and Sons, New York, 2002, pp. 369–442.
4. C. D. Klaassen and D. L. Eaton, in *Casarett and Doull's Toxicology: The Basic Science of Poisons*, 4th edn, ed. M. O. Amdur, J. Doull and C. D. Klaassen, Pergamon Press, New York, 1991, pp. 12–49.

5. T. Poole, Happy animals make good science, *Lab. Anim.*, 1997, **31**, 116–124.
6. K. Bayne, Environmental enrichment: Potential for unintended consequences and research results, *ILAR J.*, 2005, **46**, 129–139.
7. NRC (National Research Council, Committee on Toxicity Testing and Assessment of Environmental Agents), *Toxicity Testing in the 21st Century: A Vision and a Strategy*, The National Academies Press, Washington, DC, 2007.
8. M. Hall, On experiments in physiology as a question of medical ethics, *The Lancet*, 1847, **49**(1220), 58–60.
9. W. M. S. Russell and R. L. Burch, *The Principles of Humane Experimental Technique*, Methuen & Co Ltd, London, 1959.
10. J. Richmond, Criteria for humane endpoints, in *Humane Endpoints in Animal Experiments for Biomedical Research*, ed. C. F. M. Hendriksen and D. B. Morton, Royal Society of Medicine Press, London, 1999, pp. 26–32.
11. OECD, *Guidance Document on the Recognition, Assessment, and Use of Clinical Signs as Humane Endpoints for Experimental Animals Used in Safety Evaluation*, OECD, Paris, 2000.
12. J. Richmond, The Three Rs: a journey or a destination?, *Alt. Lab. Anim.*, 2000, **28**, 761–773.
13. J. Richmond, Refinement, reduction and replacement of animal use for regulatory testing; future improvements and implementation within the regulatory framework, *Inst. Lab. Anim. Res. J.*, 2002, **43**(Suppl), S63–S68.
14. W. Lane-Petter, *Provision of Laboratory Animals for Research*, Elsevier Publishing Co., Amsterdam, 1961.
15. A. K. Hansen, A. Kornerup, P. Sandøe, O. Svendsen, B. Forsman and P. Thomsen, The need to refine the notion of reduction. In: *Humane endpoints in animal experiments for biomedical research, Humane Endpoints in Animal Experiments for Biomedical Research, Proceedings of the International Conference, 22–25 November 1998, Zeist, The Netherlands*, ed. C. F. M. Hendriksen and D. Morton, The Royal Society of Medicine Press, London, 1999, pp. 139–144.
16. P. A. Flecknell, Refinement of animal use – assessment and alleviation of pain and distress, *Lab. Anim.*, 1994, **28**, 222–231.
17. C. F. M. Hendriksen and D. B. Morton (eds), *Humane Endpoints in Animal Experiments for Biomedical Research. Proceedings of the International Conference, 22–25 November 1998; Zeist, The Netherlands*, The Royal Society of Medicine Press, London, 1999.
18. J. A. Mench, Environmental enrichment and exploration, *Lab. Anim.*, 1994, **23**, 38–41.
19. C. E. Manser, D. M. Broom, P. Overend and T. H. Morris, Investigations into the preferences of laboratory rats for nest-boxes and nesting materials, *Lab. Anim.*, 1998, **32**(1), 23–35.
20. C. M. Sherwin, Preferences of individually housed TO strain laboratory mice for loose substrate or tubes for sleeping, *Lab. Anim.*, 1996, **30**, 245–251.

21. H. A. van de Weerd, P. L. P. Van Loo, L. F. M. Van Zutphen, J. M. Koolhaas and V. Baumans, Strength of preference for nesting material as environmental enrichment for laboratory mice, *Appl. Anim. Behav. Sci.*, 1998, **55**, 169–382.

22. G. Mason, D. McFarland and J. Garner, A demanding task: Using economic techniques to assess animal priorities, *Anim. Behav.*, 1998, **55**, 1071–1075.

23. C. M. Sherwin, The influences of standard laboratory cages on rodents and the validity of research data, *Anim. Welfare*, 2004, **13**(Suppl), 9–15.

24. R. D. Kirkden and E. A. Pajor, Using preference, motivation and aversion tests to ask scientific questions about animals' feelings, *Appl. Anim. Behav. Sci.*, 2006, **100**(1–2), 29–47.

25. J. Van Rooijen, The value of choice tests in assessing welfare of domestic animals, *Appl. Anim. Ethol.*, 1982, **8**, 295–299.

26. I. J. H. Duncan, The interpretation of preference tests in animal behaviour, *Appl. Anim. Ethol.*, 1978, **4**, 197–200.

27. W. Scharmann, in Physiological and ethological aspects of the assessment of pain, distress and suffering, *Humane Endpoints in Animal Experiments for Biomedical Research, Proceedings of the International Conference, 22–25 November 1998, Zeist, The Netherlands*, ed. C. F. M. Hendriksen and D. B. Morton, The Royal Society of Medicine Press, London, 1999, pp. 33–39.

28. I. J. H. Duncan, Science-based assessment of animal welfare: Farm animals, *OIE Sci. Tech. Rev.*, 2005, **24**(2), 483–492.

29. National Research Council, *Recognition and Alleviation of Distress in Laboratory Animals*, The National Academies Press, Washington, DC, 2008.

30. T. Saranteas, C. Mourouzis, C. Dannis, C. Alexopoulos, E. Lollis and C. Tesseromatis, Effect of various stress models on lidocaine pharmacokinctic properties in the mandible after masseter injection, *J. Oral Maxillofcaial Surgery*, 2004, **62**(7), 858–862.

31. R. R. Johnson, T. W. Prentice, P. Bridegam, C. R. Young, A. J. Steelman, T. H. Welsh, C. J. Welsh and M. W. Meagher, Social stress alters the severity and onset of the chronic phase of Theiler's virus infection, *J. Neuroimmunol.*, 2006, **175**(1–2), 39–51.

32. S. F. Maier and L. R. Watkins, Stressor controllability, anxiety and serotonin, *Cognitive Therapy Res.*, 1998, **22**, 595–613.

33. J. S. Mogil, S. G. Wilson, K. Bon, S. E. Lee, K. Chung, P. Raber, J. O. Pieper, H. S. Hain, J. K. Belknap, L. Hubert, G. I. Elmer, J. M. Chung and M. Devor, Heritability of nociception I: Responses of 11 inbred mouse strains on 12 measures of nociception, *Pain*, 1999, **80**(1–2), 67–82.

34. J. S. Mogil, S. B. Smith, M. K. O'Reilly and G. Plourde, Influence of nociception and stress-induced antinociception on genetic variation in isoflurane anesthetic potency among mouse strains, *Anesthesiology*, 2005, **103**, 751–758.

35. J. Paul-Murphy, J. W. Ludders, S. A. Robertson, J. S. Gaynor, P. W. Hellyer and P. L. Wong, The need for a cross-species approach to the study of pain in animals, *J. Am. Vet. Med. Assoc.*, 2004, **224**, 692–697.
36. J. M. Terner, L. M. Lomas, E. S. Smith, A. C. Barrett and M. J. Picker, Pharmacogenetic analysis of sex differences in opioid antinociception in rats, *Pain*, 2003, **106**, 381–391.
37. S. G. Wilson, S. B. Smith, E. J. Chesler, K. A. Melton, J. J. Haas, B. A. Mitton, K. Strasburg, L. Hubert, S. L. Rodriguez-Zas and J. S. Mogil, The heritability of antinociception: common pharmacogenetic mediation of five neurochemically distinct analgesics, *J. Pharmacol. Exp. Ther.*, 2003, **304**, 547–559.
38. J. V. Roughan and P. A. Flecknell, Behavioural effects of laparotomy and analgesics effects of ketoprofen and carprofen in rats, *Pain*, 2001, **90**, 65–74.
39. J. V. Roughan and P. A. Flecknell, Evaluation of a short duration behaviour-based post-operative pain scoring system in rats, *Eur. J. Pain*, 2003, **7**(5), 397–406.
40. A. Karas, P. Danneman and J. Cadillac, Strategies for Assessing and Minimizing Pain, in *Anesthesia and Analgesia in Laboratory Animals*, ed. R. Fish, M. Brown, P. Danneman and A. Karas, Academic Press, San Diego, 2008, pp. 195–218.
41. P. A. Flecknell, H. E. Orr, J. V. Roughan and R. Stewart, Comparison of the effects of oral or subcutaneous carprofen or ketoprofen in rats undergoing laparotomy, *The Veterinary Record*, 1999, **144**(3), 65–67.
42. C. J. Harvey-Clark, K. Gillespie and K. W. Riggs, Transdermal fentanyl compared with parenteral buprenorphine in post-surgical pain in swine: A case study, *Lab. Anim.*, 2000, **34**(4), 386–398.
43. P. Hawkins, Recognizing and assessing pain, suffering and distress in laboratory animals: A survey of current practice in the UK with recommendations, *Lab. Anim.*, 2002, **36**(4), 378–395.
44. D. J. Langford, A. L. Bailey, M. L. Chanda, S. E. Clarke, T. E. Drummond, S. Echols, S. Glick, J. Ingrao, T. Klassen-Ross, M. L. LaCroix-Fralish, L. Matsumiya, R. E. Sorge, S. B. Sotocinal, J. M. Tabaka, D. Wong, D, A. M. J. M. van den Maagdenberg, M. D. Ferrari, K. D. Craig and J. S. Mogil, Coding of facial expressions of pain in the laboratory mouse, *Nature Methods*, 2010, **7**, 447–449.
45. S. L. Wright-Williams, J. P. Courade, C. A. Richardson, J. V. Roughan and P. A. Flecknell, Effects of vasectomy surgery and meloxicam treatment on faecal corticosterone levels and behaviour in two strains of laboratory mouse, *Pain*, 2007, **130**, 108–118.
46. S. G. Sotocinal, R. E. Sorge, A. Zaloum, A. H. Tuttle, L. J. Martin, J. S. Wieskopf, J. C. Mapplebeck, P. Wei, S. Zhan, S. Zhang, J. J. McDougall, O. D. King and J. S. Mogil, The Rat Grimace Scale: A partially automated method for quantifying pain in the laboratory rat via facial expressions, *Molecular Pain*, 2011, **7**, 55.
47. M. C. Leach, S. Allweiler, C. Richardson, J. V. Roughan, R. Narbe and P. A. Flecknell, Behavioural effects of ovariohysterectomy and oral

administration of meloxicam in laboratory housed rabbits, *Res. Vet. Sci.*, 2009, **87**, 336–347.

48. L. S. Slingsby and A. E. Waterman-Pearson, Comparison of pethidine, buprenorphine and ketoprofen for postoperative analgesia after ovariohysterectomy in the cat, *Vet. Record*, 1998, **143**, 185–189.
49. A. M. Firth and S. L. Haldane, Development of a scale to evaluate postoperative pain in dogs, *J. Am. Vet. Med. Assoc.*, 1999, **214**, 651–659.
50. L. L. Holton, J. Reid, E. M. Scott, P. Pawson and A. M. Nolan, Development of a behaviour-based scale to measure acute pain in dogs, *Vet. Record*, 2001, **148**, 525–531.
51. M. L. Wiseman, A. M. Nolan, J. Reid and E. M. Scott, Preliminary study on owner-reported behaviour changes associated with chronic pain in dogs, *Vet. Record*, 2001, **149**, 423–424.
52. M. J. Graham, J. E. Kent and V. Molony, Effects of four analgesic treatments on the behavioural and cortisol responses of 3-week-old lambs to tail docking, *Vet. J.*, 1997, **153**, 87–97.
53. V. Molony, J. E. Kent and I. S. Robertson, Assessment of acute and chronic pain after different methods of castration of calves, *Appl. Anim. Behav. Sci.*, 1995, **26**, 33–48.
54. J. Price, S. Catriona, E. M. Welsh and N. K. Waran, Preliminary evaluation of a behaviour-based system for assessment of post-operative pain in horses following arthroscopic surgery, *Vet. Anaesthesia Analgesia*, 2003, **30**, 124–137.
55. M. C. Leach, C. A. Coulter, C. A. Richardson and P. A. Flecknell, Are we looking in the wrong place? Implications for behavioural-based pain assessment in rabbits (Oryctolagus cuniculi) and beyond, *PLoS ONE*, 2011, **6**, e13347.
56. J. S. Soothill, D. B. Morton and A. Ahmad, The HID50 (Hypothermia Inducing Dose 50): an alternative to the LD50 for the measurement of bacterial virulence, *Int. J. Exp. Pathol.*, 1992, **75**, 95–98.
57. D. B. Morton and P. H. Griffiths, Guidelines on the recognition of pain, distress and discomfort in experimental animals and an hypothesis for assessment, *Vet. Record*, 1985, **116**, 431–436.
58. NRC (National Research Council), *Recognition and Alleviation of Pain and Distress in Laboratory Animals*, National Academy Press, Washington, DC, 1992.
59. W. Tong, W. J. Welsh, L. Shi, H. Fang and R. Perkins, Structure-activity relationship approaches and applications, *Environ. Toxicol. Chem.*, 2003, **22**, 1680–1695.
60. ICCVAM (Interagency Coordinating Committee on the Validation of Alternative Methods), *ICCVAM Test Method Evaluation Report: Current Validation Status of a Proposed In Vitro Testing Strategy for U.S. Environmental Protection Agency Ocular Hazard Classification and Labeling of Antimicrobial Cleaning Products*, NIH Publication No. 10–7513, National Institute of Environmental Health Sciences, Research Triangle Park, NC, 2010.

61. ICCVAM (Interagency Coordinating Committee on the Validation of Alternative Methods), *Guidance Document on Using In Vitro Data to Estimate In Vivo Starting Doses for Acute Toxicity*, NIH Publication No. 01–4500, National Institute of Environmental Health Sciences, Research Triangle Park, NC, 2010.

62. N. G. Carmichael, H. A. Barton, A. R. Boobis, R. L. Cooper, V. L. Dellarco, N. G. Doerrer, P. A. Fenner-Crisp, J. E. Doe, J. C. Lamb and T. P. Pastoor, Agricultural chemical safety assessment: a multi-sector approach to the modernization of human safety requirements, *Crit. Rev. Toxicol.*, 2006, **36**, 1–7.

63. J. E. Doe, A. R. Boobis, A. Blacker, V. L. Dellarco, N. G. Doerrer, C. Franklin, J. I. Goodman, J. M. Kronenberg, R. Lewis, E. E. McConnell, T. Mercier, A. Moretto, C. Nolan, S. Padilla, W. Phang, R. Solecki, L. Tilbury, B. van Ravenswaay and D. C. Wolf, A tiered approach to systemic toxicity testing for agricultural chemical safety assessment, *Crit. Rev. Toxicol.*, 2006, **36**, 37–68.

64. G. D. Charles, In vitro models in endocrine disruptor screening, *ILAR J.*, 2004, **45**, 494–501.

65. H. Spielmann, Predicting the risk of developmental toxicity from in vitro assays, *Toxicol. Appl. Pharmacol.*, 2005, **207**, S375–S380.

66. D. R. Cox, *Planning Experiments*, John Wiley, New York, 1958.

67. M. F. W. Festing and D. G. Altman, Guidelines for the design and statistical analysis of experiments using laboratory animals, *ILAR J.*, 2002, **43**, 244–258.

68. A. M. Strasak, Q. Zaman, P. P. Karl, G. Gobel and H. Ulmer, Statistical errors in medical research-a review of common pitfalls, *Swiss Medical Weekly*, 2007, **137**, 44–49.

69. I. McCance, Assessment of statistical procedures used in papers in the Australian Veterinary Journal, *Aust. Vet. J.*, 1995, **72**, 322–328.

70. W. F. W. Festing, Reduction of animal use: Experimental design and quality of experiments, *Lab. Anim.*, 1994, **28**, 212–221.

71. R. Mead, *The Design of Experiments*, Cambridge University Press, Cambridge, UK, 1988.

72. S. Hoffmann and T. Hartung, Towards an evidence-based toxicology, *Human Exp. Toxicol.*, 2006, **25**, 497–513.

73. S. S. Sparrow, S. Robinson, S. Bolam, C. Bruce, A. Danks, D. Everett, S. Fulcher, R. E. Hill, H. Palmer, E. W. Scott and K. L. Chapman, Opportunities to minimise animal use in pharmaceutical regulatory general toxicology: a cross-company review, *Regulatory Toxicol. Pharmacol.*, 2011, **61**(2), 222–229.

74. J. Richmond, in *The UFAW Handbook on the Care and Management of Laboratory and Other Research Animals*, 8th edn, ed. R. Hubrecht and J. Kirkwood, Wiley-Blackwell, Oxford, UK, 2010.

75. J. B. Barley, Animal experimentation, the scientist and ethics, *Anim. Technol.*, 1999, **50**, 1–10.

76. J. Levin and F. B. Bang, A description of cellular coagulation in the limulus, *Bulletin of the Johns Hopkins Hospital*, 1964, **115**, 337–345.
77. G. L. Kedderis and J. C. Lipscomb, Application of in vitro bio-transformation data and pharmacokinetic modeling to risk assessment, *Toxicol. Ind. Health*, 2001, **17**(5–10), 315–321.
78. F. Sams-Dodd, Drug discovery: selecting the optimal approach, *Drug Discovery Today*, 2006, **11**, 465–472.
79. D. J. Wells, L. C. Playle, W. E. Enser, P. A. Flecknell, M. A. Garndiner, J. Holland, B. R. Howard, R. Hubrecht, K. R. Humphreys, I. J. Jackson, N. Lane, M. Maconochie, G. Mason, D. B. Morton, R Raymond, V. Robinson, J. A. Smith and N. Watt, *Lab. Anim.*, 2006, **40**, 111–114.
80. M. F. W. Festing and A. G. Peters, in *The UFAW Handbook on the Care and Management of Laboratory Animals*, Vol. 1, ed. T. Poole, Blackwell Science, Oxford, UK, 1999.
81. D. Hartl, *A Primer of Population Genetics*, 3rd edn, Sinauer Associates, Sunderland, MA, 2000.
82. M. F. W. Festing, in *The Laboratory Rat*, ed. H. J. Baker, J. R. Lindsey and S. H. Weisbroth, Academic Press, New York, 1979.
83. M. F. W. Festing, Origins and characteristics of inbred strains of mice, *Mouse Genome*, 1993, **91**, 393–509.
84. M. F. W. Festing, The need for better experimental design, *Trends Pharmacol. Sci.*, 2003, **24**, 341–345.
85. M. F. W. Festing and E. M. C. Fisher, Clarence Little's brainwave gave biomedical researchers their best friend..., *Nature*, 2000, **404**, 815.
86. M. F. W. Festing, The case for isogenic strains in toxicological screening, *Arch. Toxicol.*, 1986, **Suppl. 9**, 127–137.
87. R. Chia, F. Achilli, M. F. W. Festing and E. M. C. Fischer, The origins and uses of mouse outbred stocks, *Nat. Genet.*, 2005, **37**(11), 1181–1186.
88. W. Nicklas, F. R. Homberger, B. Illgen-Wilcke, K. Jacobi, V. Kraft, I. Kunstyr, M. Maehler, H. Meyer and G Pohlmeyer-Esch, Implications of infectious agents on results of animal experiments: Report of the Working Group on Hygiene of the Gesellschaft für Versuchstierkunde-Society for Laboratory Animal Science (GV-SOLAS), *Lab. Anim.*, 1999, **33**(Suppl. 1), 39–87.
89. A. K. Hansen, *Handbook of Laboratory Animal Bacteriology*, CRC Press, Boca Raton, 2000.
90. W. Nicklas, P. Baneux, R. Boot, T. Decelle, A. A. Deeny, M. Fumanelli and B. Illgen-Wilcke, Recommendations for the health monitoring of rodent and rabbit colonies in breeding and experimental units, *Lab. Anim.*, 2002, **36**, 20–42.
91. C. B. Clifford and J. Watson, Old Enemies, Still with Us after All These Years, *ILAR J.*, 2008, **49**, 291–302.
92. A. C. Benefiel, W. K. Dong and W. T. Greenough, Mandatory "Enriched" Housing of Laboratory Animals: The Need for Evidence-based Evaluation, *ILAR J.*, 2005, **46**(2), 95–105.

93. C. M. Sherwin, The influences of standard laboratory cages on rodents and the validity of research data, *Anim Welfare*, 2004, **13**(Suppl), 9–15.

94. M. R. A. Chance, The contribution of environment to uniformity, *Laboratory Animals Bureau Collected Papers*, 1957, **6**, 59–73.

95. M. R. A. Chance and W. M. S. Russell, The benefits of giving experimental animals the best possible environment, in *Comfortable Quarters for Laboratory Animals*, 8th edn, ed. V. Reinhardt, Animal Welfare Institute, Washington, DC, 1997.

96. J. Garner and G. Mason, Abnormal behavioural persistence underlies the stereotypies of laboratory-housed rodents, *Behav. Brain Res.*, 2002, **136**, 83–92.

97. J. P. Garner, G. J. Mason and R. Smith, Stereotypic route-tracing in experimentally-caged songbirds correlates with general behavioural disinhibition, *Anim. Behav.*, 2003, **66**, 711–727.

98. H. Würbel, Behaviour and the standardisation fallacy, *Nat. Genet.*, 2000, **26**, 263.

99. R. Hubrecht and J. Kirkwood (eds.), *The UFAW Handbook on the Care and Management of Laboratory and Other Research Animals*, 8th edn, Blackwell Science, Oxford, UK, 2010.

100. N. Henderson, Genetic influences on the behaviour of mice can be obscured by laboratory rearing, *J. Comparative Physiol. Psychol.*, 1970, **72**, 505–511.

101. CCAC (Canadian Council on Animal Care), *Guide to the Care and Use of Experimental Animals*, Vol. 2, CCAC, Ontario, Canada, 1984.

102. CCAC (Canadian Council on Animal Care), *Guide to the Care and Use of Experimental Animals*, Vol. 1, 2nd edn, CCAC, Ontario, Canada, 1993.

103. FELASA Working Group on Accreditation of Laboratory Animal Science Education and Training: T. Nevalainen (Convenor), H. J. M. Blom, A. Guaitani, P. Hardy, B. R. Howard and P. Vergara, FELASA recommendations for the accreditation of laboratory animal science education and training, *Lab. Anim.*, 2002, **36** (4), 373–377.

104. NRC (National Research Council, Institute of Laboratory Animal Resources, National Academy of Sciences), *Guide for the Care and Use of Laboratory Animals*, 8th edn, National Academy Press, Washington, DC, 2011.

105. S. W. Dean, Environmental enrichment of laboratory animals used in regulatory toxicology studies, *Lab. Anim.*, 1999, **33**(4), 309–327.

106. P. V. Turner, K. L. Smiler, M. Hargaden and M. A. Koch, Refinements in the care and use of animals in toxicology studies—regulation, validation, and progress, *Contemporary Topics Lab. Anim. Sci.*, 2003, **42**(6), 8–15.

107. I. A. S. Olsson and K. Dahlborn, Improving housing conditions for laboratory mice: A review of "environmental enrichment", *Lab. Anim.*, 2002, **36**, 243–270.

108. D. C. Blanchard and R. J. Blanchard, What can animal aggression research tell us about human aggression?, *Hormones and Behavior*, 2003, **44**(3), 171–177.

109. C. Brinkman, Toys for the Boys: Environmental Enrichment for Singly Housed Adult Male Macaques (Macaca fascicularis), *Lab. Prim. News*, 1996, **35**, 5–9.
110. N. Latham and G. Mason, From house mouse to mouse house: the behavioral biology of free-living Mus musculus and its implications in the laboratory, *Anim. Behav.*, 2004, **86**, 261–289.
111. C. M. Sherwin, in *Comfortable Quarters for Laboratory Animals*, 9th edn, ed. V. Reinhardt and A. Reinhardt, Animal Welfare Institute, Washington, DC, 2002, pp. 6–17.
112. G. P. Moberg, in *The Biology of Animal Stress*, ed. G. P. Moberg and J. A. Mench, CAB International, Wallingford, UK, 2000, pp. 1–21.
113. A. F. Fraser and D. M. Broom, *Farm Animal Behaviour and Welfare*, Sanders, New York, 1990.
114. S. G. Kingston and L. Hoffman-Goetz, Effect of environmental enrichment and housing density on immune system reactivity to acute exercise stress, *Physiol. Behav.*, 1996, **60**, 145–150.
115. L. Bassett, H. M. Buchanan-Smith, J. McKinley and T. E. Smith, Effects of training on stress-related behavior of the common marmoset (Callithrix jacchus) in relation to coping with routine husbandry procedures, *J. Appl. Anim. Welfare Sci.*, 2003, **6**, 221–233.
116. M. J. Gentle and S. A. Corr, Endogenous analgesia in the chicken, *Neurosci. Lett.*, 1995, **201**, 211–214.
117. M. J. Gentle and V. L. Tilston, Reduction in peripheral inflammation by changes in attention, *Physiol. Behav.*, 1999, **66**, 289–292.
118. J. J. Cooper, F. Ödberg and C. J. Nicol, Limitations on the effectiveness of environmental improvement in reducing stereotypic behaviour in bank voles (Clethrionomys glareolus), *Appl. Anim. Behav. Sci.*, 1996, **48**, 237–248.
119. V. Riley, Psychoneuroendocrine influences on immunocompetence and neoplasia, *Science*, 1981, **212**, 1100–1109.
120. R. Gattermann and R. Weinandy, Time of day and stress response to different stressors in experimental animals, *J. Exp. Anim. Sci.*, 1996, **38**, 66–76.
121. J. D. Clark, D. R. Rager and J. P. Calpin, Animal well-being I–IV: General considerations, *Lab. Anim. Sci.*, 1997, **47**(6), 564–597.
122. M. Hubert, P. Laroque, J. Gillet and K. P. Keenan, The Effects of Diet, ad Libitum Feeding, and Moderate and Severe Dietary Restriction on Body Weight, Survival, Clinical Pathology Parameters, and Cause of Death in Control Sprague-Dawley Rats, *Toxicol. Sci.*, 2000, **58**(1), 195–207.
123. W. T. Allaben, A. Turturro, J. E. Leakey, J. E. Seng and R. W. Hart, FDA pointsto-consider documents: the need for dietary control for the reduction of experimental variability within animal assays and the use of dietary restriction to achieve dietary control, *Toxicol. Pathol.*, 1996, **24**(6), 776–781.
124. M. J. Dirx, M. P. Zeegers, P. C. Dagnelie, T. Van Den Bogaard and P. A. Van Den Brandt, Energy restriction and the risk of spontaneous

mammary tumors in mice: a meta-analysis, *Int. J. Cancer*, 2003, **106**, 766–770.

125. K. P. Keenan, P. Laroque, G. C. Ballam, K. A. Soper, R. Dixit, B. A. Mattson, S. P. Adams and J. B. Coleman, The effects of diet, ad libitum overfeeding, and moderate dietary restriction on the rodent bioassay: The uncontrolled variable in safety assessment, *Toxicol. Pathol.*, 1996, **24**, 757–768.

126. S. Goto, R. Takahashi, S. Araki and H. Nakamoto, Dietary restriction initiated in late adulthood can reverse age-related alterations of protein and protein metabolism, *Ann. N. Y. Acad. Sci.*, 2002, **959**, 50–56.

127. B. Martin, J. Sunggoan, S. Maudsley and M. P. Mattson, "Control" Laboratory rodents are metabolically morbid: Why it matters, *Proc. Natl. Acad. Sci.*, 2010, **107**(14), 6127–6133.

128. M. J. Prescott, V. J. Brown, P. A. Flecknell, D. Gaffan, K. Garrod, R. N. Lemon, A. J. Parker, K. Ryder, W. Schultz, L. Scott, J. Watson and L. Whitfield, Refinement of the use of food and fluid control as motivational tools for macaques used in behavioural neuroscience research: Report of a Working Group of the NC3Rs, *J. Neurosci. Methods*, 2010, **193**(2), 167–188.

129. R. W. Hart, D. A. Neumann and R. T. Robertson, *Dietary Restriction: Implications for the Design and Interpretation of Toxicity and Carcinogenicity Studies*, ILSI Press, Washington, DC, 1995.

130. H. H. Cornish, A. B. Morrison and John M. Nelson, Effect of fasting on the toxicity and serum concentration of aminopyrine in rats, *Toxicol. Appl. Pharmacol.*, 1970, **17**(1), 217–222.

131. A. Kast and J. Nishikawa, The effects of fasting on oral acute toxicity of drugs in rats and mice, *Lab. Anim.*, 1981, **15**, 359–364.

132. R. J. Irvine, J. White and R. Chan, The influence of restraint on blood pressure in the rat, *J. Pharmacol. Toxicol. Methods*, 1997, **38**, 157–162.

133. A. Harkin, T. J. Connor, J. M. O'Donnell and J. P. Kelly, Physiological and behavioural responses to stress: what does a rat find stressful?, *Lab. Anim. Eur.*, 2002, **2**, 32–40.

134. J. L. Sharp, T. G. Zammit, T. A. Azar and D. M. Lawson, Are 'bystander' female Sprague-Dawley rats affected by experimental procedures?, *Contemp. Topics Lab. Anim. Sci.*, 2003, **42**, 19–27.

135. I. Ushijima, Y. Mizuki and M. Yamada, Development of stress-induced gastric lesions involves central adenosine A1-receptor stimulation, *Brain Res.*, 1985, **339**(2), 351–355.

136. S. Wolfensohn and P. Honess, *Handbook of Primate Husbandry and Welfare*, Blackwell Publishing, Oxford, UK, 2005.

137. M. J. Prescott and H. M. Buchanan-Smith, Training nonhuman primates using positive reinforcement techniques: Guest editors' introduction, *J. Appl. Anim. Welfare Sci.*, 2003, **6**, 157–161.

138. J. McKinley, H. M. Buchanan-Smith, L. Bassett and K. Morris, Training common marmosets (Callithrix jacchus) to cooperate during routine

laboratory procedures: ease of training and time investment, *J. Appl. Anim. Welfare Sci.*, 2003, **6**, 209–220.

139. V. Reinhardt, Working with rather than against macaques during blood collection, *J. Appl. Anim. Welfare Sci.*, 2003, **6**, 189–197.

140. A. E. Rennie and H. M. Buchanan-Smith, Refinement of the use of Non-human Primates in Scientific Research. Part III: Refinement of procedures, *Anim. Welfare*, 2006, **15**, 239–261.

141. R. P. Chilcott, B. Stubbs and Z. Ashley, Habituating pigs for in-pen, non-invasive biophysical skin analysis, *Lab. Anim.*, 2001, **35**, 30–35.

142. V. Claassen, in *Techniques in the Behavioural and Neural Sciences*, Vol. 12, ed. J. P. Huston, Elsevier, Amsterdam, 1994, pp. 5–94.

143. M. K. Meijer, B. M. Spruijt, L. F. van Zutphen and V. Baumans, Effect of restraint and injection methods on heart rate and body temperature in mice, *Lab. Anim.*, 2006, **40**, 382–391.

144. D. B. Morton, D. Abbot, R. Barcley, B. S. Close, R. Ewbank, D. Gask, M. Heath, S. Mattic, T. Poole, J. Seamer, J. Southee, A. Thompson, B. Trussel, C. West and M. Jennings, Removal of blood from laboratory mammals and birds. FIRST REPORT OF THE BVA/FRAME/RSPCA/UFA W JOINT WORKING GROUP ON REFINEMENT, *Lab. Anim.*, 1993, **27**(1), 1–22.

145. K. H. Diehl, R. Hull, D. Morton, R. Pfister, Y. Rabemampianina, D. Smith, J. M. Vidal and C. van de Vorstenbosch, A good practice guide to the administration of substances and removal of blood, including routes and volumes, *J. Appl. Toxicol.*, 2001, **21**(1), 15–23.

146. D. B. Morton, M. Jennings, A. Buckwell, R. Ewbank, C. Godfrey, B. Holgate, I. Inglis, R. James, C. Page, I. Sharman, R. Verschoyle, L. Westall and A. B. Wilson, Refining procedures for the administration of substances, *Lab. Anim.*, 2001, **35**(1), 1–41.

147. S. Wolfensohn and M. Lloyd, *Handbook of Laboratory Animal Management and Welfare*, Oxford University Press, Oxford, 1994.

148. S. J. Murphy, P. Smith, A. B. Shaivitz, M. I. Rossberg and P. D. Hurn, The effect of brief halothane anesthesia during daily gavage on complications and body weight in rats, *Contemp. Topics Lab. Anim. Sci.*, 2001, **40**, 9–12.

149. G. N. Rao, T. A. Peace and D. E. Hoskins, Training could prevent deaths due to rodent gavage procedure, *Contemp. Topics Lab. Anim. Sci.*, 2001, **40**, 7–8.

150. K. Okva, E. Tamoseviciute, A. Ciziute, P. Pokk, O. Ruksenas and T. Nevalainen, Refinements for intragastric gavage in rats, *Scand. J. Lab. Anim. Sci.*, 2006, **33**, 243–252.

151. A. P. Brown, N. Dinger and B. S. Levine, Stress produced by gavage administration in the rat, *Contemp. Topics Lab. Anim. Sci.*, 2000, **39**, 17–21.

152. I. M. Kapetanovic, M. Muzzio, Z. Huang, T. N. Thompson and D. L. McCormick, Pharmacokinetics, oral bioavailability, and metabolic profile of resveratrol and its dimethylether analog, pterostilbene, in rats, *Cancer Chemother. Pharmacol.*, 2011, **68**(3), 593–601.

153. L. Penicaud and J. Le Magnen, Recovery of body weight following starvation or food restriction in rats, *Neurosci. Biobehav. Rev.*, 1980, **4**, 47–52.
154. L. A. Toth and T. Gardiner, Food and Water Restriction Protocols: Physiological and Behavioral Considerations, *Contemp. Topics Lab. Anim. Sci.*, 2000, **39**, 9–17.
155. J. K. Vermeulen, A. De Vries, F. Schlingmann and R. Remie, Food deprivation: common sense or nonsense?, *Anim. Technol.*, 1997, **48**, 45–54.
156. G. Lukas, S. D. Brindle and P. Greengard, The route of absorption of intraperitoneally administered compounds, *J. Pharmacol. Exp. Ther.*, 1971, **178**(3), 562–564.
157. G. M. Cohen, O. M. Bakke and D. S. Davies, First-pass metabolism of paracetamol in rat liver, *J. Pharm. Pharmacol.*, 1974, **26**(5), 348–351.
158. M. M. Van der Graaff, N. P. Vermeulen and D. D. Breimer, Route- and dose-dependent pharmacokinetics of hexobarbitone in the rat: a re-evaluation of the use of sleeping times in metabolic studies, *J. Pharm. Pharmacol.*, 1985, **37**(8), 550–554.
159. J. C. Wingfield and A. Kitaysky, Endocrine responses to unpredictable environmental events: Stress or anti-stress hormones?, *Integrative Comp. Biol.*, 2002, **42**, 600–609.
160. S. Chiappin, G. Antonelli, R. Gatti and E. F. De Palo, Saliva specimen: A new laboratory tool for diagnostic and basic investigation, *Clin. Chim. Acta*, 2007, **383**, 30–40.
161. P. A. Flecknell, *Laboratory Animal Anaesthesia: an Introduction for Research Workers and Technicians*, 2nd edn, Academic Press, London, 1996.
162. P. Flecknell and A. Waterman-Pearson, *Pain Management in Animals*, W. B. Saunders, London, 2000.
163. J. V. Roughan and P. A. Flecknell, Behaviour-based assessment of the duration of laparotomy-induced abdominal pain and the analgesic effects of carprofen and buprenorphine in rats, *Behav. Pharmacol.*, 2004, **15**, 461–472.
164. J. H. Liles and P. A. Flecknell, The use of non-steroidal anti-inflammatory drugs for the relief of pain in laboratory rodents and rabbits, *Lab. Anim.*, 1992, **26**, 241–255.
165. D. P. Lovell, Variation in pentobarbitone sleeping time in mice 1. Strain and sex differences, *Lab. Anim.*, 1986, **20**, 85–90.
166. J. I. Alexander and H. G. Hill, *Postoperative Pain Control*, Blackwell Scientific Publications, Oxford, 1987.
167. R. A. Hughes and K. J. Sufka, Morphine hyperalgesic effects on the formalin test in domestic fowl (*Gallus gallus*), *Pharmacol. Biochem. Behav.*, 1991, **38**, 247–251.
168. J. V. Roughan and P. A. Flecknell, Effects of surgery and analgesic administration on spontaneous behaviour in singly housed rats, *Res. Vet. Sci.*, 2000, **69**, 283–288.
169. J. Sharp, T. Zammit, T. Azar and D. Lawson, Recovery of male rats from major abdominal surgery after treatment with various analgesics, *Contemp. Topics Lab. Anim. Sci.*, 2003, **42**, 22–7.

170. M. I. Gonzalez, M. J. Field, S. Bramwell, S. McCleary and L. Singh, Ovariohysterectomy in the rat: A model of surgical pain for evaluation of pre-emptive analgesia?, *Pain*, 2000, **88**(1), 79–88.

171. B. D. Lascelles, A. E. Waterman, P. J. Cripps, A. Livingston and G. Henderson, Central sensitization as a result of surgical pain: Investigation of the preemptive value of pethidine for ovariohysterectomy in the rat, *Pain*, 1995, **62**(2), 201–212.

172. J. A. Reichert, R. S. Daughters, R. Rivard and D. A. Simone, Peripheral and preemptive opioid antinociception in a mouse visceral pain model, *Pain*, 2001, **89**(2–3), 221–227.

173. G. A. Peyman, M. H. Rahimy and M. L. Fernandes, Effects of morphine on corneal sensitivity and epithelial wound healing: implications for topical ophthalmic analgesia, *Br. J. Ophthalmol.*, 1994, **78**, 138–141.

174. K. D. Vlach, J. W. Boles and B. G. Stiles, Telemetric evaluation of body temperature and physical activity as predictors of mortality in a murine model of staphylococcal enterotoxic shock, *Comparative Med.*, 2000, **50**, 160–166.

175. K. Kramer, *Applications and Evaluation of Radio-Telemetry in Small Laboratory Animals*, PhD Thesis, University of Utrecht, Utrecht, The Netherlands, 2000.

176. K. Kramer, L. Kinter, B. P. Brockway, H. P. Voss, R. Remie and L. F. M. Van Zutphen, The use of radiotelemetry in small laboratory animals: recent advances, *Contemp. Topics Lab. Anim. Sci.*, 2001, **40**, 8–16.

177. K. Kramer and L. B. Kinter, Evaluation and applications of radio-telemetry in small laboratory animals, *Physiol. Genomics*, 2003, **13**, 197–205.

178. N. H. Anderson, A. M. Devlin, D. Graham, J. J. Morton, C. A. Hamilton, J. L. Reid, N. J. Schork and A. F. Dominiczak, Telemetry for cardiovascular monitoring in a pharmacological study. New approaches to data analysis, *Hypertension*, 1999, **33**, 248–255.

179. A. Harkin, T. J. Connor, J. M. O'Donnell and J. P. Kelly, Physiological and behavioural responses to stress: What does a rat find stressful?, *Lab. Anim.*, 2002, **31**, 42–50.

180. S. A. B. E. Van Acker, K. Kramer, J. A. Grimbergen, J. Zhang, W. J. F. Van der Vijgh and A. Bast, A new model to test potential protectors. Doxorubicin-induced cardiotoxicity monitored by ECG in freely moving mice, *Cancer Chemother. Pharmacol.*, 1996, **38**, 95–101.

181. L. B. Kinter, Cardiovascular Telemetry and Laboratory Animals Welfare: New Reduction and Refinement Alternatives (Abstract), in *General Pharmacology/Safety Pharmacology Meeting*, Safety Pharmacology Society, Philadelphia, PA, 1996.

182. American Veterinary Medical Association, *AVMA Guidelines on Euthanasia*, AVMA, Schaumber, Illinois, 2007. http://www.avma.org/issues/animal_welfare/euthanasia.pdf.

183. B. Close, K. Banister, V. Baumans, E. M. Bernoth, N. Bromage, J. Bunyan, W. Erhardt, P. Flecknell, N. Gregory, H. Hackbarth,

D. Morton and C. Warwick, Recommendations for euthanasia of experimental animals: Part 1, *Lab. Anim.*, 1996, **30**, 293–316.

184. B. Close, K. Banister, V. Baumans, E. M. Bernoth, N. Bromage, J. Bunyan, W. Erhardt, P. Flecknell, N. Gregory, H. Hackbarth, D. Morton and C. Warwick, Recommendations for euthanasia of experimental animals: Part 2, *Lab. Anim.*, 1997, **31**, 1–32.

185. EFSA (European Food Safety Authority – Animal Health and Welfare Panel), Scientific report: Aspects of the biology and welfare of animals used for experimental and other scientific purposes, *Annex to the EFSA Journal*, 2005, **292**, 1–136.

186. I. Kimber, The local lymph node assay, in *Dermatotoxicology*, 5th edn, ed. F. N. Marzulli and H. I. Maibach, Taylor & Francis, Washington, DC, 1996, pp. 469–475.

187. A. J. Roi and B. Grune, *The ECVAM Search Guide – GOOD SEARCH PRACTICE on Animal Alternatives*, Publications Office of the European Union, Luxembourg, 2011.

CHAPTER 5

Computers Instead of Cells: Computational Modeling of Chemical Toxicity

HAO ZHU,*[a,b] MARLENE KIM,[a] LIYING ZHANG[c] AND ALEXANDER SEDYKH[d]

[a] The Rutgers Center for Computational and Integrative Biology, Camden, NJ 08102, USA; [b] Department of Chemistry, Rutgers University, Camden, NJ 08102, USA; [c] Pfizer Global Research and Development at Groton, Groton, CT 06340, USA; [d] Division of Medicinal Chemistry and Natural Products, School of Pharmacy, University of North Carolina at Chapel Hill, Chapel Hill, North Carolina 27599, USA
*Email: hao.zhu99@rutgers.edu

5.1 Challenges of Modern Chemical Toxicity Evaluations

The history of chemical toxicology could be considered as a process of understanding the mechanisms of toxicity induced by specific chemical structures. Although this is an area that has attracted great scientific effort since the Society of Toxicology was founded in 1961, chemical toxicology and safety has been regarded as the major reason for attrition of new drugs in the past decade.[1] The regulatory evaluation of chemical toxicity *in vivo* and through all the animal models used by pharmaceutical companies at the early stage of the drug discovery process is expensive and time consuming. In 2010, a joint RFA-RM-10-006 from the US Food and Drug Administration (FDA) and the US

Issues in Toxicology No. 19
Reducing, Refining and Replacing the Use of Animals in Toxicity Testing
Edited by David G. Allen and Michael D. Waters
© The Royal Society of Chemistry 2014
Published by the Royal Society of Chemistry, www.rsc.org

National Institutes of Health (NIH), entitled "Advancing Regulatory Science through Novel Research and Science-Based Technologies," emphasized the need for developing both experimental and computational tools for regulatory evaluation of drug safety. In 2011, a RFA-FD-11-017 from the FDA, entitled "Development and Qualification of Alternative Testing Methodologies for Reproductive Toxicology" indicated the urgent requirement of alternatives to animal testing in the area of developmental and reproductive toxicity. In a critical effort to advance toxicity testing in the 21st century, the US National Research Council outlined a new vision and strategies for the increased use of *in vitro* and computational technologies for chemical risk assessment.[2] Similarly, driven by the urgent need to evaluate potential health risks associated with human exposure to chemicals, the Organisation for Economic Development and Co-operation (OECD) in Europe funded the development of the Quantitative Structure Activity Relationship (QSAR) Application Toolbox.[3] The toxicity models developed by this program are "intended to be used by governments, chemical industry and other stakeholders in filling gaps in (eco)toxicity data needed for assessing the hazards of chemicals." The goal of the above recently funded programs is to emphasize the need of alternatives to traditional animal models and increase public acceptance of non-animal models for regulatory evaluation of chemical toxicity.

To find alternatives to traditional animal toxicity testing and to understand the relevant toxicological mechanisms, many *in vitro* toxicity screens and computational toxicity models have been developed and implemented.[4–11] For example, in 2006 the US EPA started a research program named "ToxCast". The goal of this program was to develop methods for utilizing *in vitro* toxicity tests and various toxicogenomics technologies to quickly evaluate the toxic potential of chemicals and to prioritize candidates for future animal testing.[7] In the first phase of the program, 320 well-characterized chemicals (primarily pesticides) were screened in around several hundred high-throughput screening (HTS) assays. All the assays were developed by different biological companies and have been used for toxicity screening tests for many years. All endpoints were presented as inhibition concentration by 50% (IC_{50}), lowest effective level (LEL) or lethal concentration by 50% (LC_{50}). Furthermore, there are other well developed *in vitro* screening programs. In 2004 the National Toxicology Program (NTP) proposed a new Roadmap as "A National Toxicology Program for the 21st Century."[12] This Roadmap program has three major goals: "refining traditional toxicology assays, developing rapid mechanism-based predictive screens and improving the overall utility of data for making public health decisions."[13] As a result, several screening centers were developed for the use of HTS techniques in chemical toxicity evaluations, e.g. the NIH Chemical Genomics Center (NCGC).[14]

The progress of HTS techniques has generated a vast quantity of *in vitro* data, which is expected to continue increasing in coming years. Thus, the current HTS studies provide the community with rich toxicology information that has the potential to be integrated into computational toxicity modeling. For example, PubChem provides the largest primary HTS dataset obtained

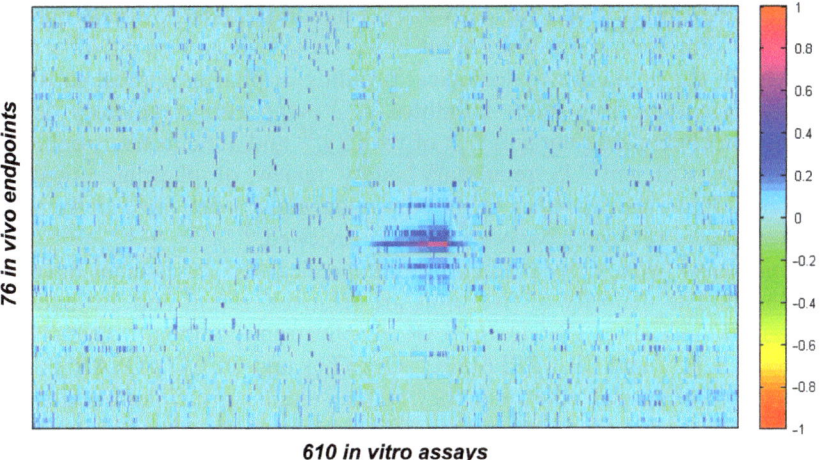

Figure 5.1   The correlation (R) heatmap between 610 *in vitro* assay and 76 *in vivo* endpoint results for 320 ToxCast compounds.

from various types of toxicity bioassays. By the end of 2012, about five million PubChem compounds, although not all of them, have been tested against around 650 000 bioassays and there are 120 million data points resulting from these bioassay experiments. However, current available *in vitro* toxicity screens have limitations: (1) they require physical samples (which may be very expensive) of compounds for testing; and (2) despite significant technical advances, they can be time consuming and resource intensive. Moreover, the relationship between the available *in vitro* assay data and the animal toxicity testing results is not always clear. For example, the results indicate that the ToxCast *in vitro* assays (mostly based on human cell lines) have no correlation with the animal *in vivo* endpoints from ToxRefDB for the same 320 compounds (Figure 5.1). Similar conditions could be found for most of the currently available *in vitro* assays data and animal toxicity for the same compounds, except in the case of genotoxicity testing (see Chapter 11 of this volume).

5.2   Conventional Cheminformatics Toxicity Models

Compared to any experimental models, the unique advantage of using a computational model in risk analysis is that a chemical could be evaluated for its toxicity potential even before being synthesized. Quantitative structure–activity relationship (QSAR) analysis is a widely used computational method to generate models and predict the toxicity of chemicals. In most QSAR studies, QSAR modeling is used as a tool to analyze how the change of functional groups affects biological activity for a generic set of compounds. However, in computational toxicology, QSAR modeling focuses on the derivation of the relationships that could link complex chemical information (normally as chemical descriptor values for a diverse set of compounds) to whole organism toxicity. Computational toxicity tools based on QSAR models have been used to assist in the predictive toxicological profiling of pharmaceutical substances for understanding drug safety liabilities,[9,15–17] supporting regulatory decision making on chemical safety and risk of toxicity,[18] and are effectively enhancing an already stringent US regulatory safety review of pharmaceutical substances.[19] Previous efforts of using computational models in toxicity studies and regulatory reviews have been recently reviewed.[20–23]

Many tools for toxicity prediction and chemical risk assessment have been developed. The majority of free tools are relatively recent and are under development, including OECD (Q)SAR Application Toolbox,[24] OpenTOX,[25] EPA T.E.S.T.,[26] and EPA EPI Suite.[27] OpenTox and OECD's QSAR Toolbox projects are of special note since they employ similar principles of data modeling, transparency, and public availability. Most of the popular commercial tools were developed a decade ago, including DEREK,[28] HazardExpert,[29] Leadscope,[30] and MCASE.[31] They rely on internally curated databases that are frequently updated, but the underlying methodologies essentially have remained unchanged since they were developed. The lack of transparency and commercial nature of these tools precludes their widespread use and validation by the computational toxicology community.

5.3   Future Directions of Computational Toxicology

In 2008, Collins *et al.* summarized the current state of "Transforming Toxicology" as viewed by the National Research Council (NRC), NTP, and other regulatory agencies.[32] In this review of the toxicology research landscape, predictive computational toxicology is described as: (1) yielding data predictive of results from animal toxicity; (2) allowing prioritization of chemicals for further testing; and (3) assisting in the prediction of risk to humans.[32] Although predictive QSAR modeling of toxicity is starting to be used to evaluate the toxicity potential for pharmaceutical companies and environmental agencies,[15,33] most of the previous studies show that currently available QSAR models do not work well in evaluating *in vivo* toxicity potentials, especially for novel compounds whose analogs do not exist in the training data.[34,35] For a new compound with a chemical structure similar to another compound in the training set but with different toxicity effects (a so-called "activity cliff" situation), traditional QSAR models will likely generate incorrect predictions. For example, metolachlor (CAS RN# 51218-45-1) and acetochlor (CAS RN# 34256-82-1) are two similar compounds in the ToxCast Phase I dataset. However, their animal toxicity results are different (Table 5.1). Compared to the high chemical similarity of these two compounds, they have different biological profiles based on the ToxCast assay results. Accordingly, the conventional computational toxicity models based purely on the chemical information usually cannot handle these cases. A novel modeling technique is needed that goes beyond the scope of conventional QSAR approaches to take

Table 5.1 Two ToxCast compounds with similar chemical structures but different *in vitro* and *in vivo* toxicity results.

	In vivo toxicity responses*				In vitro assay data**				
	Mouse kidney	Rat skelet-axial	MGR rat liver	MGR rat kidney	ACEA IC$_{50}$	Cellloss 24 h	P53act 24 h	NVS-ADME-P3A2	Solidus P450
Metolachlor	0	0	0	0	0	0	0	1	0
Acetochlor	1	1	1	1	1	1	1	0	1

*The animal toxicity results were obtained from ToxRefDB. The toxicity endpoints were defined based on the long-term animal toxicity experiments performed in NTP.

**The toxicity bioassay data were obtained from the ToxCast program. The names of the bioassays were relevant to the nature of the test and the institutes (or companies) that performed the test.

advantage of the abundant *in vitro* toxicity screening (especially the HTS) results to develop enhanced toxicity models.

5.4 Availability of Large Toxicity Databases from Data-sharing Projects

5.4.1 The Available Resources for *In Vitro* Toxicity Data

As a result of innovative technologies that enable rapid synthesis and high-throughput screening of large libraries of compounds in toxicity studies, there has been a huge increase in the number of compounds and associated testing data in different *in vitro* screens. Table 5.2 shows some examples of these data collections distributed through various data-sharing programs. PubChem is a public repository for chemical structures and their biological properties.[36,37] Bioactivity data in PubChem are contributed by hundreds of institutes, research laboratories, and specifically those screening centers under the NIH Molecular Libraries Program (MLP).[12] For example, the NIH Chemical Genomics Center (NCGC) was created in 2005 as a comprehensive screening center in the NIH MLP.[14] The mission of the NCGC is to apply the tools of small molecule screening and discovery to toxicology studies. Every year the NCGC generates millions of toxicity bioassay data points by testing thousands of diverse compounds and shares all the data with the research community via PubChem. The unique quantitative high-throughput screening (qHTS) technique developed and optimized by the NCGC generates data in a high quality and a standardized form.[4] Another large group of toxicity bioassay data in PubChem is from the European Bioinformatics Institute (EBI).[38] The EBI's goal is to provide freely available data and bioinformatics services to all branches of the scientific community. As a part of this goal, the ChEMBL database was constructed for screening data of both chemical toxicity and absorption, distribution, metabolism, and excursion (ADME) properties. ChEMBL version 11 (ChEMBL_11) was launched in 2011. It includes 3.3 million bioassay measurements covering 629 943 compounds.[39] This was obtained from curating over 42 500 scientific publications. Table 5.1 shows two examples of the NCGC and ChEMBL toxicity bioassay data currently available in PubChem.

The ToxCast program has been briefly introduced in the above section. It should be pointed out that the objective of ToxCast is to find alternatives to animal models. For this reason, this program intentionally tested compounds with rich animal toxicity information, which limits its scope to those few hundreds of compounds that have been historically tested in animal models. Compared to the ToxCast program, ToxNET contains and allows navigation through 16 separate databases of much more diverse chemicals.[40] ToxNET was developed by the National Library of Medicines' (NLM) Division of Specialized Information Services (SIS). By grouping the databases together, ToxNET allows for all information to be accessed from one query form. Although there are 14 separate databases, some are very similar and are grouped together in the example report.

Table 5.2 Public databases of *in vitro* toxicity endpoints.

NCGC qHTS cytotoxicity data[4] available through PUBCHEM[72] (via PubChem AID#)	Concentration-response profiles of 1408 substances screened for their effects on cell viability are available through PubChem for 13 cell lines: HepG2 (human hepatoma; AID #433), H-4-II-E (rat hepatoma; AID #543), BJ (human foreskin fibroblast; AID #421), Jurkat (clone E6-1, human acute T cell leukemia; AID #426), HEK293 (transformed human embryonic kidney cell; AID #427), MRC-5 (human lung fibroblast; AID #434), SK-N-SH (human neuroblastoma; AID #435), N2a (mouse neuroblastoma; AID #540), NIH 3T3 (mouse embryonic fibroblast; AID #541), HUV-EC-C (human vascular endothelial cell; AID #542), SH-SY-5Y (human neuroblastoma, subclone of SK-N-SH; AID #544), Renal Proximal Tubule (rat kidney cell; AID #545) and Mesenchymal (human renal glomeruli cell; AID #546). Each compound was tested at 14 concentrations ranging from 0 006 to 92 µM and the response was measured as % change in cell viability as compared to vehicle control at each concentration.
ChEMBLdb[38] available through PUBCHEM[72] (via PubChem AID#)	A database of bioactive drug-like small molecules abstracted and curated from the primary scientific literature. Bioactivities are represented by binding constants, pharmacology, and ADMET data. ChEMBL assays are available through PubChem. Human toxicity related endpoints are primarily from *in vitro* data, such as: cytotoxicity on SNU-354 cells (hepatoma cell line, AID #200819), antiproliferative action on L02 cells (normal hepatocytes, AID #416061), growth inhibition of SK-Hep1 cells (liver adenocarcinoma cell line, AID #201649), cytotoxicity and anticancer activity on HepG2 cells (AID #86696, 340104, 421266), etc.
ToxCast[73]	Phase I (August 7, 2009 update) provided 304 unique compounds characterized in over 600 HTS endpoints. The endpoints include biochemical assays of protein function, cell-based transcriptional reporter and gene expression, cell line and primary cell functional and developmental endpoints in zebrafish embryos, and embryonic stem cells. Additionally, mapping of these assays to 315 genes and 438 pathways was made publicly available. Phase II will complete screens of an additional 700 compounds, HTS data on nearly 10000 chemicals will be available through Tox21 collaboration in 2010.
ToxNET[40]	A data network covering toxicology, hazardous chemicals, environmental health, and related areas. Managed by US National Library of Medicine.
Chemical Effects in Biological Systems	The CEBS is a public resource that provides data to environmental health scientist. CEBS is designed to display data in the context of biology and study design, and to permit data integration across various studies.
Comparative Toxicogenomics Database	A database supporting cross-species comparative studies of toxicologically important genes and proteins and their interactions with chemicals.

Another rapidly growing area of interdisciplinary research is tox-icogenomics. Toxicogenomics is a field of toxicology that addresses information concerning gene, protein, and metabolite changes within a particular cell or tissue of an organism in response to chemicals. Many modern *in vitro* toxicity studies result in outcomes via relevant toxicity mechanisms and these findings can be translated into biomarkers that could be applied to human exposure studies.[41] Toxicogenomics investigations generate enormous amounts of "omics" data that are meant to predict toxicity or genetic susceptibility induced by chemicals. The Chemical Effects in Biological Systems (CEBS) database developed by the NIEHS is now the public repository for all NTP conventional toxicology and carcinogenicity data as well as NCGC HTS data[42] and, along with the Comparative Toxicogenomics Database (CTD) at Mount Desert Island Biological Laboratory, aims to promote comparative studies of genes and proteins across species.[43–46] Currently, CEBS is available at http://www.niehs.nih.gov/research/resources/databases/cebs/index.cfm. CTD data is searchable through the ToxNET portal.

In 2007, the NTP, NCGC, and the US EPA initiated a collaborative program called "Tox21" and were soon thereafter joined by the US FDA.[2] The objectives of the Tox21 partners were to develop, validate, and translate innovative *in vitro* testing methods that characterized toxicity pathways. The goal of the program is to prioritize those chemicals that needed more extensive toxicological evaluation from a library of 10 000 chemicals of environmental interest. Currently the NCGC is using qHTS techniques to screen these compounds.[13] Tox21 is expected to aid in developing models that can be used to more effectively predict how chemicals realize their toxicity in animals or human beings.

5.4.2 The Available Animal Toxicity Data

Compared to *in vitro* toxicity data, the availability of animal toxicity testing results is still limited. There are several individual efforts that have tried to test, accumulate, and share animal toxicity data in specific areas. For example, the Columbia Environmental Research Center (CERC) is a toxicity center which focuses on acute chemical toxicity using various fish models (Table 5.3).[47] Another example is the Carcinogenicity Potency Database (CPDB) that was developed by the Carcinogenic Potency Project at the University of California, Berkeley, and the Lawrence Berkeley National Laboratory. This database focuses on chronic, long-term animal cancer tests, both positive and negative for chemical carcinogenicity (Table 5.3).[48] However, most animal toxicity data are in the archives of regulatory agencies and pharmaceutical companies and unavailable to the public. Due to recent efforts of data curation and sharing, some historical animal toxicity data, which were obtained from animal testing over the past 50 years or longer, have become available to the research community. Table 5.3 shows some examples of the current publicly available animal toxicity databases.

The Toxicological Reference Database (ToxRefDB) is a database that includes the pesticide registration toxicity data from the US EPA over the past

Table 5.3 Publicly available databases of *in vivo* toxicity endpoints.

CERC[47]	Summarizes the results from aquatic acute toxicity tests which provide a relative starting point for hazard assessment of contaminants and is required for federal chemical registration programs. The current database contains 410 chemicals testing by 66 species of freshwater animals.
CPDB[48]	Contains reports of animal cancer tests on 1547 chemicals with tumor target site incidence and TD_{50} potencies. All the compounds were tested using standardized two year animal models.
ToxRefDB[49,74]	Captures toxicological endpoints, critical effects, and relevant dose-response data from the EPA's Office of Pesticide Programs into a relational database using a standardized data field structure and vocabulary. Chemicals included in the database represent over 800 conventional pesticide active ingredients.
DSSTox[50]	7000 chemical substances tested in rats, mice, hamsters, dogs, and other animal species; data reviewed and compiled from literature and NTP studies.
ACToR[53]	This portal provides the links of about 1100 different toxicity data resources for more than 750 000 environmental compounds. The datasets include 2-year, two species, toxicology and carcinogenesis studies collected by the NTP and other regulatory historical animal toxicity datasets.

30 years.[49] The original toxicity data were stored in the US EPA's Office of Pesticide Program. The animal toxicity data includes sub-chronic toxicity endpoints (rodents and non-rodents), prenatal developmental toxicity (rat and rabbit), reproductive and fertility effects (two-generation studies), immuno-toxicity, developmental neurotoxicity, chronic toxicity (rat, mouse, dog), and 2-year carcinogenicity bioassays (rat and mouse). The ToxRefDB standardized the toxicity data of 76 different animal toxicity endpoints for about 800 pesticides. Through this process, quality control (QC) was performed by cross-checking historical data to ensure consistency.[49] Another important toxicity data-sharing program of the US EPA is the Distributed Structure-Searchable Toxicity database network (DSSTox).[50,51] The primary goal of the DSSTox project is to share the structure-annotated toxicity data files for QSAR model development.[52] Compared to other toxicity datasets (e.g., ToxRefDB) that primarily focus on the read across of different experimental toxicity data fields, the DSSTox project has contributed the most to the standardization of chemical information space. Besides the extensive QC process of chemical annotation and representation, the DSSTox dataset also includes the information of the difference between nominal chemical structures and the "actual tested substances." For example, the chemical structure assigned to a compound in a toxicity database may only represent the major component of the tested substance, which could be a mixture, a complex formulation, or a macromolecule. The current DSSTox dataset in year 2012 contains over 7000 unique substances with relevant animal toxicity results.[52]

Starting in 2007, the US EPA's National Center for Computational Toxicology (NCCT) program initiated a unique toxicity data search program, named Aggregated Computational Toxicology Resource (ACToR).[53,54] The mission of the ACToR program was to develop a central database with links to a set of toxicity databases to bring together many types and sources of toxicity data for a large environmental chemical library. Aside from the results of *in vitro* bioassays, the current ACToR portal has the links to over 100 different animal toxicity data sources, including ToxRefDB and DSSTox.[54]

5.5 The use of Quantitative Structure *In vitro–In vivo* Relationship (QSIIR) to Develop Enhanced Computational Animal Toxicity Models

5.5.1 The Use of Experimental Parameters in the Computational Modeling

In previous computational toxicology studies, additional physicochemical properties, such as water partition coefficient (log P),[55] water solubility,[56] and melting point[57] were used successfully to augment computed chemical descriptors and improve the predictive power of QSAR models. These studies suggest that using experimental results as descriptors in QSAR modeling can prove beneficial. On the other hand, there are many previous studies that utilize *in vitro* data to correlate with specific *in vivo* toxicity results for the same compounds.[58–63] Most of these studies are performing simple read across analysis between the *in vitro* and *in vivo* data. Fielden *et al.*, however, used *in vitro* data as a new type of descriptor in animal toxicity model development.[64] The limitation of his study is that the number of compounds in the modeling set is less than 100, so the applicability of the resulting models for predicting new compounds will be likely low.

5.5.2 Advance the Traditional QSAR Approaches to QSIIR Strategy

Figure 5.2 summarizes the relationship between chemicals and their relevant *in vitro–in vivo* responses. To stress the strong appeal of the conventional QSAR approach, it should be made clear that, from a statistical viewpoint, QSAR modeling is a special case of general statistical data mining and data modeling where the data is formatted to represent objects described by multiple descriptors and the robust correlation between descriptors and a target property is sought. In computational toxicology studies, due to the recent progress of HTS and other *in vitro* assay-screening techniques, the content and interpretation of attributes and the resulting models could be different when compared to traditional QSAR toxicity modeling. Besides the descriptors that are obtained from chemical structures, the relevant bioassay data can be viewed as additional, biological information characterizing the compounds. The currently available and still rapidly growing HTS data for large and diverse chemical libraries make it possible to extend the scope of conventional QSAR in toxicity

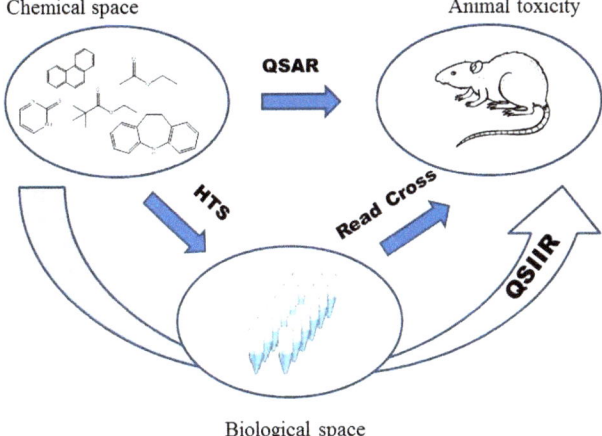

Figure 5.2 Diagram of structure *in vitro–in vivo* relationships.

studies by organizing a biological information space to complement the usual chemical descriptor space.

Based on previous reports, Zhu *et al.* proposed a new modeling technique called Quantitative Structure *In vitro–In vivo* Relationship (QSIIR) and used it in animal toxicity modeling studies.[65–67] The target properties of QSIIR modeling are still various animal toxicities, but the content and interpretation of "descriptors" and the resulting models will be different compared to traditional QSAR modeling. The QSIIR modeling approach is to treat the information from both chemical structures and biological data (normally the HTS data) as new "hybrid" descriptors and to develop enhanced animal toxicity models (Figure 5.2). The *in vitro–in vivo* correlations normally constitute the basc of this type of modeling studies.

The QSIIR modeling strategy has been applied to develop enhanced predictive models for acute toxicity,[65,67] carcinogenicity,[66] and hepatotoxicity.[68] There are reports from other labs with similar strategies for toxicity models.[69] Rusyn *et al.* summarizes the on-going efforts in this area in a recently published review article.[70] However, it should be noted that the hybrid models based on the combination of chemical and biological information does not guarantee an improvement of predictivity compared to the models based on only one type of descriptors. In a recent study, the model only with biological descriptors that were generated by toxicogenomics data demonstrated similar predictivity compared to the hybrid model.[68]

5.6 The Pitfalls of QSIIR Modeling and Potential Solutions

Data limitations, including the lack of animal toxicity testing and the relatively poor quality of current HTS data, are still major obstacles to QSIIR studies.

Although there are many toxicity databases available, as mentioned above, the experimental testing data may not be easily accessed by modelers or the quality is questionable.[71] For example, we examined the quality of the ToxCast bioassay data in our recent study. Among 320 ToxCast Phase I molecules, there are eight compounds that have duplicates or triplicates. The duplicates and triplicates were sent to the screening companies blindly for QC purposes. We compared the bioassay results of these eight compounds with those of their duplicates or triplicates (Figure 5.3a). The low correlations noted between the results of a compound and its replicate brings noise into the modeling process. The transformation from original dose response data (e.g. lowest response level (LEL)) to the category results (e.g. toxic or non-toxic) reduces such noise (but also sacrifices meaningful data variation), the correlations for the same eight compounds and their replicates become significantly higher (Figure 5.3b).

In another project, Sedykh *et al.* studied the HTS data of 13 cell viability bioassays generated by NCGC for 1408 molecules.[67] The uniqueness of this dataset is that each compound was tested at 15 different concentrations in every bioassay. However, the original NCGC-HTS data extracted from PubChem still contained high levels of noise. Figure 5.4a shows some examples of the original NCGC dose response curves for several compounds. By developing innovative data treatment procedures, we could remove the noise of low dose responses (Figure 5.4b), transform the response into classification scale to reduce data errors (Figure 5.4c), as was done in the earlier study, or convert the curve into biological fingerprints (Figure 5.4d). We used the original HTS data and the transformed data after removing the noise to develop a QSIIR model of acute animal toxicity. The results indicate that the QSIIR model based on the original HTS data (as shown in Figure 5.4a) has no improvement compared to the QSAR model based only on chemical descriptors. By using the processed data as the biological descriptors (as shown in Figure 5.4 b–d), the predictivity

	Assay->	Total	ACEA	ATG	BSK	Cellu	CLZD	NVS	NCGC	Total	ACEA	ATG	BSK	Cellu	CLZD	NVS	NCGC
	n=	610	7	73	174	57	42	239	18	610	7	73	174	57	42	239	18
CAS	Chemical	R correlation coefficients (-log [Conc.])								R correlation coefficients (binary data)							
55406-53-6	3-Iodo-2-propynyl-	0.70	0.91	NA	0.57	0.40	0.44	0.93	NA	0.93	0.86	0.99	0.86	0.74	0.93	0.98	1.00
741-58-2	Bensulide	0.87	0.38	0.97	0.16	0.35	0.47	0.91	1.00	0.94	0.79	0.99	0.90	0.75	0.87	0.99	1.00
64902-72-3	Chlorsulfuron	0.67	NA	NA	-0.03	NA	0.00	1.00	NA	0.97	0.71	0.99	0.91	0.98	0.95	1.00	1.00
84-74-2	Dibutyl phthalate	0.74	NA	0.60	0.02	-0.02	0.02	0.95	NA	0.94	0.57	0.90	0.89	0.95	0.88	0.99	1.00
51338-27-3	Diclofop-methyl	0.88	1.00	0.97	0.22	0.45	0.31	0.26	1.00	0.95	1.00	1.00	0.86	0.92	0.85	0.99	1.00
759-94-4	EPTC	0.89	NA	NA	-0.01	NA	0.98	0.00	NA	0.98	0.86	1.00	0.92	1.00	1.00	1.00	1.00
66441-23-4	Fenoxaprop-ethyl	0.85	NA	0.33	-0.02	0.10	0.00	0.03	NA	0.95	0.71	0.95	0.90	0.89	0.98	0.98	1.00
69806-50-4	Fluazifop-butyl*	0.86	NA	0.50	-0.01	0.52	0.34	0.00	NA	0.97	1.00	0.97	0.95	0.93	0.86	1.00	0.94
94125-34-5	Prosulfuron	0.86	NA	0.66	-0.02	NA	0.71	0.50	NA	0.98	1.00	0.99	0.93	0.98	0.94	1.00	1.00

Figure 5.3 QC study of ToxCast bioassay data with the results of eight compounds and their duplicates. (a) The correlation coefficients were generated by using the original data; (b) the correlation coefficients were generated by transforming the original dose-related response data into binary classifications.
*The abbreviations in the first row indicate the names of the companies that performed the testing; "n" in the second row indicates the number of bioassays; "NA" in the results indicates that there is no positive response so no correlation could be developed.

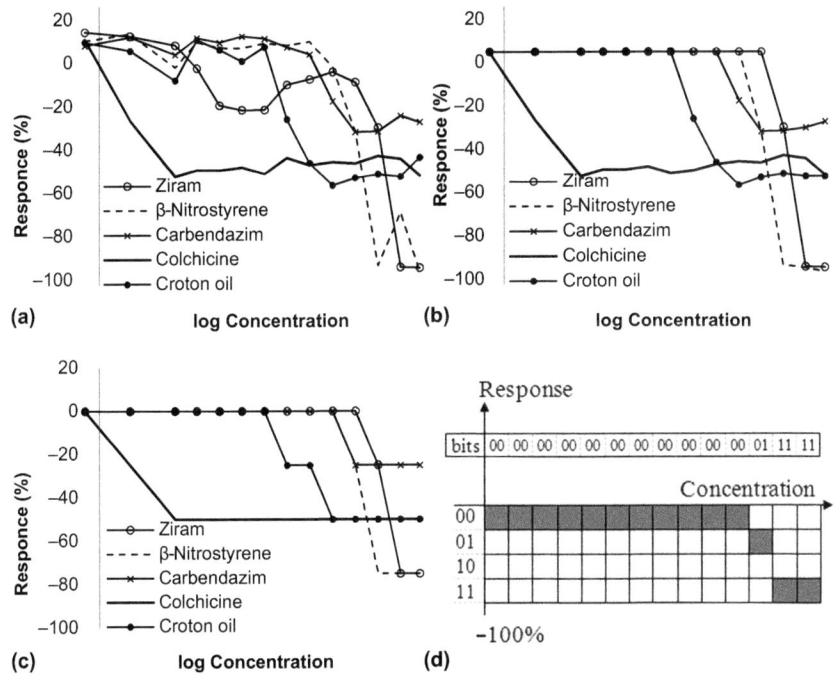

Figure 5.4 Original NCGC-HTS toxicity data (a) obtained from PubChem and three possible ways to transform it into useable formats (b)–(d).

of the QSIIR increased significantly. This research illustrates the necessity for data curation to remove experimental noise from QIIR model development.

Another problem that needs to be solved is that the QSIIR models need the bioassay data as the biological descriptors of chemicals. Although *in vitro* bioassays, especially those using IITS techniques, are relatively less expensive and time consuming compared to animal models, the predictions of compounds by QSIIR models normally require the test compounds to be synthesized and tested using the bioassays used for modeling. It is not a viable option if a large chemical library needs to be predicted. In a recent study, we used a hierarchical modeling workflow to solve this issue (Figure 5.5).[65] This workflow is based on the *in vitro–in vivo* correlation profiles of all compounds in the modeling set and the clusters (subsets) of the compounds based on the correlations. Then QSAR models are developed for individual subsets. Hence, in this modeling process, the modeling set compounds can be identified into two or more subset classes and we could develop a classification model to identify the class of a new compound. By using this workflow, the experimental assay is no longer needed when testing external compounds by using the resulting models. In this study, all chemicals are partitioned into two groups based on the relationship between IC_{50} and LD_{50} values using a novel moving regression approach. One group is formed by compounds with linear IC_{50} vs. LD_{50} relationship, and another group consists of the remaining compounds. Second, conventional binary

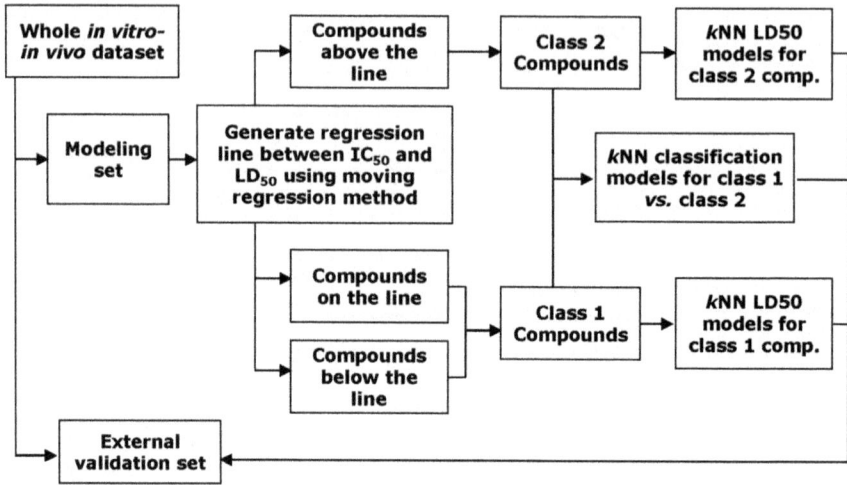

Figure 5.5 Two-step hierarchical k nearest neighbor (kNN) QSIIR workflow to develop enhanced rat acute toxicity (LD_{50}) models by using cytotoxicity data (IC_{50}) as the biological profile.

classification QSAR models are built to predict the group affiliation based on chemical descriptors only. Third, kNN continuous QSAR models are developed for each subclass individually to predict LD_{50} from chemical descriptors (Figure 5.5). Alternatively, individual QSAR models of each *in vitro* bioassay can be developed to serve as input for QSIIR and to obviate further testing if resulting model's quality permits that.[75]

Finally, for the successful QIIR, chemical space should not be under-represented as it tends to be in some toxicogenomics projects: small yet diverse sets of chemicals often lack common structural features for statistical generalization, at the same time these chemicals are extensively characterized in the biological space,[64,68] biasing prospective models toward purely biological interpretation.

5.7   Conclusions

Due to the innovative technologies that enable rapid synthesis and high-throughput screening of large libraries of compounds, relevant toxicity bioassay data is accumulating rapidly on a daily basis. Compared to well-developed traditional cheminformatics approaches (e.g., QSAR), QSIIR modeling affords an opportunity to fully explore the current chemical *in vitro–in vivo* informatics landscape in modern toxicology studies. Although there are many on-going data generation and sharing programs, the limited availability of high-quality toxicity data is still a major obstacle for relevant modeling studies. A proper data curation before modeling process would enhance the value of the available *in vitro* toxicity testing results, especially the HTS data, in

QSIIR modeling and greatly benefit the resulting models. Meanwhile, the improvement from QSAR to QSIIR brings new challenge of developing novel approaches to take advantage of the massive amount of available HTS data in the modeling process. The research efforts discussed in this chapter have proved that integrating biological data as descriptors in computational model development can be beneficial for the resulting toxicity models, especially for some specific animal toxicity endpoints. Using computational models instead of cell and/or animal testing in the early stages of developing a new compound (e.g., a new drug) could be a realistic way of minimizing expenses and saving resources. It is expected that modern computational toxicity models could be able to prioritize the compounds that are mostly likely to be toxic for experimental studies.

Acknowledgments

This work was supported, in part, by Colgate-Palmolive Alternative Research Grant (grant number 11-0897) and the Johns Hopkins Center for Alternatives to Animal Testing (2010-17). The authors want to thank Dr. Alexander Tropsha of University of North Carolina at Chapel Hill for his help in the past 5 years. Several research projects that were reviewed in this chapter were finished with his help and guidance. The authors also want to thank Kimberlee Moran of Rutgers-Camden for her help with the chapter preparation.

References

1. I. Kola and J. Landis, Can the pharmaceutical industry reduce attrition rates?, *Nat. Rev. Drug Discov.*, 2004, **3**(8), 711 715.
2. National Research Council (NRC), *Toxicity Testing in the 21st Century: A Vision and a Strategy*; The National Academies Press, Washington, DC, 2007.
3. OECD (Q)SAR Project, 2009. http://www.oecd.org/env/ehs/risk-assessment/oecdquantitativestructure-activityrelationshipsprojectqsars.htm.
4. J. Inglese, D. S. Auld, A. Jadhav, R. L. Johnson, A. Simeonov, A. Yasgar, W. Zheng and C. P. Austin, Quantitative high-throughput screening: a titration-based approach that efficiently identifies biological activities in large chemical libraries, *Proc. Natl. Acad. Sci. USA*, 2006, **103**(31), 11473–11478.
5. M. A. Cheeseman, Thresholds as a unifying theme in regulatory toxicology, *Food Addit. Contam.*, 2005, **22**(10), 900–906.
6. R. J. Riley and J. G. Kenna, Cellular models for ADMET predictions and evaluation of drug–drug interactions, *Curr. Opin. Drug Discov. Devel.*, 2004, **7**(1), 86–99.
7. D. J. Dix, K. A. Houck, M. T. Martin, A. M. Richard, R. W. Setzer and R. J. Kavlock, The ToxCast program for prioritizing toxicity testing of environmental chemicals, *Toxicol. Sci.*, 2007, **95**(1), 5–12.

8. C. Yang, L. G. Valerio, Jr. and K. B. Arvidson, Computational toxicology approaches at the US Food and Drug Administration, *Altern. Lab. Anim.*, 2009, **37**(5), 523–531.

9. L. G. Valerio, Jr., In silico toxicology for the pharmaceutical sciences, *Toxicol. Appl. Pharmacol.*, 2009, **241**(3), 356–370.

10. A. Dash, W. Inman, K. Hoffmaster, S. Sevidal, J. Kelly, R. S. Obach, L. G. Griffith and S. R. Tannenbaum, Liver tissue engineering in the evaluation of drug safety, *Expert Opin. Drug Metab. Toxicol.*, 2009, **5**(10), 1159–1174.

11. M. V. Park, D. P. Lankveld, H. van Loveren and W. H. van de Jong, The status of in vitro toxicity studies in the risk assessment of nanomaterials, *Nanomedicine (Lond.)*, 2009, **4**(6), 669–685.

12. C. P. Austin, L. S. Brady, T. R. Insel and F. S. Collins, NIH Molecular Libraries Initiative, *Science*, 2004, **306**(5699), 1138–1139.

13. S. J. Shukla, R. Huang, C. P. Austin and M. Xia, The future of toxicity testing: a focus on in vitro methods using a quantitative high-throughput screening platform, *Drug Discov. Today*, 2010, **15**(23–24), 997–1007.

14. C. J. Thomas, D. S. Auld, R. Huang, W. Huang, A. Jadhav, R. L. Johnson, W. Leister, D. J. Maloney, J. J. Marugan, S. Michael, A. Simeonov, N. Southall, M. Xia, W. Zheng, J. Inglese and C. P. Austin, The pilot phase of the NIH Chemical Genomics Center, *Curr. Top. Med. Chem.*, 2009, **9**(13), 1181–1193.

15. S. K. Durham and G. M. Pearl, Computational methods to predict drug safety liabilities, *Curr. Opin. Drug Discov. Devel.*, 2001, **4**(1), 110–115.

16. D. Jacobson-Kram and J. F. Contrera, Genetic toxicity assessment: employing the best science for human safety evaluation. Part I: Early screening for potential human mutagens, *Toxicol. Sci.*, 2007, **96**(1), 16–20.

17. W. Muster, A. Breidenbach, H. Fischer, S. Kirchner, L. Muller and A. Pahler, Computational toxicology in drug development, *Drug Discov. Today*, 2008, **13**(7–8), 303–310.

18. A. B. Bailey, R. Chanderbhan, N. Collazo-Braier, M. A. Cheeseman and M. L. Twaroski, The use of structure–activity relationship analysis in the food contact notification program, *Regul. Toxicol. Pharmacol.*, 2005, **42**(2), 225–235.

19. L. Valerio, Jr., Tools for evidence-based toxicology: computational-based strategies as a viable modality for decision support in chemical safety evaluation and risk assessment, *Hum. Exp. Toxicol.*, 2008, **27**(10), 757–760.

20. N. L. Kruhlak, R. D. Benz, H. Zhou and T. J. Colatsky, (Q)SAR modeling and safety assessment in regulatory review, *Clin. Pharmacol. Ther.*, 2012, **91**(3), 529–534.

21. R. J. Kavlock, G. Ankley, J. Blancato, M. Breen, R. Conolly, D. Dix, K. Houck, E. Hubal, R. Judson, J. Rabinowitz, A. Richard, R. W. Setzer, I. Shah, D. Villeneuve and E. Weber, Computational toxicology – a state of the science mini review, *Toxicol. Sci.*, 2008, **103**(1), 14–27.

22. M. T. Khan, Predictions of the ADMET properties of candidate drug molecules utilizing different QSAR/QSPR modelling approaches, *Curr. Drug Metab.*, 2010, **11**(4), 285–295.

23. A. M. Voutchkova, T. G. Osimitz and P. T. Anastas, Toward a comprehensive molecular design framework for reduced hazard, *Chem. Rev.*, 2010, **110**(10), 5845–5882.
24. OECD (Q)SAR Application Toolbox, 2012. http://www.qsartoolbox.org/index.html.
25. OpenTox, 2012. www.opentox.org.
26. EPA QSAR, 2013. http://www.epa.gov/nrmrl/std/qsar/qsar.html.
27. EPA EPI, 2010. www.epa.gov/oppt/exposure/pubs/episuite.htm.
28. DEREK, 2013. http://www.lhasalimited.org/products/derek-nexus.htm.
29. M. P. Smithing and F. Darvas, HazardExpert: an expert system for predicting chemical toxicity, in *Food Safety Assessment*, ed. J. W. Finlay, S. F. Robinson, D. J. Armstrong, American Chemical Society, Washington, DC, 1992, pp. 191–200.
30. Leadscopem 2010. www.leadscope.com/product_info.php?products_id = 78.
31. MultiCASE, 2010. www.multicase.com.
32. F. S. Collins, G. M. Gray and J. R. Bucher, Toxicology – Transforming environmental health protection, *Science*, 2008, **319**(5865), 906–907.
33. R. D. Snyder, An update on the genotoxicity and carcinogenicity of marketed pharmaceuticals with reference to in silico predictivity, *Environ. Mol. Mutagen.*, 2009, **50**(6), 435–450.
34. E. Zvinavashe, A. J. Murk and I. M. Rietjens, On the number of EINECS compounds that can be covered by (Q)SAR models for acute toxicity, *Toxicol. Lett.*, 2009, **184**(1), 67–72.
35. E. Zvinavashe, A. J. Murk and I. M. Rietjens, Promises and pitfalls of quantitative structure–activity relationship approaches for predicting metabolism and toxicity, *Chem. Res. Toxicol*, 2008, **112**(2), 385–393.
36. Y. Wang, J. Xiao, T. O. Suzek, J. Zhang, J. Wang and S. H. Bryant, PubChem: a public information system for analyzing bioactivities of small molecules, *Nucleic Acids Res.*, 2009, **37**(Web Server issue), W623–W633.
37. Y. Wang, E. Bolton, S. Dracheva, K. Karapetyan, B. A. Shoemaker, T. O. Suzek, J. Wang, J. Xiao, J. Zhang and S. H. Bryant, An overview of the PubChem BioAssay resource, *Nucleic Acids Res.*, 2010, **38**(Database issue), D255–D266.
38. ChEMBL, 2010. www.ebi.ac.uk/chembldb/index/.
39. A. Gaulton, L. J. Bellis, A. P. Bento, J. Chambers, M. Davies, A. Hersey, Y. Light, S. McGlinchey, D. Michalovich, B. Al-Lazikani and J. P. Overington, ChEMBL: a large-scale bioactivity database for drug discovery, *Nucleic Acids Res.*, 2012, **40**(Database issue), D1100–D1107.
40. G. C. Fonger, D. Stroup, P. L. Thomas and P. Wexler, TOXNET: A computerized collection of toxicological and environmental health information, *Toxicol. Ind. Health*, 2000, **16**(1), 4–6.
41. C. M. McHale, L. Zhang, A. E. Hubbard and M. T. Smith, Toxicogenomic profiling of chemically exposed humans in risk assessment, *Mutat. Res.*, 2010, **705**(3), 172–183.

42. M. Waters, S. Stasiewicz, B. A. Merrick, K. Tomer, P. Bushel, R. Paules, N. Stegman, G. Nehls, K. J. Yost, C. H. Johnson, S. F. Gustafson, S. Xirasagar, N. Xiao, C. C. Huang, P. Boyer, D. D. Chan, Q. Pan, H. Gong, J. Taylor, D. Choi, A. Rashid, A. Ahmed, R. Howle, J. Selkirk, R. Tennant and J. Fostel, CEBS – Chemical Effects in Biological Systems: a public data repository integrating study design and toxicity data with microarray and proteomics data, *Nucleic Acids Res.*, 2008, **36**(Database issue), D892–D900.

43. C. J. Mattingly, M. C. Rosenstein, G. T. Colby, J. N. Forrest, Jr. and J. L. Boyer, The Comparative Toxicogenomics Database (CTD): a resource for comparative toxicological studies, *J. Exp. Zool. A Comp. Exp. Biol.*, 2006, **305**(9), 689–692.

44. C. J. Mattingly, M. C. Rosenstein, A. P. Davis, G. T. Colby, J. N. Forrest, Jr. and J. L. Boyer, The comparative toxicogenomics database: a cross-species resource for building chemical–gene interaction networks, *Toxicol. Sci.*, 2006, **92**(2), 587–595.

45. C. J. Mattingly, G. T. Colby, M. C. Rosenstein, J. N. Forrest, Jr. and J. L. Boyer, Promoting comparative molecular studies in environmental health research: an overview of the comparative toxicogenomics database (CTD), *Pharmacogenomics J.*, 2004, **4**(1), 5–8.

46. C. J. Mattingly, G. T. Colby, J. N. Forrest and J. L. Boyer, The Comparative Toxicogenomics Database (CTD), *Environ. Health Perspect.*, 2003, **111**(6), 793–795.

47. F. L. Mayer and M. R. Ellersieck, *Manual of Acute Toxicity: Interpretation and Database for 410 Chemicals and 66 Species of Freshwater Animals*, US Fish and Wildlife Service Resource Publ., Washington, DC, 1986.

48. L. S. Gold, T. H. Slone, N. B. Manley, G. B. Garfinkel, E. S. Hudes, L. Rohrbach and B. N. Ames, The Carcinogenic Potency Database: analyses of 4000 chronic animal cancer experiments published in the general literature and by the US National Cancer Institute/National Toxicology Program, *Environ. Health Perspect.*, 1991, **96**, 11–15.

49. M. T. Martin, R. S. Judson, D. M. Reif, R. J. Kavlock and D. J. Dix, Profiling chemicals based on chronic toxicity results from the US EPA ToxRef Database, *Environ. Health Perspect.*, 2009, **117**(3), 392–399.

50. A. M. Richard and C. R. Williams, Distributed structure-searchable toxicity (DSSTox) public database network: a proposal, *Mutat. Res.*, 2002, **499**(1), 27–52.

51. A. M. Richard, C. R. Williams and N. F. Cariello, Improving structure-linked access to publicly available chemical toxicity information, *Curr. Opin. Drug Discov. Devel.*, 2002, **5**(1), 136–143.

52. A. M. Richard, C. Yang and R. S. Judson, Toxicity data informatics: supporting a new paradigm for toxicity prediction, *Toxicol. Mech. Methods*, 2008, **18**(2–3), 103–118.

53. R. S. Judson, M. T. Martin, P. Egeghy, S. Gangwal, D. M. Reif, P. Kothiya, M. Wolf, T. Cathey, T. Transue, D. Smith, J. Vail, A. Frame,

S. Mosher, E. A. Cohen Hubal and A. M. Richard, Aggregating data for computational toxicology applications: The US Environmental Protection Agency (EPA) aggregated computational toxicology resource (ACToR) system, *Int. J. Mol. Sci.*, 2012, **13** (2), 1805–1831.

54. R. Judson, A. Richard, D. Dix, K. Houck, F. Elloumi, M. Martin, T. Cathey, T. R. Transue, R. Spencer and M. Wolf, ACToR – Aggregated computational toxicology resource, *Toxicol. Appl. Pharmacol.*, 2008, **233**(1), 7–13.

55. G. Klopman, H. Zhu, G. Ecker and P. Chiba, MCASE study of the multidrug resistance reversal activity of propafenone analogs, *J. Comput. Aided Mol. Des.*, 2003, **17**(5–6), 291–297.

56. C. L. Stoner, E. Gifford, C. Stankovic, C. S. Lepsy, J. Brodfuehrer, J. V. N. V. Prasad and N. Surendran, Implementation of an ADME enabling selection and visualization tool for drug discovery, *J. Pharm. Sci.*, 2004, **93**(5), 1131–1141.

57. P. Mayer and F. Reichenberg, Can highly hydrophobic organic substances cause aquatic baseline toxicity and can they contribute to mixture toxicity?, *Environ. Toxicol. Chem.*, 2006, **25**(10), 2639–2644.

58. R. S. Thomas, L. Pluta, L. Yang and T. A. Halsey, Application of genomic biomarkers to predict increased lung tumor incidence in 2-year rodent cancer bioassays, *Toxicol. Sci.*, 2007, **97**(1), 55–64.

59. R. S. Thomas, T. M. O'Connell, L. Pluta, R. D. Wolfinger, L. Yang and T. J. Page, A comparison of transcriptomic and metabonomic technologies for identifying biomarkers predictive of two-year rodent cancer bioassays, *Toxicol. Sci.*, 2007, **96**(1), 40–46.

60. T. J. Flynn and M. S. Ferguson, Multiendpoint mechanistic profiling of hepatotoxicants in HepG2/C3A human hepatoma cells and novel statistical approaches for development of a prediction model for acute hepatotoxicity, *Toxicol. In Vitro*, 2008, **22**(6), 1618–1631.

61. P. J. O'Brien, W. Irwin, D. Diaz, E. Howard-Cofield, C. M. Krejsa, M. R. Slaughter, B. Gao, N. Kaludercic, A. Angeline, P. Bernardi, P. Brain and C. Hougham, High concordance of drug-induced human hepatotoxicity with in vitro cytotoxicity measured in a novel cell-based model using high content screening, *Arch. Toxicol.*, 2006, **80**(9), 580–604.

62. T. Uehara, A. Ono, T. Maruyama, I. Kato, H. Yamada, Y. Ohno and T. Urushidani, The Japanese toxicogenomics project: application of toxicogenomics, *Mol. Nutr. Food Res.*, 2010, **54**(2), 218–227.

63. U. Bernauer, A. Oberemm, S. Madle and U. Gundert-Remy, The use of in vitro data in risk assessment, *Basic Clin. Pharmacol. Toxicol.*, 2005, **96**(3), 176–181.

64. M. R. Fielden, R. Brennan and J. Gollub, A gene expression biomarker provides early prediction and mechanistic assessment of hepatic tumor induction by nongenotoxic chemicals, *Toxicol. Sci.*, 2007, **99**(1), 90–100.

65. H. Zhu, L. Ye, A. Richard, A. Golbraikh, F. A. Wright, I. Rusyn and A. Tropsha, A novel two-step hierarchical quantitative structure–activity

relationship modeling work flow for predicting acute toxicity of chemicals in rodents, *Environ. Health Perspect.*, 2009, **117**(8), 1257–1264.

66. H. Zhu, I. Rusyn, A. M. Richard and A. Tropsha, Use of cell viability assay data improves the prediction accuracy of conventional quantitative structure activity relationship models of animal carcinogenicity, *Environ. Health Perspect.*, 2008, **116**(4), 506–513.

67. A. Sedykh, H. Zhu, H. Tang, L. Zhang, A. Richard, I. Rusyn and A. Tropsha, Use of in vitro HTS-derived concentration-response data as biological descriptors improves the accuracy of QSAR models of in vivo toxicity, *Environ. Health Perspect.*, 2011, **119**(3), 364–370.

68. Y. Low, T. Uehara, Y. Minowa, H. Yamada, Y. Ohno, T. Urushidani, A. Sedykh, E. Muratov, V. Kuz'min, D. Fourches, H. Zhu, I. Rusyn and A. Tropsha, Predicting drug-induced hepatotoxicity using QSAR and toxicogenomics approaches, *Chem. Res. Toxicol.*, 2011, **24**(8), 1251–1262.

69. W. Tong, H. Fang, Q. Xie, H. Hong, L. Shi, R. Perkins, U. Scherf, F. Goodsaid and F. Frueh, Gaining confidence on molecular classification through consensus modeling and validation, *Toxicol. Mech. Methods*, 2006, **16**(2–3), 59–68.

70. I. Rusyn, A. Sedykh, Y. Low, K. Z. Guyton and A. Tropsha, Predictive modeling of chemical hazard by integrating numerical descriptors of chemical structures and short-term toxicity assay data, *Toxicol. Sci.*, 2012, **127**(1), 1–9.

71. C. Yang, R. D. Benz and M. A. Cheeseman, Landscape of current toxicity databases and database standards, *Curr. Opin. Drug Discov. Devel.*, 2006, **9**(1), 124–133.

72. PubChem, 2008. http://pubchem.ncbi.nlm.nih.gov/.

73. ToxCast, 2010. www.epa.gov/comptox/toxcast/.

74. T. B. Knudsen, M. T. Martin, R. J. Kavlock, R. S. Judson, D. J. Dix and A. V. Singh, Profiling developmental toxicity of 387 environmental chemicals using EPA's toxicity reference database (ToxRefDB), *Birth Defects Research Part A – Clinical and Molecular Teratology*, 2009, **85**(5), 406.

75. R. Huang, N. Southall, M. Xia, M. H. Cho, A. Jadhav, D. T. Nguyen, J. Inglese, R. R. Tice and C. P. Austin, Weighted feature significance: a simple, interpretable model of compound toxicity based on the statistical enrichment of structural features, *Toxicol. Sci.*, 2009, **112**(2), 385–393.

CHAPTER 6

Acute Systemic Toxicity: Oral, Dermal and Inhalation Exposures

DAVID J. ANDREW

TSGE LLP, Concordia House, St James Business Park, Knaresborough
HG5 8QB, UK
Email: david.andrew@tsgeurope.com

6.1 Definition of Acute Toxicity

Acute toxicity can be defined as the adverse effects resulting from the administration of a single dose of, or single exposure to, a chemical substance or mixture.[1] Acute toxicity studies performed to regulatory guidelines typically investigate toxicity for 14 days following dosing or exposure.

6.1.1 Use of Acute Toxicity Data

Acute toxicity data are used primarily for the purposes of classification and labelling according to national or international regulations,[1,2] i.e. to categorize chemicals according to their apparent hazard and to assign appropriate labelling for the purposes of safe use, storage and transport. Data from acute toxicity studies may also be used as the basis for risk assessments relating to the use of chemicals in the home or workplace, or for emergency planning measures following accidental release. Observations made in acute toxicity studies are generally restricted to mortality and clinical signs and are of limited use in

Issues in Toxicology No. 19
Reducing, Refining and Replacing the Use of Animals in Toxicity Testing
Edited by David G. Allen and Michael D. Waters

predicting the target organ or likely effects of overdose or accidental exposure or in guiding the medical treatment of exposed individuals. In the context of a more comprehensive toxicological assessment, acute toxicity studies may be used to inform the selection of dose levels for repeated dose toxicity studies or for other studies of toxicity *in vivo*.

6.1.2 Historical Context

The endpoint most frequently associated with studies of acute toxicity is lethality. The assessment of lethality was formalized in 1927 through introduction of the concept of the median lethal dose[3] (the LD_{50}), which is defined as the amount of a chemical shown to or predicted to cause the death of 50% of a population of animals. This concept was introduced as a measure of toxic potency for substances intended to be used in humans.[4] Early studies of acute toxicity included large numbers of animals and multiple dose groups in order to measure the LD_{50} with a high degree of statistical precision and studies using well in excess of 100 animals (typically rats) were not uncommon. Furthermore, in order to identify LD_{50} values for substances of very low toxicity, dose levels used in some of these older studies were excessively high. The expression of acute oral toxicity is affected by numerous potential variables including species, sex, strain, nutritional status and dosing vehicle; therefore even repeating the study in the same laboratory using the same dose levels could result in a different outcome.[1] The statistical accuracy with which LD_{50} values were described (using techniques including probit analysis, moving average or maximum likelihood estimation methods) therefore implies a high degree of certainty which is largely spurious. As a consequence, the 'classic' LD_{50} procedure has been criticized by a number of authors for scientific reasons and also due to concerns over animal welfare.[2,5]

6.2 Predicting Acute Toxicity *In Vitro*

6.2.1 Cytotoxicity

Lethality in acute toxicity studies is considered, for the majority of chemicals, to be the result of non-specific interference of the chemical with structures or processes vital for cellular survival, proliferation or function. These non-specific interactions with basic functions common to all cells lead to organ- or tissue-specific toxicity and/or subsequent lethality.[6] As a consequence, it may be possible to predict the acute systemic toxicity of most chemicals in studies in cells *in vitro*, using non-specific effects as endpoints. While it is recognised that studies of cytotoxicity in cultured cells *in vitro* may be useful in predicting toxicity *in vivo*, it is also clear that predictions will not be reliable for all chemicals, and that a proportion of chemicals will produce cell-specific toxicity at lower concentrations than those eliciting cytotoxicity in other cell types. The use of a battery of *in vitro* assays in order to more reliably predict acute systemic toxicity has therefore been recognized by many authors.[6–9] Building on this concept, three main types of cytotoxicity can be defined:[7]

- general (basal) cytotoxicity, in which a chemical adversely affects structures or functions common to all cell types;
- selective cytotoxicity, in which some types of cell are more sensitive to the effects of a chemical than others; and
- cell-specific function cytotoxicity, in which a chemical affects structures or functions not critical to the cells themselves (e.g. cell–cell communication or specific transport proteins), but which may be critical to the intact organism.[7]

6.2.2 Toxicokinetic Factors

The toxicity of a chemical *in vivo* is ultimately governed by its toxicokinetic properties (i.e. the time-dependent and interrelated processes of absorption, distribution, metabolism and excretion). Given the critical influence of these processes on the expression of toxicity *in vivo*, consideration of toxicokinetic parameters must also be incorporated into any battery of *in vitro* tests intended to reliably predict acute systemic toxicity. Toxicokinetic data are most frequently generated *in vivo*; however, useful information can also be obtained using other approaches. Importantly, data on the likely oral absorption of a chemical *in vivo* can be inferred from basic data on its physicochemical properties (molecular weight, water solubility, partition coefficient), from *in vitro* assays (e.g. in CaCo-2 cells or isolated rat intestine) or from PBPK modelling. Metabolites of a chemical can also be predicted to some extent based on chemical structure. Due to the critical role of the liver in the detoxification (and potential bioactivation) of chemicals, the use of metabolically competent liver-derived cells has been investigated extensively. Assays in liver (or other metabolically competent) cells are likely to be a critical part of any *in vitro* testing battery.

6.2.3 *In Vitro* Testing Strategies

Any non-animal method intended to replace the currently accepted regulatory studies of acute systemic toxicity *in vivo* must therefore provide information on a number of factors potentially influencing the expression of toxicity *in vivo*. Successful testing strategies will need to involve the assessment not only of basal cytotoxicity, cell-specific toxicity and cell-specific function,[7] but will also need to take into account complex biological processes such as toxicokinetics. It is likely, therefore, that in addition to a battery (or batteries) of *in vitro* tests, a tiered assessment including consideration of structural and physicochemical properties will also be required in order to provide a realistic estimate of acute systemic toxicity.

6.3 Development of *In Vitro* Testing Strategies

The potential use of *in vitro* basal cytotoxicity data to predict acute lethality in animals has been recognized for many years.[10–12] It is recognized that *in vitro* basal cytotoxicity data are more reliable as a predictor of low acute systemic toxicity and are less reliable for chemicals causing high acute systemic toxicity, most likely as a consequence of toxicokinetic or cell-specific effects. Recent initiatives have therefore sought to identity *in vitro* test batteries which assess the potential for cell-specific toxicity and which also take into account the potential influence of toxicokinetic factors in order to improve predictivity. These initiatives have sought to reduce the numbers of animals used in studies of acute toxicity *in vivo* though the refinement of study design, but also have the eventual aim of replacing *in vivo* studies entirely. Considerable funds and effort have been expended into identifying suitable testing strategies over recent years and a number of major initiatives in this area are described below.

6.3.1   Registry of Cytotoxicity

The German Centre for Documentation and Evaluation of Alternatives to Animal Use (ZEBET) have generated an invaluable database of published cytotoxicity values and corresponding rodent LD_{50} values for nearly 350 chemical substances - the Registry of Cytotoxicity.[13,14] These data have been used to construct a regression model to enable the estimation of acute oral LD_{50} values from *in vitro* cytotoxicity data. Using this model, predictions of acute oral toxicity were shown to be accurate (within a defined dose range) for 73% of the substances contained in the database.

6.3.2   Multicenter Evaluation of *In Vitro* Cytotoxicity (MEIC) Program

The MEIC program (established in 1989) assessed the relevance of *in vitro* cytotoxicity data for human lethality rather than the prediction of LD_{50} values in experimental animals.[9,15] This program identified 50 reference chemicals for which reliable data on lethal human blood concentrations were available from case reports of human poisoning. Nearly 100 participating MEIC laboratories tested the 50 reference chemicals in more than 60 different assays *in vitro* using human and animal primary cell cultures and cell lines. One outcome of the MEIC project was the identification of a limited battery of three 24-hour assays in human cell lines which showed good predictivity for lethal human blood concentrations (Table 6.1). While it is notable that this limited battery showed a better prediction of lethal human blood concentrations than the rodent LD_{50} values,[9] this level of prediction was not confirmed using an independent dataset.

Although it was considered that metabolic factors were likely to play a critical role in acute toxicity *in vivo* in only a relatively small proportion of cases, further improvements to the predictivity of the testing battery were

Table 6.1   MEIC: Cytotoxicity Assay Battery.

HepG2 human liver cell: total protein content	Human lymphocyte HL60: ATP content	Chang human liver cells: morphology and pH change

recommended, including additional assays for kinetic factors potentially influencing toxicity, including blood–brain barrier permeability, oral absorption, distribution and metabolism. The predictive capability of the MEIC training set was further improved by the addition of assessments for protein binding and partition coefficient;[16–18] however, this improved level of prediction has not been extended to an independent dataset.

6.3.3 ACuteTox

The ACuteTox Project (2005–2010), which included 35 partners from a large number of European countries, had the ambitious objective of developing an *in silico/in vitro* testing strategy for acute toxicity which was sufficiently robust to completely replace acute toxicity testing *in vivo* for the purposes of hazard classification. ACuteTox built on previous initiatives including the Registry of Cytotoxicity, a US validation study[19] coordinated by in the US by the Interagency Coordinating Committee on the Validation of Alternative Methods (ICCVAM) and the National Toxicology Program (NTP) Interagency Center for the Evaluation of Alternative Toxicological Methods (NICEATM) and the MEIC program. ACuteTox aimed to identify factors that could optimize the correlation between *in vitro* and *in vivo* results and therefore minimise the under- or over-classification of chemicals. The project compiled high-quality data on a set of reference chemicals to facilitate comparative analyses, and aimed to identify those factors influencing the correlation of *in vitro* and *in vivo* results. Importantly, ACuteTox also had an eye to future developments, in that any proposed testing strategy should be amenable to incorporation into HTS platforms. The project used a set of 97 reference chemicals, which were selected based on the availability of human poisoning cases (with reliable measurement of blood levels), rodent acute LD_{50} values and representation of all hazard classification categories and different generic use classes.[20] Cytotoxicity data for the reference chemicals were generated using a battery of six assays in human and animal cell lines. Those chemicals not showing a good *in vitro–in vivo* correlation were identified and subjected to further testing in a large number of additional *in vitro* assays. The additional assays included assessment of toxicokinetic factors and organ-specific toxicity endpoints in order to identify the reasons for the outliers. PBPK modelling was also used to predict those kinetic factors resulting in outliers; the extent of oral absorption, lipophilicity, clearance and protein binding were identified as the most important parameters. Incorporating corrections for kinetic parameters was shown to further improve the correlation between cytotoxicity *in vitro* and acute systemic toxicity. The ACuteTox project culminated in a pre-validation study using a total of nine endpoints in seven different assay systems (Table 6.2). This combination of assay demonstrated a 69% correct classification rate and underestimated toxicity for only two of the 32 substances tested.

One of the principle reasons for performing studies of acute toxicity is to assign a chemical substance or product into a classification category based on the LD_{50} value. ACuteTox therefore included further detailed analyses to identify potential *in vitro/in silico* strategies to allow assignment into the

Table 6.2   ACuteTox: Cytotoxicity Pre-Validation Assay Battery.

CFU-GM/human cord blood cell assay	Primary rat brain aggregates: lowest gene expression	HepG2 cell line: MMP assay
3T3/NRU assay	Primary rat brain aggregates: HSP-32 expression	Rat hepatocyte: MTT assay
SH-SY5Y cell line	Primary rat brain aggregates: NF-H expression	Whole blood: IL-1 release

appropriate acute toxicity classification categories under GHS (the UN Globally Harmonized System of Classification and Labelling of Chemicals) or CLP (Regulation (EC) No. 1272/2008 on the Classification, Labelling and Packaging of Substances and Mixtures).

6.4   Acute Oral Toxicity

The concept of the LD_{50} as a comparator of acute toxic potency was incorporated into the early test guidelines of the Organisation for Economic Co-operation and Development (OECD). However OECD Test Guideline 401 (1981) did not require the statistical precision associated with classical LD_{50} studies and hence was not associated with the same level of animal usage. This guideline specified the use of a more limited number of dose levels (typically three, with five animals of each sex per group) spanning the LD_{50} value and intended to provide an approximate estimate. OECD 401 also introduced the concept of the limit dose, an upper dose level (5000 mg/kg bw) not to be exceeded, thereby avoiding the use of excessively high dose levels. For substances of low toxicity, the use of a single group of animals at a high dose level (a limit test) was therefore possible. Subsequent revision of OECD 401 in 1987 brought refinements with testing possible in animals of one sex (with confirmation of the absence of sex differences in toxicity by testing in a single group in the second sex) and a reduction of the limit dose to 2000 mg/kg bw; the study typically used up to 25 animals. OECD 401 was deleted in 2002 following the development and introduction of three alternative methods to assess acute oral toxicity.

6.4.1   Acute Oral Toxicity: Current *In Vivo* Methods

The current alternative *in vivo* methods for the assessment of acute oral toxicity are the Fixed Dose Procedure (OECD 420; 1992, revised 2001); the Acute Toxic Class method (OECD 423; 1996, revised 2001); and the Up-and-Down Procedure (OECD 425;1998, revised 2001 and 2008). Each of these methods represents a refinement or reduction approach to testing compared to the deleted OECD 401 test method while remaining adequately predictive.

6.4.1.1   Fixed Dose Procedure

Following a concept introduced by the British Toxicology Society,[21] the Fixed Dose Procedure (FDP) involves the stepwise dosing of small groups of animals of a single sex with a series of predetermined dose levels and relies on the observation of 'clear signs of toxicity' rather than lethality. As such, this method uses fewer animals (typically between 5 and 9) and causes less suffering,[4] while allowing the ranking of substances according to their acute oral toxicity in a similar manner to other acute toxicity testing methods.

6.4.1.2 Acute Toxic Class Method

The Acute Toxic Class Method (ATC) similarly uses the concept of stepwise testing of small groups of animals of a single sex at fixed dose levels, progression to the next dose level being informed by the absence or presence of lethality. This method retains lethality as the principal endpoint of interest but uses fewer animals (typically 7) than OECD 401.[4]

6.4.1.3 Up-and-Down Procedure

The Up-and-Down Procedure (UDP) estimates the LD_{50} by testing single animals of one sex (typically females) sequentially at a series of set dose levels, with the outcome (i.e. mortality/moribundity) for the previous animal determining the choice of dose level for the next animal tested.[22] This test method relies on lethality as an endpoint but generally uses fewer animals (as few as three, typically 4 to 6 and up to a total of 15) than required for OECD 401.

6.4.2 QSAR

QSAR models for the prediction of acute oral toxicity are available (e.g. TOPKAT, MCASE, ACD/Tox Suite and others) and may be useful for predicting acute toxicity in terms of broad categories. The quantitative prediction of LD_{50} values is less reliable, most likely due to the multiple mechanisms involved in acute systemic toxicity. QSAR predictions are, however, useful in highlighting structural groups associated with specific modes of toxicity. Existing QSAR models for the prediction of acute systemic toxicity are not sufficiently predictive for regulatory acceptance, but are an important part of any battery of tests used in a tiered testing strategy.

6.4.3 Acute Oral Toxicity: *In Vitro* Assays

In vitro assays of cytotoxicity are used to predict systemic toxicity *in vivo* and are therefore potentially of relevance to all exposure routes. Assays have been identified and developed primarily for the purposes of predicting acute oral toxicity and classification, but are also of relevance to other routes. The assessment of acute toxicity following dermal and inhalation exposure involves the consideration of route-specific factors which are outlined below.

6.4.3.1   Acute Oral Toxicity In Vitro: Basal Cytotoxicity Assays

A large number of *in vitro* endpoints are suitable for the assessment of cyto-toxicity in various cell lines (Table 6.3). The majority of workers have focussed on endpoints relevant to basal cytotoxicity such as cell number, cell viability, cell proliferation, membrane integrity or metabolic status and assays may be performed in any number of animal or human primary cell cultures or cell lines. The use of rodent (rat or mouse) cell lines to generate basal cytotoxicity data has been used extensively in order to demonstrate correlation with acute systemic toxicity in rat and mouse studies for the purposes of method validation. It is arguable that the ideal correlation would be between cytotoxicity data from cells of human origin and human poisoning data; however, while the use of a human cell line may be preferable, differences in response at a cellular level are likely to be relatively infrequent.

6.4.3.2   Acute Oral Toxicity In Vitro: Specific Cytotoxicity Assays

While basal cytotoxicity data show reasonable predictivity for acute systemic toxicity, the degree of predictivity may be improved through the additional use of assays designed to detect cytotoxicity in other organs or tissues potentially more sensitive to acute chemical toxicity (specific cytotoxicity). Assays investigating the potential for organ- or tissue-specific effects must therefore be included in any effective testing battery designed to replace *in vivo* studies. While it is obviously not feasible to include the assessment of cytotoxicity in all

Table 6.3   Endpoints in Cytotoxicity Assays.

Metabolic activity	MTT reduction	XTT reduction
	MTS reduction	WST reduction
	Alamar blue reduction	Creatine kinase activity
	Glucose consumption	ATP content
	Fluorescein diacetate hydrolysis	Cellular pH change
Membrane integrity	Lactate dehydrogenase (LDH) release	Trypan Blue Exclusion
	51Cr-release	Tritiated thymidine release
	ATP leakage	Transepithelial resistance
Cell number	Neutral Red Uptake (NRU)	Cell counting
Cell morphology	Visual assessment	Flow cytometry
Apoptosis induction	Caspase gene expression	DNA laddering
Cellular stress	Heat Shock Protein expression	ROS production
	Antioxidant status	Glutathione status
Mitochondrial potential	Rhodamine 123 staining	JC-1 staining
Total protein content	Sulphorodamine B staining	Coomassie blue staining
	Kenacid blue staining	Lowry staining
DNA content	Gentian violet staining	Ethidium bromide staining

potential cellular targets, the inclusion of assays in cells from those organs or tissues most likely to be targets (and therefore resulting in the expression of toxicity *in vivo*) is critical. Common or sensitive targets of toxicity include the nervous system, kidneys and liver; assays may investigate relative cytotoxicity or cell-specific functional endpoints. The choice of specific cytotoxicity assay(s) must also be guided by information on structural or metabolic alerts and relationship to substances with demonstrated target organ toxicity. The incorporation of other specific cytotoxicity assays should be considered for chemicals known to target other organs or tissues.

6.4.3.2.1 Specific Cytotoxicity in Nervous System Cells. Neurotoxicity is perhaps the most obvious example of organ- or tissue-specific toxicity resulting in acute systemic toxicity; chemicals showing high acute toxicity through a specific interaction with nerve cell function are unlikely to be identified in assays for basal cytotoxicity. While the complexity of the central and peripheral nervous systems are such that they cannot be fully represented by *in vitro* models, a large number of assays of varying complexity including single neuronal-derived cell lines, mixed cell cultures, organotypic explants or reaggregated brain cells have been used to investigate multiple endpoints of relevance to the assessment of specific neurotoxicity.[23–26] Significant efforts have therefore been made into identifying suitable screening assays for neurotoxicity *in vitro* and a tiered assessment strategy incorporating both general and specific indicators of neurotoxicity has been recommended by a number of authors.[23–25] Assessment of neurotoxicity potentially includes assays of specific neuronal and glial cell toxicity, oxidative stress, calcium homeostasis, demyelination, excitoxicity and effects on neurotransmission. Any assessment must also incorporate a prediction of blood–brain barrier permeation and the potential role of metabolic activation in the expression of neurotoxicity.[27] A large number of potential assays of neurotoxicity were assessed by the ACuteTox project[26] for inclusion into an *in vitro* testing battery (Table 6.4).

ACuteTox concluded that, while assays did show good predictivity, no single assay was identified as a suitable stand-alone for the prediction of neurotoxicity *in vivo*. This conclusion reflects not only the complexity of the nervous system, but also the numerous mechanisms potentially involved in the expression of neurotoxicity *in vivo*. A more limited battery of neurotoxicity assays was identified as the most promising candidates for incorporation into a tiered *in vitro* testing strategy. This 'optimised battery' (Table 6.5) was shown to provide improved detection of chemicals causing specific neurotoxicity; i.e. those substances with a poor correlation between *in vitro* cytotoxicity in the mouse fibroblast 3T3 NRU assay for basal cytotoxicity and *in vivo* data in rats and humans.

6.4.3.2.2 Specific Cytotoxicity in Kidney Cells. Specific renal toxicity is recognised as a relatively frequent mechanism of acute toxicity;[28] consequently the development of models of nephrotoxicity for incorporation into *in vitro* testing batteries has been seen as a priority for a number of years.[29] The

Table 6.4 Specific Neurotoxicity assays (ACuteTox).[26]

Model	Endpoint	Assay
Mature rat brain cell aggregates	Cytotoxicity	LDH activity: photometric
	Cell function: enzyme activity	Choline acetyltransferase activity: radiometric
		Glutamic acid decarboxylase activity: radiometric
		Glutamine synthetase activity: radiometric
		Acetylcholinesterase activity: radiometric
		2′,3′-cyclic nucleotide 3′-phosphohydrolase (CNP) activity: photometric
	Glycolytic activity	^{3}H-2-deoxyglucose activity
	Protein synthesis	^{14}C-uridine incorporation
	Neuronal neurofilament mRNA expression	RT-PCR, SYBR green DNA binding
	Astrocytic GFAP mRNA expression	RT-PCR, SYBR green DNA binding
	Myelin basic protein (MBP) mRNA expression	RT-PCR, SYBR green DNA binding
	Oxidative stress: inducible nitric oxide synthetase (iNOS) expression	RT-PCR, SYBR green DNA binding
Rat cerebellar granule cells	Cell viability	Alamar blue metabolism
	Neuronal cell viability	NF-H (neurofilament protein) ELISA
	Glial cell viability	GFAP ELISA
	Mitochondrial membrane potential	JC-1 dye permeability (fluorescence)
	Cell viability	MTT metabolism: photometric
	Calcium homeostasis	Calcium influx: fluorescence
	Ionotropic glutamate receptor activation	Calcium influx: fluorescence
	Oxidative stress: reactive oxygen species (ROS) generation	DCF oxidation: flow cytometry
	Membrane integrity	LDH leakage: photometric
	Neurotoxicity target gene identification	Microarray; RT-PCR, TaqMan
	Caspase-3 mRNA expression	RT-PCR, TaqMan
Mouse cerebellar granule cells	Cell function: transmitter uptake	^{3}H-aspartate uptake
	Cell function: transmitter release	Glutamate release (HPLC)
Mouse cortical neurones	mRNA expression	Quantitative RT-PCR/ TaqMan
	GABA-A receptor inhibition	^{36}Cl-uptake
	GABA transporter inhibition	^{3}H-GABA uptake
	Cell viability (membrane permeability)	LDH leakage (photometric)
	Membrane potential	Fluorescence
	Cell viability	LDH activity (photometric)

Table 6.4   (*Continued*)

Model	Endpoint	Assay
Mouse brain slice	Cell viability	LDH activity (photometric)
SH-SY5Y human	Cell viability	LDH activity (photometric)
neuroblastoma cells	Voltage-operated ion channel function	K^+-induced cytotoxicity
	Acetylcholinesterase activity	Photometric
	Cell function: transmitter uptake	^{3}H-noradrenaline uptake
	Cell viability (membrane permeability)	LDH leakage (photometric)
Differentiated SH-SY5Y human neuroblastoma cells	Membrane potential	Fluorescence
	Acetylcholine receptor function	Carbachol-induced calcium influx: fluorescence
	Voltage-operated ion channel function	Potassium-induced calcium influx: fluorescence
Purified bovine acetylcholinesterase [cell-free assay]	Acetylcholinesterase inhibition	Acetylcholinesterase activity (photometric)

Table 6.5   Optimised neurotoxicity battery (ACuteTox).[26]

Model	Endpoint	Assay
Cerebellar granule cells	Apoptosis	Caspase-3 gene expression
Reaggregate cultures	Neuronal toxicity	NF-H gene expression
	Astrocyte toxicity	GFAP gene expression
	Oligodendrocyte toxicity	MBP gene expression
	Cellular stress	Heat Shock Protein 32 (HSP32) gene expression
	Cell viability	Total RNA synthesis
	Cell metabolism	Glucose uptake
SH-SY5Y cells	Neurotransmission	Acetylcholine esterase (AChE) activity
	Neurotransmission	Cell membrane potential (CMP)
Primary cortical neurons	Neurotransmission	GABA-A receptor inhibition

most common renal effect seen following acute chemical exposure is acute tubular necrosis and, in the large majority of cases, the primary target is the proximal tubule where cells have high metabolic activity and significant concentrating function.[30,31]

In vitro models of renal toxicity (Table 6.6) vary in complexity from cultured cells of renal origin though to the isolated perfused kidney.[31,32] The primary advantages of the isolated perfused kidney as a model of acute toxicity are the retention of tissue architecture, cell function and cell–cell contact; however, the use of this model in practice is low for reasons of technical complexity, limited viability and animal usage. Kidney slices are readily prepared and enable the simultaneous assessment of toxicity in various renal cell types; tissue architecture, cell morphology and function (including transport and drug

Table 6.6 Specific Nephrotoxicity Assays.

Model	Endpoint	Assay
Precision-cut kidney slices	Membrane integrity	LDH leakage[38,39]
	Cellular energy	ATP depletion[39]
	Transport function	Anion/cation accumulation[38]
	Membrane integrity	ALP, AST, ALP leakage[39]
	Oxidative stress	Glutathione depletion[38,39]
LLC-PK1 cell line	Cellular metabolism	Alamar blue reduction[40]
	Cellular energy	ATP depletion[40]
	Metabolic activity	MTT metabolism[40]
	Oxidative stress	ROS production[41]
	Metabolic activity	WST-1 metabolism[41]
	Membrane integrity	Trypan blue exclusion[42]
	Cellular metabolism	Histochemical[42]
	Cell viability	Trans-epithelial resistance[37,43]
MDCK cell	Membrane integrity	Trypan blue exclusion[42]
		Trans-epithelial resistance[42,43]
	Cell proliferation	BrdU incorporation[43]
	Cellular energy	ATP depletion[43]
	Metabolic activity	MTT metabolism[44]
	Cellular metabolism	Histochemical[44]
Primary human proximal tubular epithelial kidney cells	Membrane integrity	NAG leakage[45]
	Cellular energy	ATP depletion[45]
	Cellular injury	Kidney Injury Molecule 1 (KIM-1) expression[45]
	Oxidative stress	ROS production[41]
	Metabolic activity	WST-1 metabolism[41]
IP15 human glomerular mesangial cell	Oxidative stress	ROS production[41]
	Metabolic activity	WST-1 metabolism[41]

metabolism) are retained but, as for the isolated organ, viability may be limited.[32,33] Cultures of primary animal and human renal cells are representative of the cell *in vivo*; cultured cells show typical morphological and functional features, particularly when grown on porous membranes, although some differentiated characteristics may be lost in long-term cultures.[34] The use of transformed cell lines of renal origin is technically less demanding than primary cultures; however, cell lines may lack critical morphological, functional and biochemical characteristics of renal cells and are therefore less representative. The maintenance of polar architecture, solute and water transport systems and xenobiotic uptake are largely absent from most renal cell lines[35] which may be of questionable relevance as models of specific renal toxicity. The commonly used LLC-PK1 porcine proximal tubule cell line is, however, shown to retain characteristic and well-differentiated epithelial morphology with a brush border and transport processes. Cells grow in a monolayer and show good barrier properties.[36] The LLC-PK1 cell line has been assessed as a screen of specific renal toxicity using endpoints of both general viability (Alamar blue reduction) and specific function. While the assessment of viability was not shown to offer any greater sensitivity over the assessment of basal cytotoxicity in the 3T3-NRU assay, the functional endpoint (TER) was shown to provide a sensitive indication of nephrotoxic potential.[37] LLC-PK1 cells have limited xenobiotic metabolic capacity and therefore this assay alone will not detect chemicals causing renal toxicity following metabolic activation. MDCK cells are also a commonly used model; these cells originate from the canine distal tubule and maintain some transport features,[34] but are potentially of less relevance to the assessment of renal toxicity than the (proximal tubule-derived) LLC-PK1 cell line.

6.4.3.2.3 Specific Cytotoxicity in Liver Cells. The liver is also a common target of acute toxicity *in vivo* for a number of reasons relating to anatomy, physiology and biochemistry. The liver is the site of first contact following the absorption of chemicals from the gastrointestinal tract and toxic metabolites may be generated by xenobiotic metabolizing enzymes present at high levels in hepatocytes. The role of the liver in xenobiotic excretion may result in the presence of relatively high levels of metabolites in bile and consequent local effects. Additionally, the liver has a large number of specialized functions, perturbation of which may manifest as acute toxicity. Assays performed in metabolically competent liver cells may also be also potentially informative as preferential cytotoxicity may indicate the role of metabolic activation in toxicity. The incorporation of a specific assay of liver cytotoxicity in any *in vitro* testing battery is therefore likely to improve the prediction of acute systemic toxicity.

Models of liver toxicity *in vitro* range in complexity from cultured cells to the isolated perfused organ.[46–49] Studies in the isolated perfused rat liver have the obvious advantage of preserving the anatomical and biochemical characteristics of the organ, including phase I and (Table 6.7) phase II metabolism, but have obvious drawbacks in terms of animal usage, technical complexity and viability limited to a few hours and are therefore little used routinely. Additionally, organs from small laboratory animals may not be as representative of the human liver as those from larger animals. Precision-cut liver slices maintain the lobular structure of the organ, tissue organisation and cell–cell interaction; thin slices (<250 µM) offer reasonable viability (2–3 days) due to better oxygen and nutrient diffusion, but lack bile collection functions.[32,47,49,50] The most frequently used *in vitro* models of hepatotoxicity utilise isolated primary hepatocytes or cell lines of hepatic origin.[47,51] Primary hepatocytes are typically obtained following collagenase treatment of the liver (although this method of isolation may be associated with cell damage) and are cultured in a submerged monolayer. A critical aspect of the assessment of specific hepatotoxicity relates to the potential generation of toxic metabolites. It is essential, therefore, that any *in vitro* model of hepatotoxicity retains xenobiotic metabolic capacity. Cultured primary hepatocytes are phenotypically unstable but, depending on the culture conditions, may retain this metabolic capacity at least in the short-term.[52] Commonly used cell line models of hepatotoxicity include the HepG2

Table 6.7   Examples of Specific Hepatotoxicity Assays.

Model	Endpoint	Assay
Primary hepatocyte spheroids	Cellular stress	Heath shock protein induction[61]
	Cellular energetics	ATP status[61]
BRL 3A rat liver cell	Membrane integrity	LDH leakage[63]
	Metabolic activity	MTT reduction[63]
	Oxidative stress	ROS production (fluorescence)[63]
	Mitochondrial potential	Rhodamine 123 staining[63]
	Glutathione status	Glutathione level (colorimetric)[63]
Chang liver cell	Cell viability	Morphology[64]
		pH change[64]
HepG2 cell line	Mitochondrial activity	Alamar blue[53]
	Protein content	Lowry staining[64]
	Metabolic activity	MTT reduction[64]
	Protein content	Sulphorhodamine B staining[64]
	Cellular stress	Heath shock protein induction[53]
	Apoptosis	Caspase activity (fluorescence)[53,54]
	Necrosis	Adenylate kinase release (fluorescence)[53]
	Antioxidant status	Glutathione, malondialdehyde, cysteine levels[53,54]
		Catalase activity[53,54]
		Glutathione reductase activity[53,54]
HepaRG cell line	Antioxidant status	Glutathione status[55]
	Membrane integrity	LDH leakage[55]
	Mitochondrial potential	Rhodamine 123 staining[55]

human liver and the human hepatoma-derived HepaRG cell lines.[53–55] HepaRG cells are phenotypically and functionally more similar to primary hepatocytes, retain metabolic capacity[56] and show some advantages to HepG2 cells for the assessment of toxicity due to a higher metabolic competence.[47,57]

Disadvantages of cell culture models include the loss of tissue architecture and cell–cell interaction; however, for most chemicals these are likely to play a less important role in toxicity than metabolic capacity. More recent developments include 'sandwich' cultures, in which hepatocytes are grown between two layers of collagen or a similar gel matrix. The collagen matrix is shown to stimulate the repolarization of cells and also triggers the extensive formation of biliary canaliculi.[47] Sandwich cultures therefore permit the assessment of more specific markers of liver toxicity in addition to non-specific markers of basal cytotoxicity. Primary rat hepatocytes co-cultured with rat liver epithelial cells also maintain more characteristic morphology and biochemical features. The culture of liver cells on three-dimensional substrates has been shown to improve viability, facilitate cell growth and result in the formation of multiple cell layers with more typical morphological characteristics. The use of HepaRG cells in a three-dimensional model has also recently been proposed as a potentially valuable tool for hepatotoxicity screening.[47,58–62]

6.5 Acute Dermal Toxicity

In vivo data on toxicity following acute dermal exposure are frequently required for chemicals or products for which skin contact is likely or possible; an acute dermal toxicity study is therefore a standard data requirement under regulatory regimes covering industrial chemicals and pesticides. An *in vitro* system capable of reliably predicting acute dermal toxicity faces essentially the same issues as those assessing acute oral toxicity, but with a small number of important differences relating to the potential for local effects and the influence of dermal absorption on the expression of systemic toxicity.

6.5.1 Acute Dermal Toxicity: Current *In Vivo* Methods

The current standard method for the assessment of acute dermal toxicity *in vivo* (OECD 402; 1987) specifies the application of a chemical substance to at least 10% of the skin surface area of groups of five animals of a single sex, typically rats or rabbits. The study is performed with at least three dose levels; alternatively (and most frequently) a limit test may alternatively be performed with a single group of five animals of each sex. Exposure is terminated after 24 hours by washing the application site and animals are observed for a 14-day period, followed by limited post-mortem observation. Lethality remains the primary endpoint and the results of the study provide limited data on the likely mechanism of toxicity.

6.5.2 Acute Dermal Toxicity: Refined *In Vivo* Methods

The design and statistical evaluation of a fixed concentration procedure for acute dermal toxicity (similar to the acute oral FDP) has been described;[65] this method specifies the use of dose levels causing only moderate toxicity and avoids the administration of dose levels expected to be lethal or to cause pain or distress though marked local effects. A starting dose, predicted not to cause severe toxicity or mortality, is administered to one group of five animals; testing continues using a series of set dose levels until 'evident toxicity' is observed. The test method does not therefore rely on lethality as the primary endpoint and represents a refinement method as well as potentially a reduction in animal numbers. This method is currently available in draft form as OECD Test Guideline 434.

6.5.3 Acute Dermal Toxicity: Intelligent Testing Strategies

The generation of acute dermal toxicity data is rarely required in isolation; studies are most frequently performed as part of a package performed to satisfy regulatory requirements and also include studies of acute oral toxicity and skin irritation. Consequently it may be possible to meet the regulatory requirement for information on acute dermal toxicity potential using a stepwise and more informed approach, maximising the use of existing data. While there are clear physiological reasons why chemicals may potentially demonstrate greater acute systemic toxicity following dermal exposure than following oral exposure (e.g. the absence of first-pass detoxification following dermal administration), it is also likely for the large majority of chemicals that dermal penetration is less rapid and less extensive than absorption from the gastrointestinal tract following oral administration.

Where a dermal toxicity study is required for the purposes of hazard classification and labelling, the redundancy of testing by this route where data on acute oral toxicity are available has been suggested.[66–69] This concept has been challenged in an analysis;[70] which suggests a higher degree of 'false positives' (over-classification) and 'false negatives' (under-classification) by applying oral classifications to the dermal route, particularly when comparing oral classifications derived in the rat with dermal classifications derived in the rabbit.

For industrial chemicals where both acute oral and dermal toxicity data are available, the majority of substances are not classified (according to EU CLP criteria) or are classified in Category 5 (GHS criteria) for both acute oral and dermal toxicity. For a smaller proportion of chemicals classified for acute oral toxicity (i.e. in CLP/GHS Categories 1–4), classification for acute dermal toxicity is predominantly in a less severe hazard category. A small proportion of chemicals are classified in the same hazard category for acute oral and acute dermal toxicity; these chemicals appear to be those which would be predicted based on physicochemical properties (low molecular weight, log P_{ow}) or QSAR analysis to have high rates of dermal penetration.[71] A small proportion of substances are classified in a more severe category for acute dermal toxicity; however, analyses indicate that these substances are typically those causing severe local effects (i.e. corrosivity), for which testing would not be required in any case.[66] These data suggest that testing for acute dermal toxicity provides little useful additional information for the majority of chemicals which are not classified (EU CLP) or are classified in Category 5 (GHS) for acute oral toxicity. These chemicals are very likely to be similarly classified for acute dermal toxicity. Investigation of the acute dermal toxicity of substances classified in Category 4 may be required only if data indicate extensive dermal absorption. For these substances, information on the extent of dermal absorption can be taken from physicochemical properties or from studies *in vitro*. Further assessment of acute dermal toxicity will be required for substances classified in Categories 1–3 for acute oral toxicity, analyses indicate that allocation of the chemical to the same classification category for acute oral and acute dermal toxicity would be a precautionary approach but it is acknowledged that this may be overly conservative. It is clearly demonstrated that those physiological factors potentially leading to increased acute toxicity by the dermal route are counteracted by the less rapid and less extensive absorption likely following dermal exposure. Data therefore suggest that there is no need to test the majority of chemicals for acute dermal toxicity *in vivo* where there are data available on acute oral toxicity; an intelligent strategy (Figure 6.1) should be used to inform the need for testing for acute dermal toxicity *in vivo*.

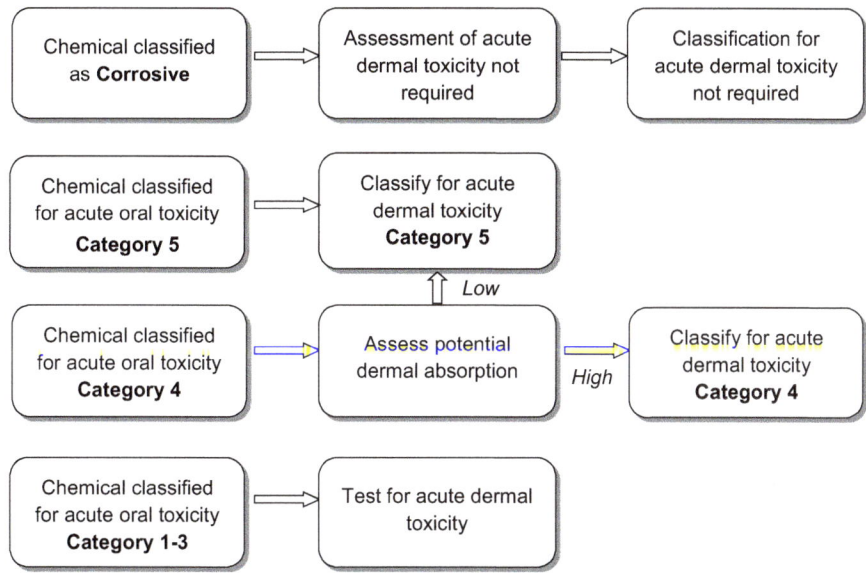

Figure 6.1 Acute dermal toxicity: Intelligent Testing Strategy.

6.5.4 Acute Dermal Toxicity: Testing *In Vitro*

Lethality in studies of acute dermal toxicity *in vivo* may result from systemic effects, local (i.e. site of contact) effects, or through a combination of systemic and local effects. Lethality as a consequence of site of contact effects is relevant for corrosive substances, but corrosivity can readily be predicted either from the physicochemical properties of a chemical (e.g. extremes of pH), or through the use of a validated *in vitro* assay for skin corrosivity (e.g. EpiDerm™ or EPISKIN™, models specified in OECD 431; Corrositex, specified in OECD 435; or the rat skin transcutaneous electrical resistance (TER) test). The testing of corrosive substances for acute dermal toxicity is therefore not required; such chemicals will be identified as corrosive, classified as such and conditions of use specified to minimise the likelihood of dermal exposure.

Systemic toxicity in an acute dermal toxicity study requires dermal penetration, a property which may be predicted to some extent by QSAR or based on physicochemical properties (physical form, molecular weight, water solubility, volatility, partition coefficient). Dermal penetration may be excluded for a small proportion of substances (e.g. those of very high molecular weight and/ or low solubility) but, for the majority of chemicals, dermal absorption to some extent will be predicted using these approaches. A more useful and quantitative prediction of dermal absorption can be obtained using validated *in vitro* studies in skin membranes (e.g. OECD 428). There is also clearly the potential for the assessment of dermal absorption in reconstructed human epidermis (RHE) models such as SkinEthic, EpiDerm and EpiSkin; these models are not

Table 6.8 Suggested *In Vitro* Tiered Testing Battery for Acute Dermal Toxicity.

Stage	Endpoint	Assay	Comment
1	Local effects	BCOP test (OECD 437) EpiDerm™, EPISKIN™ (OECD 431) for irritation/corrosivity	Further assessment may not be required if severe local effects are predicted
2	Dermal penetration	Dermal penetration in skin membranes (OECD 429) Dermal penetration in Reconstructed Human Epithelium (RHE) model	Further assessment may not be required if dermal penetration is not predicted
3	Basal cytotoxicity	Neutral Red Uptake (NRU) assay	Further assessment not required if there is data on acute oral toxicity
4	Role of metabolic activation in toxicity	Comparative cytotoxicity in primary rat hepatocytes Comparative cytotoxicity in a metabolically competent cell line	Possible generation of toxic metabolites assessed by QSAR
5	Potential for specific cytotoxicity	Comparative toxicity in cells from other organs/tissues	Choice of cell guided by chemical structure/QSAR

currently validated for this purpose, but may provide a useful indication of dermal absorption.[72,73]

Where acute oral toxicity data are not available to inform the assessment acute dermal toxicity, this may be predicted using a battery of assays assessing endpoints including severe local effects (corrosivity), the extent of dermal penetration and basal cytotoxicity. The role of metabolic activation in systemic toxicity may also be investigated using a comparative assay in metabolically competent cells. A more comprehensive assessment of systemic toxicity can be achieved through the use of additional cytotoxicity assays utilising cells from potentially sensitive organs, as described for the assessment of acute oral toxicity. The US NTP has evaluated an *in vitro* testing battery in order to reduce the requirement for 14-day range-finding dermal toxicity studies;[74] this battery includes assays for local effects, dermal penetration, basal cytotoxicity and also includes an assessment of metabolic activation. The prediction of acute dermal toxicity *in vitro* may therefore be undertaken using a similar battery of assays, as suggested below (Table 6.8).

6.6 Acute Inhalation Toxicity

Assessment of acute inhalation toxicity is generally required for chemicals which are gases, volatile liquids or fine powders; or for chemicals where the use pattern indicates the potential for inhalation exposure (e.g. those used by spraying). The physicochemical properties of the chemical influence not only the likelihood and extent of inhalation exposure, but also the site(s) of exposure. Inhalation is perhaps the most difficult exposure route for which to predict acute toxicity. Unlike acute dermal toxicity, acute inhalation toxicity cannot be reliably predicted from data on acute oral toxicity. Reasons for this lack of predictivity may include the potential for rapid systemic exposure compared to other routes of exposure, the lack of a first-pass effect and the marked potential for local effects (e.g. for chemicals with irritant or surfactant properties) to cause lethality. Additionally, the chemical and physicochemical properties of a substance strongly influence the areas of the respiratory tract likely to be exposed and therefore also dictate the cell type(s) in which toxicity may be elicited.

6.6.1 Acute Inhalation Toxicity: Current *In Vivo* Methods

The standard test method (OECD 403) for the assessment of acute inhalation toxicity specifies the exposure of three groups of rats (5/sex) to different concentrations of a chemical (as a gas, vapour, dust or mist) or in order to provide an estimation of the median lethal concentration (LC_{50}). A limit test may be performed for chemicals of low toxicity. An alternative concentration × time (C×T) protocol may be used when there is a specific need to assess toxicity over different exposure durations for emergency response planning. The current version of OECD 403 (2009) allows a reduction in animal numbers by testing in animals of a single (susceptible) sex. An Acute Toxic Class method for inhalation toxicity has also been adopted (OECD 436). In common with OECD 423 (for acute oral toxicity), this guideline allows the use of a series of fixed exposure concentrations in order to estimate the LC_{50} value. This guideline involves the use of fewer animals, although lethality remains the key endpoint.

6.6.2 Acute Inhalation Toxicity: Future *In Vivo* Methods

A draft Fixed Concentration Procedure for acute inhalation toxicity (OECD 433) is currently under discussion. In common with the Fixed Dose Procedure for acute oral toxicity, this method allows the estimation of the LC_{50} value in a single sex (females) using a series of fixed exposure concentrations and does not rely on lethality as the exclusive or intended endpoint. The aim of this test method is that exposure of rats to concentrations expected to be lethal is avoided; equally, exposure to concentrations causing corrosivity or severe irritation (and consequent pain or distress) is also avoided. Instead, the method relies on signs of 'evident toxicity' as an endpoint on which to base classification of the chemical.

6.6.3 Acute Inhalation Toxicity: *In Vitro* Methods

A large number of models are used for the assessment of toxicity to the respiratory tract, ranging in complexity from cultured primary cells or cell lines (Table 6.9), through to air–liquid interface cultures, lung tissue constructs or lung slices. Models cover all major areas of the respiratory tract from nasal tissue to the deep lung.[75,76]

Table 6.9 Cellular Assays for Respiratory Tract Toxicity *In Vitro*.

Model	Endpoint	Assay
A549 cell	Cell viability	Neutral Red Uptake[79,84]
		Apoptosis induction[87]
		Morphology[87]
	Cell number	Gentian violet binding[79]
	Metabolic activity	MTT metabolism[84]
		WST-8 reduction[87]
	Oxidative stress	ROS generation (fluorescent)[87,88]
	Membrane integrity	Trypan Blue exclusion[87]
		LDH leakage[83,87]
	Cellular energetics	Mitochondrial membrane integrity[88]
	Cell proliferation	PCNA analysis[88]
BEAS-2B	Membrane integrity	LDH leakage[89]
	Apoptosis induction	DNA fragmentation[89]
		Annexin V binding[89]
	Metabolic activity	MTT metabolism[90]
	Oxidative stress	ROS generation (fluorescent)[91]
		Glutathione status[91]
Human lung cells	Membrane integrity	LDH leakage[92]
	Metabolic activity	Alamar blue reduction[92]
MucilAir	Cell viability	Alamar blue metabolism[85]
	Metabolic activity	MTT reduction[83]
	Membrane integrity	LDH leakage[83]
	Tissue integrity	TEER[83]
	Functional	Cilia beating[85]
EpiAirway	Tissue integrity	TEER[86]
	Metabolic activity	MTT[86]

6.6.3.1 Lung Slices as Models of Acute Inhalation Toxicity

Precision-cut lung slices can be obtained from a number of species and are most readily prepared following agarose instillation *ex vivo*; a number of slices (up to 20) may be taken from a single animal.[32,76] Lung slices are widely used for the assessment of respiratory tract toxicity. The major advantages of this model are the inclusion of a number of respiratory cell types, the retention of normal cell morphology and tissue architecture and metabolic competence. Drawbacks include the technical difficulties associated with preparing lung slices, inter-slice variability and a limited duration of viability in culture.[77]

6.6.3.2 Primary Cell Models of Acute Inhalation Toxicity

The most commonly used primary cell model is constructed using normal human tracheal/bronchial epithelial (NHT/BE) cells. Cells can be cultured under submerged conditions but may also be cultured on collagen-coated microporous membranes and, following confluence, raised to the air–liquid interface. Cultures may also be grown on fibroblast-seeded collagen gels. Culture of NHT/BE cells at the air–liquid interface results in the differentiation of bronchial epithelial cells, with the subsequent formation of tight junctions and a pseudostratified layer containing mixtures of basal cells, intermediate cells, ciliated columnar cells and goblet cells in representative proportions. Cells grown in these conditions can also be exposed apically to chemicals. The differentiated nature of the cells means that toxicity can be measured not only through the general assessment of viability but also through the assessment of specific respiratory cell morphology or function. Three-dimensional air–liquid interface cultures of NHT-BE cells are commercially available from a number of sources, including MatTek (EpiAirway®) and Epithelix (MucilAir™).

6.6.3.3 Cell Line Models of Acute Inhalation Toxicity

Respiratory cell lines typically do not retain many characteristic features and are not representative of the whole respiratory tract, but are more readily obtained and cultured than primary cells. Cell lines such as Calu-3 (human bronchial epithelium) cells, A549 (human alveolar epithelium) cells, 16HBE14o, BEAS-2B (human bronchial epithelium) cells and RPMI 2650 (human nasal epithelium) cells have been used in submerged culture to assess cytotoxicity.[78–81] Cells grown in submerged culture express fewer typical morphological characteristics and cannot be exposed directly to chemicals in gas or vapour form or solid particulates, which are of obvious interest in the assessment of respiratory tract toxicity. These limitations of submerged cell cultures may be overcome through the culture of cells at the air–liquid interface. Depending on the precise culture conditions, air–liquid interface cultures also tend to show the formation of tight junctions and more characteristic cell surface morphology including ciliation and biochemistry and have the distinct advantage that cells may be exposed directly (apically) to gases and vapours.[82] The prediction of acute inhalation toxicity using cytotoxicity values from monolayer cultures of A549 cells, 3T3 cells and three-dimensional culture models of the human respiratory epithelium was compared against rodent acute inhalation LC_{50} values. A generally poor predictivity was shown, but the authors concluded that the use of more specific endpoints reflecting respiratory cell function and specific mechanisms of toxicity may improve the power of the models.[83]

6.7 Alternative Approaches to Assessing Acute Systemic Toxicity

Potential alternatives to meeting regulatory data requirements for acute toxicity include making better use of existing toxicity data, introducing more intelligent testing strategies or the use of *in vitro* assays. Each of these methods can ultimately result in the reduction of animal numbers, a refinement procedure or the replacement of *in vivo* studies.

6.7.1 Use of Existing Data

Acute toxicity studies form part of the standard toxicological dataset for chemical substances and products under various regulatory regimes. There is a danger that studies may be performed in isolation without considering other potentially relevant factors, and without questioning the absolute need for a study. One example is in the pharmaceutical sector, where acute oral toxicity studies were previously required; revision of the ICH M3 guideline removed the requirement for acute toxicity studies to support the first clinical trial in humans as it as shown that this study provided limited information on signs of toxicity and was not used to set dose levels for human trials.[93] In the industrial chemicals sector, there may be a potential to identify chemicals of low acute toxicity from the results of repeated dose toxicity, without the need for separate testing in an acute toxicity study.[2] In a detailed analysis,[94] data from over 4000 chemicals were assessed in order to identify a no observed adverse effect level (NOAEL) threshold for the 28-day study which permitted the identification of the majority (63%) of acutely non-toxic chemicals. Using this approach, only 1% of harmful substances were misclassified as acutely non-toxic. It was estimated that following this approach could potentially reduce the numbers of animals used in acute oral toxicity studies by half.

For some substance groups (e.g. low manufacture tonnage industrial chemicals), acute toxicity studies may represent a significant part of the toxicological dataset and other data may be limited. For other groups, however, acute toxicity studies form a small part of a much larger dataset involving animal testing for period of up to a lifetime, or over a number of successive generations. As part of an overall testing programme, short-term range-finding or sighting studies are routinely performed to inform the choice of dose levels in longer term studies. Where such studies are already available, these can often provide a good estimation of the maximum tolerated or minimal lethal dose and therefore have the potential to be used to estimate the acute oral LD_{50}. In such cases, studies of acute oral toxicity (to GLP and regulatory guidelines) are required under the relevant legislation but may simply serve to confirm what is already known from non-standard investigations.

6.7.2   Intelligent Testing Strategies

As previously detailed, it is clear that the acute dermal toxicity study shows a large degree of redundancy for pesticides and industrial chemicals. While it may be possible to adequately address the potential for the acute dermal of a chemical using an intelligent testing strategy taking into account acute oral toxicity data, dermal penetration and the potential for local effects, the study remains a data requirement under regulatory regimes.

6.7.3   Use of *In Vitro* Data on Acute Toxicity

One of the primary aims of *in vivo* assays of acute toxicity is to assign chemical substances or chemical products to hazard classification categories. In this respect, *in vitro* measures of cytotoxicity may either be used to inform the design of *in vivo* studies (selection of the starting dose), or potentially to replace the *in vivo* study entirely.

6.7.3.1   Use of In Vitro Data to Set the Starting Dose In Vivo

Any basic information on the toxic potency of a chemical is useful in setting the starting dose for studies of acute oral toxicity *in vivo* according to current testing guidelines (OECD 420, 423, 425). Indications of potency from *in vitro* assays of basal cytotoxicity in any animal cell line can be used in this respect and therefore provide the potential to reduce the numbers of animals used in studies *in vivo*. A NICEATM/ECVAM international multi-laboratory validation study performed in 2002–2005 using two *in vitro* cytotoxicity assays (the 3T3-NRU assay and the NHK-NRU assay) in conjunction with the ZEBET regression model concluded that, while these assays were not sufficiently accurate to replace the acute oral toxicity study for the purposes of hazard classification, they may be used to predict the starting dose for *in vivo* studies[95] While it has been proposed[13] that following this approach could reduce the numbers of animals used in studies *in vivo* by 30%; recent experience in ICCVAM and one industrial laboratory indicates that, realistically, more limited reductions may be obtained.[95,96] Basal cytotoxicity data may also be used to identify those chemicals for which testing at the limit dose may be appropriate, without the need for a sighting study *in vivo*, thus leading to modest reductions in animal use.

6.7.3.2   Use of In Vitro Data as Part of an Integrated Approach

ECVAM[97] have recently recommended the use of the 3T3 NRU cytotoxicity assay as part of an integrated approach (weight of evidence analysis or integrated testing strategy) to the identification of non-classified substances, when using a classification threshold of 2000 mg/kg bw. This assay is intended to be used as part of an approach that also includes assessment of other data sources including chemical analogues (read-across), physicochemical properties, structural alerts, SAR and toxicokinetic data. While the supporting EURL ECVAM validation study demonstrated a low false-negative rate, it is cautioned that, since the assay will not detect organ-specific mechanisms of action, its use in this respect should be restricted to chemicals that have not been designed to interact with specific molecular or cellular targets. The assay is therefore of particular relevance to the assessment of industrial chemicals which are not designed to act on specific biological targets and also tend to be of low acute toxicity. Industrial chemicals which do exhibit toxicity are likely to act through non-specific mechanisms (basal cytotoxicity) affecting most cell types. It is further noted that care must be taken in interpreting the results of this assay in cases where the chemical needs metabolic activation to exert its toxicity or is detoxified. However, the high rate of false-positive results in the validation study means that positive results cannot be used as a reliable indication of acute oral toxicity *in vivo*.

6.7.3.3   Potential Use of In Vitro Data to Replace In Vivo Studies

The ultimate goal of recent initiatives is the complete replacement of *in vivo* studies of acute toxicity with batteries of *in vitro* tests. While it is clear that substantial progress has been made in this direction, there remains some way to go before such a strategy can gain regulatory acceptance. *In vitro* basal cytotoxicity data show better predictivity of toxicity for substances of low acute toxicity and can be used to guide the choice of starting dose in studies *in vivo*, but are not sufficiently accurate in isolation to enable the reliable prediction of acute systemic toxicity *in vivo* for all chemicals. In this respect, the use of a more intelligent tiered testing strategy has been proposed by a number of groups.[8] A suitable battery of validated assays (Figure 6.2) must include investigations of basal and specific cytotoxicity, assessment of metabolic activation and kinetic parameters including oral absorption and blood–brain barrier permeability.[27,98,99] *In vitro* cytotoxicity batteries will need to be complemented by reliable QSAR analysis, PBPK modelling and possibly toxicogenomic and proteomic investigation. Such a battery may, in the future, be considered sufficiently predictive of human toxicity for the purposes of classification and labelling. While it is doubtful that any totally *in vitro* approach will ever be sufficiently powerful to reliably predict systemic human toxicity for all chemicals, it is important to note that the current rodent *in vivo* models of acute

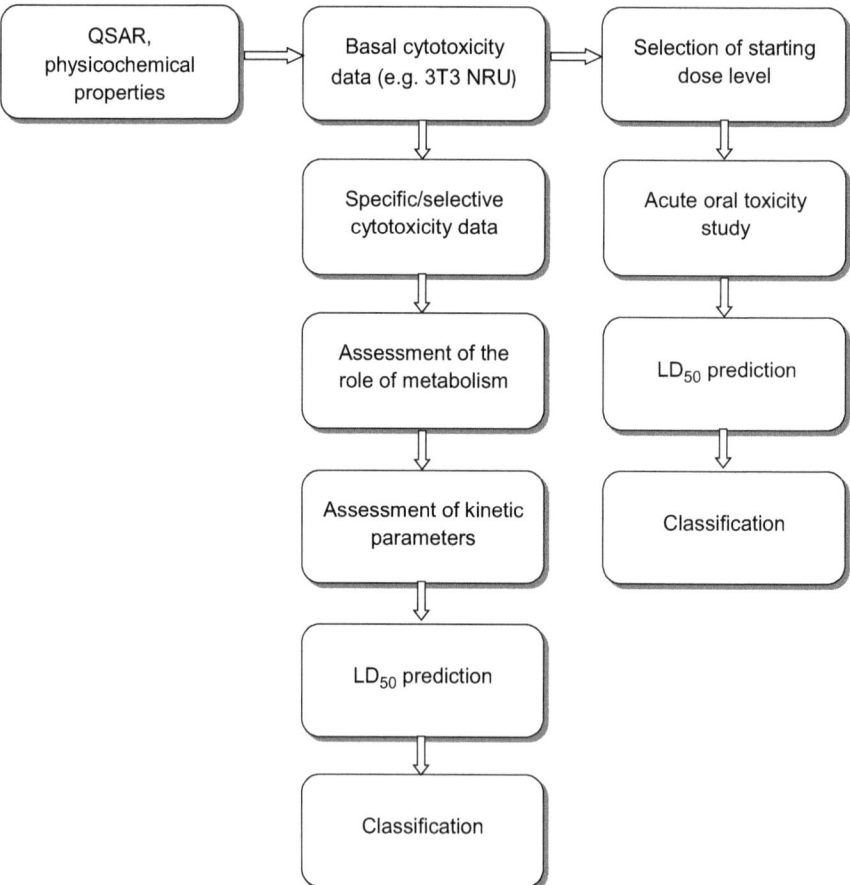

Figure 6.2   Potential use of *in vitro* data in the assessment of acute toxicity

toxicity are equally unlikely to be completely predictive of effects in humans, largely on the basis of potential metabolic differences.

References

1. D. J. Andrew, Acute toxicity, in *General and Applied Toxicology*, 3rd edn, B. Ballantyne, T. C. Marrs and T. Syversen, Wiley, 2009.
2. T. Seidle, S. Robinson, T. Holmes, S. Creton, P. Prieto, J. Scheel and M. Chlebus, Cross-sector review of drivers and available 3Rs approaches for acute systemic toxicity testing, *Toxicol. Sci.*, 2010, **116**(2), 382.
3. J. Trevan, The error of determination of toxicity, *Proc. R. Soc.*, 1927, **101B**, 483.
4. P. A. Botham, Acute systemic toxicity, *ILAR J.*, 2002, **43**, S27.
5. G. Zbinden and M. Flury-Roversi, Significance of the LD50-test for the toxicological evaluation of chemical substances, *Arch. Toxicol.*, 1981, **47**(2), 77.
6. B. Ekwall, Correlation between cytotoxicity in vitro and LD50-values, *Acta Pharmacol. Toxicol.*, 1983, **52**(S2), 80.
7. H. Siebert, M. Balls, J. H. Fentem, V. Bianchi, R. H. Clothier, P. J. Dierickx, B. Ekwall, M. J. Garle, M. J. Gómez-Lechón, L. Gribaldo, M. Gülden, M. Liebsch, E. Rasmussen, R. Roguet, R. Shrivstava and E. Walum, Acute toxicity testing *in vitro* and the classification and labelling of chemicals, *Altern. Lab. Anim.*, 1986, **24**, 499.
8. C. J. Cao, J. S. Madren-Whalley, C. Krishna and J. J. Valdes, *In vitro* methods to measure toxicity of chemicals, 2004, AMSRD-RRT-BM Report.
9. B. Ekwall, Overview of the final MEIC results: II. The in vitro ± in vivo evaluation, including the selection of a practical battery of cell tests for prediction of acute lethal blood concentrations in humans, *Toxicol. In Vitro*, 1999, **13**, 665.
10. C. M. Pomerat and C. D. Leake, Short-term cultures for drug assays: general considerations, *Ann. N.Y. Acad. Sci.*, 1954, **58**, 1110.
11. H. Eagle and G. E. Folgt, The cytotoxic action of carcinolytic agents in tissue culture, *Am. J. Med.*, 1956, **21**, 739.
12. M. J. Garle, J. H. Fentem and J. R. Fry, *In vitro* cytotoxicity tests for the prediction of acute toxicity *in vivo*, *Toxicol. In Vitro*, 1994, **8**(6), 1303.
13. W. Halle, The Registry of Cytotoxicity: toxicity testing in cell cultures to predict acute toxicity (LD50) and to reduce testing in animals, *Altern. Lab. Anim.*, 2003, **31**(2), 89.
14. M. Liebsch, B. Grune, A. Seiler, D. Butzke, M. Oelgeschläger, R. Pirow, S. Adler, C. Riebeling and A. Luch, Alternatives to animal testing: current status and future perspectives, *Arch. Toxicol.*, 2011, **85**, 841.
15. C. Clemedson and B. Ekwall, Overview of the final MEIC results: I. The *in vitro* ± *in vitro* evaluation, *Toxicol. In Vitro*, 1999, **13**, 657.
16. C. Clemedson, P. J. Dierickx and M. Sjöström, The prediction of human acute systemic toxicity by the EDIT/MEIC in vitro test battery: the

importance of protein binding and of partitioning into lipids, *Altern. Lab. Anim.*, 2003, **31**(3), 245.

17. M. Gülden, S. Mörchel and H. Seibert, Factors influencing nominal effective concentrations of chemical compounds *in vitro*: cell concentration, *Toxicol. In Vitro*, 2001, **15**, 233.

18. H. Seibert, S. Mörchel and M. Gülden, Factors influencing nominal effective concentrations of chemical compounds *in vitro*: serum albumin concentration, *Toxicol. In Vitro*, 2002, **16**, 289.

19. ICCVAM Test Method Evaluation Report, *In vitro* cytotoxicity test methods for estimating starting doses for acute oral systemic toxicity tests, Interagency Coordinating Committee on the Validation of Alternative Methods (ICCVAM) National Toxicology Program (NTP) Interagency Center for the Evaluation of Alternative Toxicological Methods (NICEATM). NIH Publication No. 07-4519, 2006.

20. S. Hoffmann, A. Kinsner-Ovaskainen, P. Prieto, I. Mangelsdorf, C. Bieler and T. Cole, Acute oral toxicity: Variability, reliability, relevance and interspecies comparison of rodent LD(50) data from literature surveyed for the ACuteTox project, *Regul. Toxicol. Pharmacol.*, 2010, **58**(3), 395.

21. British Toxicology Society Working Party on Toxicity, Special report: a new approach to the classification of substances and preparations on the basis of their acute toxicity, *Hum. Toxicol.*, 1984, **3**, 85.

22. R. D. Bruce, An up-and-down procedure for acute toxicity testing, *Fundam. Appl. Tox.*, 1985, **5**, 51.

23. J. Gartlon, A. Kinsnera, A. Bal-Price, S. Coecke and R. H. Clothier, Evaluation of a proposed *in vitro* test strategy using neuronal and non-neuronal cell systems for detecting neurotoxicity, *Toxicol. In Vitro*, 2006, **20**(8), 1569.

24. S. Coecke and A. Price, Why *in vitro* neurotoxicity approaches are not formally validated and used for regulatory purposes: the way forward, ALTTOX, 2007.

25. C. K. Atterwill, A. Bruinink, J. Drejer, E. Duarte, E. McFarlane Abdulla, C. Meredith, P. Nicotera, C. Regan, E. Rodríguez-Farré, M. G. Simpson, R. Smith, B. Veronesi, H. Vijverberg, E. Walum and D. C. Williams, In vitro neurotoxicity testing: the report and recommendations of ECVAM Workshop 3, *Altern. Lab. Anim.*, 1994, **22**, 350.

26. A. Forsby, A. K. Bal-Price, A. Camins, S. Coecke, N. Fabre, H. Gustafsson, P. Honegger, A. Kinsner-Ovaskainen, M. Pallas, V. Rimbaud, E. Rodríguez-Farré, C. Suñol, J. A. Vericat and M. G. Zurich, Neuronal in vitro models for the estimation of acute systemic toxicity, *Toxicol. In Vitro*, 2009, **23**(8), 1564.

27. S. Coecke, H. Ahr, B. J. Blaauboer, S. Bremer, S. Casati, J. Castell, R. Combes, R. Corvi, C. L. Crespi, M. L. Cunningham, G. Elaut, B. Eletti, A. Freidig, A. Gennari, J. F. Ghersi-Egea, A. Guillouzo, T. Hartung, P. Hoet, M. Ingelman-Sundberg, S. Munn, W. Janssens, B. Ladstetter, D. Leahy, A. Long, A. Meneguz, M. Monshouwer, S. Morath, F. Nagelkerke, O. Pelkonen, J. Ponti, P. Prieto, L. Richert, E. Sabbioni,

B. Schaack, W. Steiling, E. Testai, J. A. Vericat and A. Worth, Metabolism: a bottleneck in *in vitro* toxicological test development, *Altern. Lab. Anim.*, 2006, **34**, 49.

28. P. Prieto, Barriers, nephrotoxicology and chronic testing *in vitro*, *Altern. Lab. Anim*, 2002, **30**(S2), 101.

29. G. M. Hawksworth, P. H. Bach, J. F. Nagelkerke, W. Dekant, J. E. Diezi, E. Harpur, E. A. Lock, C. MacDonald, J.-P. Morin, W. Pfaller, F. A. J. J. L. Rutten, M. P. Ryan, H. J. Toutain and A. Trevisan, Nephrotoxicity testing *in vitro*: the report and recommendations of ECVAM Workshop 10, *Altern. Lab. Anim*, 1995, **23**, 713.

30. L. H. Lash, *In vitro* methods of assessing renal damage, *Toxicol. Pathol.*, 1998, **26**, 33.

31. P. Jennings, C. Koppelstätter, J. Lechner and W. Pfaller, Renal cell culture models: contribution to the understanding of nephrotoxic mechanisms, in *Clinical Nephrotoxins*, ed. M. E. de Broe and G. A. Porter, Springer, 2008.

32. P. H. Bach, A. E. M. Vickers, R. Fisher, A. Baumann, E. Brittebo, D. J. Carlile, H. J. Koster, B. G. Lake, F. Salmon, T. W. Sawyer and G. Skibinski, The use of tissue slices for pharmacotoxicology studies: the report and recommendations of ECVAM Workshop 20, *Altern. Lab. Anim.*, 1996, **24**, 893.

33. C. E. Ruegg, A. J. Gandolfi, R. B. Nagle, C. L. Krumdieck and K. Brendel, Preparation of positional renal slices for study of cell-specific toxicity, *J. Pharmacol. Methods*, 1987, **17**(2), 111.

34. P. D. Williams, The application of renal cells in culture in studying drug-induced nephrotoxicity, *In Vitro Cell. Develop. Biol.*, 1989, **25**(9), 800.

35. J.-P. Morin, M. E. De Broe, W. Pfaller and G. Schmuck, Nephrotoxicity testing *in vitro*: the current situation: ECVAM Nephrotoxicity Task Force Report 1, *Altern. Lab. Anim.*, 1997, **25**, 497.

36. R. Nielsen, H. Birn, S. K. Moestrup, M. Nielsen, P. Verroust and E. I. Christensen, Characterization of a kidney proximal tubule cell line, LLC-PK1, expressing endocytotic active megalin, *J. Am. Soc. Nephrol.*, 1998, **9**, 1767.

37. ACuteTox: Optimisation and Pre-Validation of an In Vitro Test Strategy for Predicting Human Acute Toxicity, Progress Report, 2011.

38. J. B. Traloff, R. S. Goldstein, A. C. Silver, W. R. Hewitt and J. B. Hook, Intrinsic susceptibility of the kidney to acetaminophen toxicity in middle-aged rats, *Toxicol. Lett.*, 1990, **52**(1), 101.

39. D. K. Obatomi, N. T. K. Thanh, S. Brant and P. H. Bach, The toxic mechanism and metabolic effects of atractyloside in precision-cut pig kidney and liver slices, *Arch. Toxicol.*, 1998, **72**, 524.

40. F. Massicot, H. Dutertre-Catella, C. Pham-Huy, X.-H. Liu, H. T. Duc and J.-M. Warnet, *In vitro* assessment of renal toxicity and inflammatory events of two protein phosphatase inhibitors cantharidin and nor-cantharidin, *Basic Clin. Pharmacol. Toxicol.*, 2005, **96**, 26.

41. B. L'Azou, J. Jorly, D. On, E. Sellier, F. Moisan, J. Fleury-Feith, J. Cambar, P. Brochar and C. Ohayon-Courtès, In vitro effects of nano-particles on renal cells, *Particle Fibre Toxicol.*, 2008, **5**, 22.

42. R. Rezzani, P. Angoscini, E. Borsani, L. Rodella and R. Bianchi, Cyclosporine A-induced toxicity in two renal cell culture models (LLC-PK1 and MDCK), *Histochem. J.*, 2002, **34**(1–2), 27.

43. L. B. Zimmerhackl, F. Momm, G. Wiegele and M. Brandis, Cadmium is more toxic to LLC-PK1 cells than to MDCK cells acting on the cadherin-catenin complex, *Am. J. Physiol.*, 1998, **275**, 143.

44. X. Dan, Z. Yinan, W. Yili, J. Yu and C. Yu, Cytotoxicity of gold nanorods to human Hep-2 and canine MDCK cells, *Res. J. Biotech.*, 2012, **7**(3), 34.

45. P. C. Wilga, J. M. McKim Jr., J. F. Pregenzer and D. K. Petrella, Evaluation of a novel anti-tumor drug using in vitro toxicity screening in rat hepatoma (H4IIE) cells, normal rat kidney (NRK) cells and rat primary hepatocytes, Ceetox (www.ceetox.com).

46. D. A. Groneberg, C. Grosse-Siestrup and A. Fischer, *In vitro* models to study hepatotoxicity, *Toxicol. Pathol.*, 2002, **30**, 394.

47. V. Y. Soldatow, E. L. LeCluyse, L. G. Griffiths and I. Rusyn, *In vitro* models for liver toxicity testing, *Toxicol. Res.*, 2013, **2**, 23.

48. E. L. LeCluyse, R. P. Witek, M. E. Andersen and M. J. Powers, Organotypic liver culture models: Meeting current challenges in toxicity testing, *Crit. Rev. Toxicol.*, 2012, **42**(6), 501.

49. P. H. Bach, A. E. M. Vickers, R. Fisher, A. Baumann, E. Brittebo, D. J. Carlile, H. J. Foster, B. J. Lake, F. Salmon, T. W. Sawyer and G. Skibinski, The use of tissue slices for pharmacotoxicological studies: the report and recommendations of ECVAM Workshop 20, *Altern. Lab. Anim.*, 1996, **24**, 893.

50. K. Amin, C. Ip, L. Jimenez, C. Tyson and H. Behrsing, *In vitro* detection of differential and cell-specific hepatobiliary toxicity induced by geldanamycin and 17-allylaminogeldanamycin using dog liver slices, *Tox. Sci.*, 2005, **87**(2), 442.

51. A. Guillouzo, Liver cell models in *in vitro* toxicology, *Environ. Health Persp.*, 1998, **106**(S2), 511.

52. I. Horii and H. Yamada, *In vitro* hepatotoxicity testing in the early phase of drug discovery, *AATEX*, 2008, **14**, 437.

53. S. Miret, E. M. De Groene and W. Klaffke, Comparison of *in vitro* assays of cellular toxicity in the human hepatic cell line HepG2, *J. Biomol. Screen.*, 2006, **11**, 184.

54. J. Ogony, R. Matthews, H. Anni, K. Shannon and N. J. Ercal, The mechanism of elevated toxicity in HepG2 cells due to combined exposure to ethanol and ionizing radiation, *J. Appl. Toxicol.*, 2008, **28**(3), 345.

55. M. R. McGill, H. Yan and H. Jaeschke, Acetaminophen-induced injury in HepaRG cells: a novel human cell line for studies of drug hepatotoxicity, *FASEB J.*, 2010, **24**(Suppl.), 759.

56. A. Guillouzo, A. Corlu, C. Aninat, D. Glaise, F. Morel and C. Guguen-Guillouzo, The human hepatoma HepaRG cells: A highly differentiated model for studies of liver metabolism and toxicity of xenobiotics, *Chem. Biol. Interact.*, 2007, **168**, 66.

57. M. R. McGill, H.-M. Yan, A. Ramachandran, G. J. Murray, D. E. Rollins and H. Jaeschke, HepaRG Cells: a human model to study mechanisms of acetaminophen hepatotoxicity, *Hepatology*, 2011, **53**(3), 974.

58. F. Evenou, T. Fujii and Y. Sakai, Liver cells culture on three-dimensional micropatterned polydimethylsiloxane surfaces, *AATEX*, 2008, **14**, 665.

59. M. Bokhari, R. J. Carnachan, N. R. Cameron and S. A. Przyborski, Culture of HepG2 liver cells on three dimensional polystyrene scaffolds enhances cell structure and function during toxicological challenge, *J. Anat.*, 2007, **211**, 567.

60. S. B. Leite, I. Wilk-Zasadna, J. M. Zaldivar, E. Airola, M. A. Reis-Fernandes, M. Mennecozzi, C. Guguen-Guillouzo, C. Chesne, C. Guillou, P. M. Alves and S. Coecke, Three-dimensional HepaRG model as an attractive tool for toxicity testing, *Toxicol. Sci.*, 2012, **130**(1), 106.

61. C. Dilworth, G. A. Hamilton, E. George and J. A. Timbrell, The use of liver spheroids as an *in vitro* model for studying induction of the stress response as a marker of chemical toxicity, *Toxicol. In Vitro*, 2000, **14**(2), 169.

62. G. Mazzoleni and N. Steimberg, New models for the *in vitro* study of liver toxicity: 3D culture systems and the role of bioreactors, in *The Continuum of Health Risk Assessments*, ed. M. G. Tyshenko, 2012, Intech.

63. S. M. Hussain, K. L. Hess, J. M. Gearhart, K. T. Geiss and J. J. Schlager, *In vitro* toxicity of nanoparticles in BRL 3 A rat liver cells, *Toxicol. In Vitro*, 2005, **19**, 975.

64. The Multicenter Evaluation of In Vitro Cytotoxicity (MEIC) Summary, September 2000, National Toxicology Program (NTP) Interagency Center for the Evaluation of Alternative Toxicological Methods (NICEATM).

65. N. Stallard, A. Whitehead and I. Indans, Statistical evaluation of an acute dermal toxicity test using the dermal fixed dose procedure, *Hum. Exp. Toxicol.*, 2004, **23**(8), 405.

66. N. P. Moore, D. J. Andrew, D. L. Bjerke. S. Creton, D. Dreher, T. Holmes, I. Indans, P. Prieto, T. Seidle and T. G. Rowan, Can acute dermal systemic toxicity tests be replaced with oral tests? A comparison of route-specific systemic toxicity and hazard classifications under the Globally Harmonized System of Classification and Labelling of Chemicals (GHS). *Regul. Toxicol. Pharmacol.*, 2013, **66**(1), 30.

67. S. Creton, I. C. Dewhurst, L. K. Earl, S. C. Gehen, R. Guest, J. A. Hotchkiss, I. Indans, M. R. Woolhiser and R. Billington, Acute toxicity testing of chemicals: opportunities to avoid redundant testing and use alternative approaches, *Crit. Rev. Toxicol.*, 2010, **40**, 50.

68. T. Seidle, Opportunities and barriers to the replacement of animals in acute systemic toxicity testing, 2007. www.alttox.org.

69. H. D. Thomas and I. C. Dewhurst, What does a dermal acute toxicity study add to the information on a plant protection pesticide, *Toxicology*, 2007, **231**, 104.
70. M. Paris, J. Strickland, F. Stack, D. Allen, W. Casey and W. Stokes, Analysis to determine if acute systemic toxicity data can be used to estimate and avoid acute dermal systemic toxicity testing, poster presentation, SOT meeting, 2012.
71. D. Andrew and S. Wright-Williams, Analysis of acute toxicity and irritation data submitted under the first REACH deadline, poster presentation, British Toxicology Society, 2011.
72. C. Lotte, C. Patouillet, M. Zanini, A. Messager and R. Roguet, Permeation and skin absorption: reproducibility of various industrial reconstructed human skin models, *Skin Pharmacol. Appl. Skin Physiol.*, 2002, **15**(S1), 18.
73. M. Schäfer-Korting, U. Bock, W. Diembeck, H. J. Düsing, A. Gamer, E. Haltner-Ukomadu, C. Hoffmann, M. Kaca, H. Kamp, S. Kersen, M. Kietzmann, H. C. Korting, H. U. Krächter, C. M. Lehr, M. Liebsch, A. Mehling, C. Müller-Goymann, F. Netzlaff, F. Niedorf, M. K. Rübbelke, U. Schäfer, E. Schmidt, S. Schreiber, H. Spielman, A. Vuia and M. Weimer, The use of reconstructed human epidermis for skin absorption testing: Results of the validation study, *Altern. Lab. Anim*, 2008, **36**(2), 161.
74. R. S. Chabra, N. B. Ress, J. W. Harbell and R. D. Curren, Evaluation of some *in vitro* tests to reduce and replace the sub-acute animal toxicity studies, *Altern. Lab. Anim.*, 2004, **32**, 137.
75. K. BéruBé, M. Aufderheide, D. Breheny, R. Clothier, R. Combes, R. Duffin, B. Forbes, A. Gaça, Gray, I. Hall, M. Kelly, M. Lethem, M. Liebsch, L. Merolla, J.-P. Morin, J. Seagrave, M. A. Schwartz, T. D. Tetley and M. Umachandran, *In vitro* methods of inhalation toxicity and disease: the report of a FRAME Workshop, *Altern. Lab. Anim.*, 2009, **37**, 89.
76. J. Wan, M. Johnson, J. Schilz, M. V. Djordjevic, J. R. Rice and P. G. Shields, Evaluation of *in vitro* assays for assessing the toxicity of cigarette smoke and smokeless tobacco, *Cancer Epidemiol. Biomarkers Prev.*, 2009, **18**(12), 3263.
77. C. Monteil, M. Guerbet, E. Le Prieur, J.-P. Morin, J. M. Jouany and J. P. Fillastre, Characterisation of precision-cut rat lung slices in a biphasic gas/liquid exposure system: effect of O_2, *Toxicol. In Vitro*, 1999, **13**, 467.
78. C. I. Grainger, L. L. Greenwell, D. J. Lockley, G. P. Martin and B. Forbes, Culture of Calu-3 cells at the air interface provides a representative model of the airway epithelial barrier, *Pharm. Res.*, 2006, **23**(7), 1482.
79. C. D. Lindsay and J. L. Hambrook, Protection of A549 cells against the toxic effects of sulphur mustard by hexamethylenetetramine, *Hum. Exp. Toxicol.*, 1997, **16**(2), 106.
80. D. J. Andrew and C. D. Lindsay, Protection of human upper respiratory tract cell lines against sulphur mustard toxicity by hexamethylenetetramine (HMT), *Hum. Exp. Toxicol.*, 1998, **17**, 373.

81. M. Davoren, E. Herzog, A. Casey, B. Cottineau, G. Chambers, H. J. Byrne and F. M. Lyng, *In vitro* toxicity evaluation of single walled carbon nanotubes on human A549 lung cells, *Toxicol. In Vitro*, 2007, **21**(3), 438.

82. S. Mülhopt, S. Diabaté, T. Krebs, C. Weiss and H.-R. Paur, Lung toxicity determination by *in vitro* exposure at the air liquid interface with an integrated online dose measurement, *J. Phys.*, 2009, **170**, 1.

83. U. G. Sauer, S. Vogel, A. Hess, S. N. Kolle, L. Ma-Hock, B. van Ravenzwaay and R. Landsiedel, *In vivo - in vitro* comparison of acute respiratory tract toxicity using human 3D airway epithelial models and human A549 and murine 3T3 monolayer cell systems, *Toxicol. In Vitro*, 2013, **27**(1), 174.

84. S. Lanone, F. Rogerieux, J. Geys, A. Dupont, E. Maillot-Marechal, J. Boczkowski, G. Lacroix and P. Hoet, Comparative toxicity of 24 manufactured nanoparticles in human alveolar epithelial and macrophage cell lines, *Part. Fibre Toxicol.*, 2009, **6**, 14.

85. S. Constant, The use of an *in vitro* cell model of the Human Airway Epithelium (MucilAir™) in preclinical development. www.epithelix.com.

86. D. Balharry, K. Sexton and K. A. BéruBé, An *in vitro* approach to assess the toxicity of inhaled tobacco smoke components: nicotine, cadmium, formaldehyde and urethane, *Toxicology*, 2008, **244**(1), 66.

87. Y. Chang, S.-T. Yang, J.-H. Liu, E. Dong, Y. Wang, A. Cao, Y. Liu and H. Wang, *In vitro* toxicity evaluation of graphene oxide on A549 cells, *Toxicol. Lett.*, 2011, **200**, 201.

88. P. Chairuangkitti, S. Lawanprasert, S. Roytrakul, S. Aueviriyavit, D. Phummiratch, K. Kulthong, P. Chanvorachote and R. Maniratanachote, Silver nanoparticles induce toxicity in A549 cells via ROS-dependent and ROS-independent pathways, *Toxicol. In Vitro*, 2013, **27**(1), 330.

89. W. K. Nichols, R. Mehta, K. Skordos, K. Macé, A. M. A. Pfeifer, B. A. Carr, T. Minko, S. W. Burchiel and G. S. Yost, 3-Methylindole-induced toxicity to human bronchial epithelial cell lines, *Tox. Sci.*, 2003, **71**, 229.

90. N. V. Vallabani, S. Mittal, R. K. Shukla, A. K. Pandey, S. R. Dhakate, R. Pasricha and A. Dhawan, Toxicity of graphene in normal human lung cells (BEAS-2B), *J. Biomed. Nanotechnol.*, 2011, **7**(1), 106.

91. E.-J. Park, J. Choi, Y.-K. Park and K. Park, Oxidative stress induced by cerium oxide nanoparticles in cultured BEAS-2B cells, *Toxicology*, 2008, **12**, 245.

92. S. Mülhopt, S. Diabaté, T. Krebs, C. Weiss and H.-R. Paur, Lung toxicity determination by in vitro exposure at the air liquid interface with an integrated online dose measurement, Nanosafe 2008: International Conference on Safe production and use of nanomaterials, *J. Physics*: Conference Series 170 (2009).

93. K. Chapman and S. Robinson, Challenging the regulatory requirement for acute toxicity studies in the development of new medicines: a workshop

report, 2007, National Centre for the Replacement, Refinement and Reduction of Animals in Research.

94. A. Bulgheroni, S. Kinsner-Ovaskainen, S. Hoffman, T. Hartung and P. Prieto, Estimation of acute oral toxicity using the No Observed Adverse Effect Level (NOAEL) from the 28 day repeated dose toxicity studies in rats, *Regul. Toxicol. Pharmacol.*, 2009, **53**(1), 16.

95. W. S. Stockes and L. Schectman, *In vitro* cytotoxicity test methods for estimating starting doses for acute oral systemic toxicity tests. ICCVAM Test Method Evaluation Report (2006), ICCVAM.

96. A. Schrage, K. Hempel, M. Schulz, S. N. Kolle, B. van Ravenzwaay and R. Landsiedel, Refinement and reduction of acute oral toxicity testing: a critical review of the use of cytotoxicity data, *Altern. Lab. Anim.*, 2011, **39**(3), 273.

97. EURL ECVAM recommendation on the 3T3 Neutral Red Uptake Cytotoxicity Assay for Acute Oral Toxicity testing, European Commission JRC Scientific and Policy Reports, Report EUR 25946, April 2013.

98. A. Gennari, C. van den Berghe, S. Casati, J. Castell, C. Clemedson, S. Coecke, A. Colombo, R. Curren, G. Dal Negro, A. Goldberg, C. Gosmore, T. Hartung, I. Langezaal, I. Lessigiarska, W. Maas, I. Ingelsdorf, R. Parchment, P. Prieto, J. Riego Sintes, M. Ryan, G. Schmuck, K. Sitzel, W. Stokes, J. A. Vericat and L. Gribaldo, Strategies to replace in vivo acute systemic toxicity testing: the report and recommendations of ECVAM Workshop 50, *Alt. Lab. Anim.*, 2004, **32**, 437.

99. H. Seibert, M. Balls, J. H. Fentem, V. Bianchi, R. H. Clothier, P. J. Dierickx, B. Ekwall, M. J. Garle, M. J. Gómez-Lechón, L. Gribaldo, M. Gülden, N. Liebsch, E. Rasmussen, R. Roguet, R. Shrivistava and E. Walum, Acute toxicity testing *in vitro* and the classification and labeling of chemicals: The report and recommendations of ECVAM Workshop 16, *Altern. Lab. Anim.*, 1996, **24**, 499.

CHAPTER 7

In Vitro Toxicology Models for Acute Eye and Skin Irritation Assessment

GERTRUDE-EMILIA COSTIN AND HANS RAABE*

Institute for In Vitro Sciences, Inc. (IIVS), Gaithersburg, MD, USA
*Email: hraabe@iivs.org

7.1 Introduction

With the expansion and proliferation of personal care, consumer, household, pharmaceutical, and industrial chemicals and products, there is a growing need to evaluate and assure their safe handling and use throughout the manufacturing, transportation, distribution, and end-use life cycles. There is an obvious need to be able to identify the potential hazards of chemical ingredients in personal care products and pharmaceuticals which are intentionally applied to the skin and eyes. However, accidental exposures to the eyes and skin in spill and splash events are also commonplace in the occupational and consumer arenas, and thus there is a universal need to be able to identify and communicate potential eye and skin hazards. As a historical note, in response in part to reports of eye injuries induced by the inclusion of the analine dye paraphenylenediamine in some mascaras, the US Congress passed the Federal Food, Drug, and Cosmetic Act of 1938, thus granting the US Food and Drug Administration (US FDA) regulatory authority over the use of chemical ingredients in cosmetic products.[1] Shortly thereafter, FDA toxicologists John H. Draize and Geoffrey Woodard, and division chief Herbert Calvery, reported on

Issues in Toxicology No. 19
Reducing, Refining and Replacing the Use of Animals in Toxicity Testing
Edited by David G. Allen and Michael D. Waters
© The Royal Society of Chemistry 2014
Published by the Royal Society of Chemistry, www.rsc.org

methods for evaluating the eye and skin irritation potential of cosmetic ingredients, using albino rabbits.[2] Currently, in the US, several federal regulatory
agencies (e.g., the Food and Drug Administration, Environmental Protection
Agency, Consumer Product Safety Commission, Department of Agriculture,
and the Occupational Safety and Health Administration) rely upon variations
of the Draize procedures for ocular hazard classification of potential eye irritants based upon their legislative mandate and chemicals or products under
their purview.[3]

7.2   Advent of Alternative Test Methods to the Draize Eye and Skin Irritation Tests

For almost 70 years, the basic methods of Draize and Woodard have been used
in both the regulatory and industrial sectors for assessing the eye and skin irritation potential of a wide range of chemicals and products, and are still in use
today to fulfill certain regulatory requirements. However, during much of this
time, scientists have searched for alternatives that would reduce, refine, or replace the use of animals. In fact, the first *in vitro* methods to be developed were
those methods targeted at replacing these traditional Draize eye and skin tests.
There are several factors driving this area of development, but political pressure
from animal welfare groups on the cosmetics and personal care industries to
abolish the Draize eye and skin irritation tests was perhaps the most influential.
Animal welfare organizations effectively targeted these industries by calling
into question the ethics of using animals for testing cosmetic products – categories of products that essentially provide for society's beauty and vanity needs.
Furthermore, non-animal test methods were pursued by some cosmetics and
personal care companies to meet their own corporate animal welfare mandates,
and in the absence of specific regulatory requirements for product safety testing, industry investigated how these non-animal methods may be used in the
development of safer products. The following scientific issues were also raised
regarding the use of the Draize test for testing cosmetics:

1 Considering that most cosmetics are generally of low irritation potential,
   there seemed to be little justification for subjecting animals to either the
   test materials, or the requisite positive controls, the latter of which by
   design induced substantial eye or skin irritation to demonstrate a valid
   test.
2 Concerns of the use of excessive dose volumes in the Draize test, and high
   variability in test results were voiced.[4–7]

Therefore, much of the initial research and development of alternative non-
animal test methods was conducted within the cosmetics and personal care
industry research labs, often in collaboration with the academic and contract
research laboratories they support. The subsequent sections will describe the
nature of eye and skin irritation events *in vivo*, followed by an overview of the

major groups of non-animal test methods developed for predicting these events, and culminating in a current presentation of available test methods for regulatory classification and labeling.

7.3 Acute Eye and Skin Irritation: Similarities in Exposure and Response Events

There are many factors in common between eye and skin irritation events *in vivo*, and the general methods developed for predicting these events *in vitro* warrant their presentation in the same chapter.

First, the apical corneal or cutaneous epithelia *in vivo* are primary sites of exposure to neat chemicals, chemical mixtures, and complex product formulations. Regardless of whether the exposure events are accidental (unintentional exposures in the industrial, commercial, or consumer settings) or intentional (as is the case for cosmetics and personal care products, and topically applied pharmaceuticals), these external epithelial tissues are uniquely exposed to undiluted formulations or chemical mixtures, and thus exposed to the full representation of active ingredients at the concentrations found in the formulations. In contrast, exposures of chemicals to internal organ systems after systemic uptake vary greatly owing to the effects of the differences in the pharmacokinetic factors (i.e., dose and route of exposure, selective systemic absorption of individual chemicals from complex mixtures, systems distribution and accumulation, metabolism, and excretion), and the impact of dilution of the absorbed chemicals in the body.

Secondly, since the eyes and skin are directly exposed to chemicals and formulations, the irritation responses typically occur locally at the site of chemical exposure, rather than distally from the sites of exposure. Acute skin exposures to corrosive or irritant chemicals typically result in edema, erythema, and/or overt necrosis at the immediate site of exposure. Similarly, for acute exposures to the eyes, irritant and corrosive effects tend to be observed in the conjunctivae and surrounding tissues, sclera, and cornea, but in some cases may also involve the iris and lens. The conjunctivae and sclera exhibit local erythematous and edematous responses, as well as necrotic damage not unlike that observed in the epidermal tissues. The corneal tissues result in a variety of effects, including necrosis and/or loss of corneal epithelium, coagulation, or edema of the corneal epithelium and stroma resulting in opacities, and subsequent inflammatory responses, all of which are likely to affect visual acuity.

Third, it appears that cell damage and cell death at the sites of exposure are key to both acute eye and skin irritation and corrosion, particularly for the more severe effects. In fact, Robinson, *et al.*[8] stated that cellular damage and cell death were the underlying mechanisms for chemical-induced skin irritation effects *in vivo*. Similarly, Jester, *et al.*[9] demonstrated that the primary mechanism for ocular damage was cell death, by correlating initial *in vivo* rabbit eye irritation observations (using a low dose volume variation of the Draize eye irritation test) to the subsequent live-dead staining of corneal cells in the test animals' enucleated eyes. Furthermore, Jester, *et al.*[9] and Maurer, *et al.*[10,11] suggested that the depth and the degree of the injuries in the cornea shortly after exposure generally correlate with the eventual degree and duration of the ocular lesions, and proposed that determining this information shortly after exposure could be useful for predicting long-term outcomes.

This leads us to a fourth common concept observed in irritation responses in ocular and dermal tissues, namely, the degree of tissue damage tends to be the most severe nearest to the site of exposure, and becomes progressively less severe deeper into the tissues, or distally from the focal point of exposure. The most severe or corrosive chemicals are those that are both highly cytotoxic and penetrable into the corneal and epidermal/dermal tissues, and are more likely to induce irreversible damage deep into the underlying stromal tissues. In contrast, the effects of the lesser irritant materials may be limited to the superficial epithelial layers, where rapid turnover of these tissues typically ensures repair and recovery.

Lastly, it should be stated that ocular and dermal exposures can result in systemic absorption as well, with potential distal toxic effects. In fact the skin is one of the four most common routes for systemic exposures to chemicals (in addition to inhalation, ingestion, and injection). However, the systemic absorption of chemicals via the ocular and dermal routes should be evaluated separately from the prediction of local acute ocular and dermal irritation.

7.4 Development of Alternative Predictive Models for Eye and Skin Irritation

Many of the current *in vitro* safety test protocols validated for regulatory classification and labeling for acute eye and skin irritation were originally designed to support product development and safety evaluations by the personal care, cosmetic, and household products industries for product stewardship, and by the pharmaceutical and chemical industries for industrial hygiene/worker safety goals. As presented above, in general the acute irritation effects in the eye and skin are relatively simple in comparison to the complex toxicological effects observed in organ system responses after whole body exposures. Accordingly, from a technical perspective, the development of non-animal assays and endpoints to predict acute eye and skin irritation appeared to be notably less challenging than developing assays to predict systemic toxicity and organ systems toxicology endpoints, where the complex pharmacokinetic and pharmacodynamic effects necessitate more comprehensive understanding and analyses. Consequently, alternative test methods to predict eye and skin irritation were among the first to be optimized for industry and regulatory use.

Several assays were initially developed specifically for a narrow industry niche. For example, in the early 1990s, Merck investigated the use of *ex vivo* bovine corneas for determining eye irritation potential of pharmaceutical intermediates specifically for identifying the appropriate personal protective equipment (PPE) for worker exposures.[12] The assay, the Bovine Corneal

Opacity and Permeability Assay (BCOP), targeted measuring key changes only in the cornea, with the justification that corneal effects were most relevant to impacts upon visual function and acuity. This assay, however, has since gained regulatory approval and widespread acceptance among the personal care and household products industries for product development and safety assessments, and by the chemical and pharmaceutical industries for worker safety/chemical hygiene purposes because of its ability to generally predict a wide spectrum of corneal effects.[13]

Other assays were developed for very specific regulatory purposes. The Corrositex Time Monitor assay (InVitro International, Irvine, CA) was developed for corrosive chemical packing group categorization. This assay has also gained regulatory approval and general market acceptance, but is only used for corrosive sub-categorization.[14] The Microtox assay was originally developed for rapid wastewater effluent toxin detection, and was promoted for predicting eye irritation in the late 1980s when very few alternative assays had been developed. However, the assay has not gained widespread acceptance for eye irritation screening although it was reported to be among the more promising assays to predict ocular irritation for some cosmetic and personal care product ingredients.[15]

Individual *in vitro* assay endpoints are developed to address specific events, but in many cases they do not provide the same overall series of events observed *in vivo*. For example, the Fluorescein Leakage (FL)[16] and the Short Time Exposure (STE)[17] assays are both monolayer cell-based systems that utilize dilutions of test chemicals for ocular irritation predictions. However, the FL assay measures increases in fluorescein passage as a chemical concentration-dependent loss of tight junction barrier function, considered an initiating event in corneal penetration, while the STE assay measures the chemical concentration-dependent loss of cell viability, typically associated with corneal injuries. Thus, each assay addresses a specific event in the action of ocular irritants.

The *in vitro* test systems used for eye and skin irritation predictions differ in many aspects from the *in vivo* Draize rabbit test. First, *in vitro* tests tend to be reductionist by design, by simplifying the test system, controlling the exposures, and focusing on a few key endpoints, all with the goal of answering very specific toxicological queries. Because of this simplification, they tend to be more consistent and provide more reproducible results relative to animal systems. For example, the *in vitro* toxicologist has greater control and precision over the exposure and dosing activities – dose volumes are precisely applied and retained on the test system for exactly the prescribed exposure times and rinsed using defined methods. In contrast, in the Draize tests the application and retention of chemical doses may be affected by animal responses such as blinking, grooming, and pawing at the exposure sites, all of which can affect the outcome of the test predictions. Lastly, the majority of non-animal test methods for eye and skin irritation are evaluated using objective machine-read results, while in contrast, the Draize tests are scored based on subjective observations by a technician.

Whether developed for eye or for skin irritation studies, alternative or non-animal test methods generally fall into four categories: simple *in vitro* monolayer cell systems, *ex vivo* tissue or organ systems, complex *in vitro* re-constructed cell-based tissue models, and a variety of artificial non-cell-based eye or skin models. The following sections will provide examples of each of these major categories and will present some classic applications and limitations of these models.

7.4.1 *In Vitro* Monolayer Cell Systems

Non-animal methods designed to predict eye and skin irritation events were developed with several common concepts. Among those concepts was the recognition that eye and skin irritation typically resulted from cell damage or cell death. Consequently, many of the early *in vitro* cell-based methods measured loss of cell viability as the primary endpoint. Some methods were based upon simple monoculture cell systems where test chemicals were dosed as aqueous dilutions onto the cells to determine the concentration inducing a defined loss of cell viability. Based upon the concept, materials that were known to be strong eye or skin irritants generally induced cytotoxic responses at lower doses than materials of lesser irritation potential, and concentration-based prediction models were developed to rank order and/or identify eye and skin irritants. For examples, the Neutral Red Uptake assay (NRU) and the Neutral Red Release assay (NRR), both using human primary keratinocytes, were developed and evaluated for their abilities to predict dermal and ocular irritation, respectively.[18,19]

These types of assays provided reasonable predictions for many aqueous soluble materials, but complex lipid-rich formulations were frequently problematic. Lipophilic or aqueous-insoluble ingredients prove difficult to dissolve and dilute in the aqueous media used for preparing dosing dilutions, and typically form immiscible two-phase dilutions or non-homogeneous suspensions. Consequently, the resulting dosing dilutions may not adequately present hydrophobic ingredients to the test system. In addition, monoculture dilution-based test systems do not model the barrier functions of the eye and skin, which selectively control the penetration and diffusion of chemicals in a mixture or formulation to the underlying cells. For example, the skin's stratum corneum provides an effective barrier to the penetration of hydrophilic chemicals (i.e., those with low log P values), while lipophilic chemicals with high log P values may diffuse through the stratum corneum more readily.[20,21] Thus the skin barrier may impact the composition of ingredients actually penetrating to the underlying epidermal cells. In the absence of these barriers, such as is typical for cell monolayer test systems, the modeling of exposure and availability of ingredient chemicals simply differs considerably from what is experienced *in vivo*.

Although monoculture assay systems do not adequately model the eye or skin barrier functions, these test systems do provide useful hazard assessments. Assays such as the 3T3 Neutral Red Uptake Phototoxicity Test[22] and the KeratinoSens sensitization screening assay[23] provide mechanistically-relevant platforms for predicting the potential for chemicals to have phototoxic or electrophilic activity, respectively. The ability to determine these types of hazards early in a product development setting helps product formulators to identify and sequester offending chemicals from inclusion in personal care and cosmetic products, thus precluding the need to evaluate complex formulations downstream.

7.4.2 *Ex Vivo* Tissue and Organ Models

A variety of *ex vivo* human and non-human corneal and epidermal tissues have been investigated for their applicability to predicting eye and skin irritation potential. The use of *ex vivo* tissues for research is especially appealing since they closely resemble the tissues of interest, they allow test formulations to be applied to the tissues just as they are exposed *in vivo*, they are typically inexpensive, and in general animal-sourced tissues are readily available. When harvested and prepared appropriately, the viability of *ex vivo* tissues may be maintained for the duration of most acute exposure scenarios to allow for relevant cell-based responses to chemical insults. Accordingly, the endpoints determined *in vitro* can readily reflect most of the cellular expressions of toxicity observed *in vivo*.

Ex vivo tissues closely model intact tissues *in vivo*, and include much of their native complexity. For example, excised human and mammalian skin is often used in *in vitro* percutaneous penetration studies since the innate barrier properties affecting chemical absorption and diffusion are maintained in the *in vitro* setting.[24] Excised skin includes the stratum corneum (providing the primary barrier function) and epidermal cell layers, as well as all of the accessory features of skin including hair follicles, and sweat and sebaceous glands, all of which impact the transcutaneous passage of chemicals. In contrast, no current commercially available cell models or tissue reconstructions exactly reproduce the skin features of native skin, and thus do not model all of the penetration and absorption pathways occurring *in vivo*. Similarly, corneas from various mammalian species share several common features, and in cross section are typically comprised of the anterior corneal epithelium, which includes the outer squamous epithelial layer, and progressing posteriorly, include the wing and basal epithelial cell layers. Progressing deeper into the eye, the corneal epithelial layers are typically attached to a basement membrane covering the corneal stroma (and in humans, covers Bowman's layer), the substantia propria or stroma, Descemet's membrane, and a single cell layer corneal endothelium.

A variety of animal-sourced whole globe eye and excised corneal tissues have been considered for eye irritation studies. *Ex vivo* eye irritation test methods using excised rabbit eyes (Rabbit Enucleated Eye Test, or REET) were originally developed since the excised rabbit eye represented a test system which most closely modeled the *in vivo* rabbit Draize test.[25,26] Prinsen reported on the evaluation of excised eyes from slaughterhouse animals (bovine, porcine, and chicken), and had found that all of these animal-sourced eyes provide reasonable mechanistic models for *ex vivo* eye irritation studies.[27]

In some cases, *ex vivo* tissues not of ocular or dermal origin are selected because they resemble the eye or skin tissues being modeled. For example, the chorioallantoic membrane of the fertilized hens egg has been used as a model for conjunctivae owing to the extensive vascular network immediately below the epithelial membrane. Several protocols have been used and evaluated for eye irritation testing including the Chorioallantoic Membrane Vascular Assay (CAMVA), used primarily by the personal care and cosmetics industry for modeling conjunctival effects.[28]

Ex vivo tissues allow one to apply test chemicals and formulations onto the tissue surface undiluted, generally with sufficient control over dose volumes to mimic a variety of exposure scenarios. Accordingly, the methods for dosing neat chemicals, chemical mixtures, or complex formulations in *ex vivo* corneal and epidermal tissues accurately model the exposures expected *in vivo*. For example, in the Isolated Chicken Eye test (ICE), liquid test chemicals and formulations are applied at a dose volume of 30 µL for a single 10-second exposure.[29] The dose volume was modified to reflect the dose volume to surface area ratio of the Draize test that is accepted for regulatory use, Draize test, and thus accurately models the exposure of the *in vivo* assay. These *ex vivo* tissues also allow one to adapt exposures to accurately model *in vivo* exposure scenarios. For example, a variety of rinse-off and leave-on exposures can readily be represented in *ex vivo* tissues simply by controlling the exposure times of the test formulations. Since these test systems typically are dosed topically, and do not require continuous immersion in support medium, there is no need to dilute test formulations.

Lastly, excised eyes from slaughterhouse animals are readily available to re-searchers globally, and are generally inexpensive to acquire. In the US, major metropolitan areas are served locally by abattoirs thus providing resources for the researcher. In some regions, certain species may not be readily available; however, it is possible to ship whole globe eyes overnight provided the eyes are packaged in cool physiological saline solutions (e.g., Hanks' Balanced Salt Solution containing antibiotics) and shipped appropriately. Companies such as Pel Freez Biologicals (Rogers, AR) and Sioux-Preme Packing Company (Sioux City, IA) specialize in providing research-grade whole globe eyes and corneas from abattoir-derived animals, thus eliminating the need to establish a local supplier.

Excised human tissues would be expected to provide the most relevant test systems to support studies for human safety predictions, and would be more relevant than any non-human *ex vivo* tissues. However, obtaining a reliable source for high-quality human tissues to support the various needs of the re-search community at large can be challenging at best. It is currently still pos-sible to obtain sufficient good quality skin from tissue donors (generally from elective surgery, or deceased donors) to support certain types of *in vitro/ex vivo* safety studies, including determining percutaneous penetration of chemicals; however, there is a growing competitive need for the highest quality skin for use for clinical transplant purposes. In contrast, it is not feasible to obtain high quality human eyes or corneas to support any meaningful toxicology screening programs, due to the dire demand for healthy corneas for clinical transplant.

Therefore, a variety of *ex vivo* tissues, and in particular eyes, from animals have been investigated to support product safety and toxicology programs.

Despite the apparent benefits of using *ex vivo* tissues, it is imperative that one understands the differences between the *ex vivo* model and the native *in vivo* organ being modeled, the limitations of these tissues, as well as the protocols employed, and ensure that one doesn't over interpret the results. Perhaps the most important facet is to recognize that excised tissues are isolated from the whole body, and thus no longer model the effects and support provided in the whole organism. The recruitment of immune cells is no longer possible in the isolated tissues, and thus any collateral damage due to inflammatory cell responses at the site of chemical insult will not be modeled. Furthermore, no other whole body response mechanisms are available, that could induce changes locally at the site of exposure, or distally in other organ systems after exposure and absorption. In the absence of an intact nervous system or lacrimation response mechanism in excised eyes or corneas, there are no innate blinking or tear responses. Given that excised tissues impart confidence in modeling exposures, owing to the resemblance of these tissues to the organism *in vivo*, it may be easy to forget that these aforementioned responses may not be expressed *in vitro*. One should also recognize that the specific protocols used and the endpoints evaluated directly influence the predictive capacity of the assays. Although both the REET and BCOP assays utilize *ex vivo* eyes for eye irritancy predictions, the REET assay with swelling and histological endpoints discriminated among high and low irritant shampoo formulations, and accurately predicted the *in vivo* data, while the BCOP assay was found to be less sensitive and did not readily discriminate among the various shampoo formulations except for when histological evaluation was added to the standard corneal opacity and permeability endpoints.[30] It is not clear why the BCOP assay was less sensitive than the REET assay for predicting milder shampoo formulations. However, this reinforces the concept that users of any test system must understand the nature of the test systems and the applications and limitations of the protocols used. Differences in corneal thickness, dose used, exposure method applied, and endpoints measured play key roles in determining how predictive these tools will be.

7.4.3 Three-dimensional Reconstructed Human Cell-based Tissue Models

Perhaps the best opportunities for predicting human eye and skin irritation events will come from the development and refinement of reconstructed human cell-based corneal and epidermal tissues. Based upon technologies developed for regenerative tissues for clinical applications, several commercially available models for research purposes have been validated (or are currently undergoing validation) for classification and labeling of chemicals for skin and eye corrosion and irritation potential. These tissues are generally "reconstructed" using primary or undifferentiated human cells isolated from the target tissues *in vivo*, and cultured *in vitro* using specialized media and culture techniques designed to encourage "normal" growth and differentiation, to replicate the native tissues.

The skin models typically are comprised of a three-dimensional (3D) cross section, histologically similar to human epidermis, complete with proliferative basal epidermal cells, intermediate spinous and granular cells, and most importantly a functional stratum corneum. The corneal models typically present proliferative basal epithelial cells, intermediate wing-like cells, and a squamous epithelial barrier, architecturally similar to the human corneal epithelium. They offer several distinct advantages over cells cultured in monolayers. For example, these models are more highly integrated as part of the tissue complex, they take on more of the functionality of cells in the native tissues, and have a greater degree of cell-to-cell interactions than monolayer culture models. Additionally, since the reconstructed human skin and eye models are cultured at the air–liquid interface, these models allow for topical dosing of chemicals and complex formulations in undiluted form, thus presenting these materials to the corneal and epidermal barriers in the same manner as what occurs *in vivo*. The dosing of these tissue reconstructions can be conducted similarly to the exposure protocols used for the *ex vivo* tissues presented above. Accordingly, the exposures of these *in vitro* 3D reconstructed tissues accurately model the exposure events experienced *in vivo*.

To date, reconstructed human tissues have been most successfully validated for regulatory classification and labeling of skin irritation and corrosion. There are several commercially available reconstructed human epidermis (RhE) models, complete with test method protocols, validated for skin corrosion and skin irritation according to the Globally Harmonized System (GHS) classes 1 and 2, and these methods are described in Organisation of Economic Co-operation and Development (OECD) Test Guidelines 431[31] and 439,[32] respectively. However, recent test method development and validation activities based upon reconstructed human corneal epithelial tissues for predicting mild to moderate eye irritants show great promise.[33] As these technologies mature, perhaps these 3D tissue reconstructions will supplant the use of *ex vivo* tissues for research and regulatory toxicology.

Since cell damage and cell death typically occur in both eye and skin irritation and corrosion events *in vivo*, the most common endpoint evaluated in 3D tissue reconstructions is the determination of changes in cell viability after chemical exposure to these tissues. The *de facto* standard assay for measuring cell viability in these 3D tissues is the MTT (3-[4,5-dimethylthiazol-2-yl]-2, 5-diphenyltetrazolium bromide) conversion or reduction assay, which measures the NAD(P)H-dependent microsomal enzyme reduction of MTT (and to a lesser extent, the succinate dehydrogenase reduction of MTT) to a blue formazan precipitate in the tissues.[34] All of the current skin irritation and corrosion test methods for regulatory classification and labeling purposes use the MTT viability assay as the basis for the predictions. In essence, tissues are treated with test chemical for fixed exposure times, rinsed of treatment, and are evaluated for reductions in viability using the MTT viability assay. In concept the exposure times used in the skin corrosion assays are modeled after the Draize test, where tissue viability is assessed after 3- and 60-minute exposure times. Those materials expected to be highly corrosive would reduce tissue viability rapidly within the 3- and 60-minute exposure times. Materials that are not corrosive would be tolerated for longer exposure times, and would not reduce viability significantly at these short exposure times. Protocols used to identify milder materials evaluate tissue viability after notably longer chemical exposure times, or long post-treatment expression times, before conducting the MTT viability endpoint.

In addition to the MTT viability endpoint, other cell-based endpoints have been used to evaluate ocular and dermal irritation potential in these reconstructed tissues. Measuring the release of cytokines, such as IL-1α, TNF-α, IL-8, and others, from treated tissues provides useful information on the potential for inflammatory responses after chemical exposures. These endpoints may be useful to supplement the predictions using the MTT viability endpoint, especially when the MTT endpoint does not predict overt cell damage. The skin irritation test protocol designed to predict GHS 2 skin irritants originally proposed by Cotovio[35] included the measurement of IL-1α released into the medium after test chemical treatments as a second tier test to the MTT endpoint. However, none of the currently accepted skin corrosion and skin irritation assays require these supplemental endpoints for classification and labeling.

Although all RhE models have a stratum corneum, the barrier properties vary from model to model, and at least among the commercial models do not yet exhibit the full barrier properties found in native healthy skin.[36–39] Consequently, variations in the barrier function between different RhE models might become particularly relevant when testing chemicals with different physical properties. As is the case for developing test protocols using various *ex vivo* tissues, specific test methods are optimized to the specific tissues to account for the differences in the barrier properties of the RhE models. For RhE models with robust barrier properties, exposure times are characteristically longer than the exposure times optimized for the more permeable RhE models, to assure the appropriate balance of assay sensitivity and selectivity. Additionally, dose volumes should reasonably approximate skin exposures expected *in vivo*, but may be adjusted to account for differences in tissue barrier and downstream endpoint responses.

7.4.4 Artificial Non-viable Eye or Skin Models

A fourth group of *in vitro* test systems for predicting eye and skin irritation are a collection of assays that do not rely upon viable cells or tissues. These test systems are generally based upon modeling the basic physicochemical properties of corneal and epidermal tissues, and predict eye and skin irritation events based upon interactions between test chemicals and the test systems. InVitro International (Irvine, CA) developed and markets several commercially available test kits for eye and skin irritation prediction, as well as for corrosive subcategorization for various classification and labeling purposes.

The Irritection assays are designed for predicting the skin or eye irritation potential of chemicals after treatment of the test systems, which are comprised of organized matrices of proprietary proteins, glycoproteins, lipids, and low molecular weight components designed to mimic the biochemical components of corneal and epidermal tissues.[40] The endpoints of these assays are based upon chemical-induced changes in the optical properties of the biochemical matrices, thus modeling the biochemical changes observed in cells and tissues *in vivo*.

The Corrositex assay, validated and approved for predicting and sub-categorizing corrosive materials, evaluates the time for chemicals to break through an artificial biobarrier designed to mimic epidermal tissues.[41] There are a number of practical advantages of these assay systems which make them attractive for widespread use. The kits are relatively inexpensive to purchase, they do not require specialized equipment (other than an incubator and 96-well plate spectrophotometer) or specialized expertise and training, and can be readily stored for immediate on-demand applications.

7.5 Current Assays for Acute Eye and Skin Irritation Classification and Labeling

The following section introduces currently available *in vitro* and *ex vivo* test methods for regulatory classification and labeling of eye and skin irritation hazards.

Tables 7.1 and 7.2 provide summaries of the nature of the test systems, applicability to testing and dosing chemical ingredients or complex formulations, and brief comments on the assay endpoints measured. The currently accepted non-animal test methods for regulatory classification and labeling for eye irritation potential and skin irritation and corrosion are presented in Tables 7.3 and 7.4, respectively. These tables indicate the validation status of each method, and the regulatory framework for acceptance of data. Many of the validated methods have been adopted as OECD Test Guidelines (TG) over the last four years, while several others are being proposed. The recent ardent activities to adopt these non-animal methods coincide with several legislative drivers, including:

- the European Union's (EU) Registration, Evaluation, Authorisation and Restriction of Chemicals (REACh) European Commission (EC) No 1907/2006 entered in force in June 2007 to improve the protection of human health and the environment from the risks that can be posed by chemicals;[42]
- the Regulation (EC) No 1272/2008 on classification, labeling and packaging (CLP) of substances and mixtures entered in force in January 2009 (European Chemicals Agency);[43]
- the EU Cosmetics Directive 2003/15/EC (a.k.a. the 7th Amendment to the Cosmetics Directive) which stipulated a prohibition to test finished cosmetic products and cosmetic ingredients on animals (testing ban), and a phased prohibition to market in the European Community, finished cosmetic products and ingredients included in cosmetic products which were tested on animals (marketing ban).[44]

Table 7.1 Characteristics and applications of *in vitro* ocular irritation assays.

Method	Test system description	Dosing, exposure, and endpoints	Applicability and limitations
In Vitro Monolayer cell systems			
Fluorescein Leakage Assay (FL)	MDCK cells, *in vitro*, monoculture on permeable substrate tissue culture insert Model of squamous epithelium/tight junctions Machine-read spectrophotometric determination of short exposure-time dose-related changes in passage of fluorescein through model	Aqueous dilution series of substances are prepared and dosed in medium Doses are applied for 1 minute, followed by a wash-off, and immediate determination of fluorescein passage Endpoint: the dose which induces a 20% increase in fluorescein passage (FL_{20}), relative to negative controls	Limited to aqueous-soluble substances, and single-phase homogeneous suspensions (e.g., surfactants) Limitations exclude strong acids and bases, cell and protein fixatives, and precipitating solids that may induce coagulation or other mechanisms that retard fluorescein passage Used in non-regulatory eye irritation safety assessments to determine the potential for initial disruption of tight junctions in squamous epithelium, typically for evaluating mild products Used in regulatory testing to identify severe irritants of aqueous-soluble substances
Cytosensor Microphysiometer (CM)	L929 cells, *in vitro*, monoculture on permeable substrate tissue culture insert Cell-based model to evaluate short exposure-time dose-related changes in cellular metabolism and viability Machine-read measurement of changes in rates of cellular respiration, by measuring the acidification rate of the medium surrounding the cell population	Aqueous dilution series of substances in low-buffer medium Doses are applied for 13.5 minutes, followed by a 6-minute wash-off, and immediate cellular respiration determination (acidification rate determination) Endpoint: the dose which induces a 50% decrease in metabolic rate (MRD_{50})	Limited to aqueous-soluble substances, and single-phase homogeneous suspensions (e.g., surfactants) Generally not compatible with complex formulations with lipophilic ingredients Used in non-regulatory eye irritation safety assessments to predict a full range of responses Used in regulatory testing to identify non-irritant surfactant-containing materials, and to identify severe irritants of aqueous-soluble substances Accepted by EPA OPP Antimicrobial Cleaning Products Program for EPA I and III/IV labeling

Table 7.1 (*Continued*)

Method	Test system description	Dosing, exposure, and endpoints	Applicability and limitations
Short Time Exposure Assay in SIRC cells (STE)	SIRC cells, *in vitro*, monoculture in 96-well plate format Cell-based model to evaluate short exposure-time dose-related changes in cellular viability Machine-read spectrophotometric determination of changes in vital dye (MTT) reduction	Aqueous dilution series of substances are prepared and dosed in medium. Aqueous-insoluble substances can be tested using a mineral oil vehicle Doses are applied for 5 minutes, followed by a wash-off, and immediate determination of cell viability (by MTT viability assay) Endpoint: relative viabilities at specified doses are used for irritation classifications	Limitations exclude chemicals that may readily become non-irritant with minimal dilution in aqueous vehicles (e.g., small chain solvents like ethanol, isopropanol, and acetone, and weak acids or bases) Used in non-regulatory eye irritation safety assessments to predict a full range of responses Proposed for use in regulatory testing to identify GHS categories 1, 2, and non-classified (NC) substances

Ex Vivo Tissue or Organ Systems

Method	Test system description	Dosing, exposure, and endpoints	Applicability and limitations
Bovine Corneal Opacity and Permeability Assay (BCOP)	Bovine, *ex vivo*, from abattoir activities Full thickness viable cornea with epithelial, stromal and endothelial tissues Machine-read measurement of changes in corneal opacity and barrier function (by fluorescein passage) Optional histopathology	Neat (undiluted) application of liquids at full formulation strength for 10 minutes, followed by a rinse-off and a 2-hour expression incubation Surfactant liquids and solid substances and concentrated surfactant formulations may be diluted to 10% in aqueous and tested for 10 minutes, followed by a rinse-off and a 2-hour expression incubation Solids (non-surfactants) applied as 20% aqueous dilutions for 4 hours, followed by a rinse-off Endpoints: corneal opacity, fluorescein permeability	Small chain alcohols and ketones (and likely other organic solvents) over-predicted, possibly due to impact of the infinite dose volume used in the assay Some solid materials have been identified to be under-predicted, possibly due to preparation of aqueous suspensions of hydrophobic materials Chemistries that induce delayed effects (reactive chemistries, oxidizers, alkylation, etc.) require use of alternate extended post-treatment protocols Used in non-regulatory eye irritation safety assessments to predict a range of responses from mild to severe irritants Used in regulatory testing to identify severe irritants Accepted by EPA OPP Antimicrobial Cleaning Products Program for EPA I

Isolated Chicken Eye Assay (ICE)	Chicken, *ex vivo*, from abattoir activities full thickness viable cornea with epithelial, stromal and endothelial tissues Operator-evaluated measurement of changes in corneal opacity, corneal swelling, and epithelial barrier erosion (by fluorescein retention) Optional histopathology	Neat (undiluted) application of liquids and solids at full formulation strength for 10 seconds, and then rinsed Endpoints are evaluated periodically between 30 and 240 minutes post-treatment Endpoints: corneal opacity, corneal swelling, and fluorescein retention	Small chain alcohols and ketones (and likely other organic solvents) over-predicted Some solid materials have been identified to be under-predicted Chemistries that induce delayed effects (reactive chemistries, oxidizers, etc.) may be under-predicted Used in non-regulatory eye irritation safety assessments to predict a full range of responses Used in regulatory testing to identify severe irritant substances
Hen's Egg Test on the Chorioallantoic Membrane (HET-CAM)	Chicken egg, <1/2 gestation, *in vivo*, commercially-available eggs for research Vascularized chorioallantoic membrane (CAM) to model conjunctivae and ocular epithelium Operator-evaluated determination of changes in vascular integrity and vessel function	Neat (undiluted) application of liquids and solids at full formulation strength Substances are applied to CAM for up to 5 minutes. CAM observed at 0.5, 2, and 5 minutes for appearance of vascular hemorrhage, vascular lysis, or coagulation of blood flow Endpoints: The time to appearance of vascular effects is used for irritation classification	Used in non-regulatory eye irritation safety assessments to predict a full range of responses including assessment of potential conjunctival effects Proposed for use in regulatory testing to identify GHS category 1, EPA category I severe eye irritant substances
Isolated Rabbit Eye Assay (IRE)	Rabbit, *ex vivo*, from abattoir activities Full thickness viable cornea with epithelial, stromal and endothelial tissues Operator-read measurement of changes in corneal opacity and epithelial barrier Optional histopathology	Neat (undiluted) application of liquids and solids at full formulation strength for 10 seconds, and then rinsed. Longer exposures to 60 seconds may be used to discriminate among milder substances Endpoints are evaluated periodically between 30 and 240 minutes post-treatment Endpoints: corneal opacity, corneal swelling, and fluorescein retention	Small chain alcohols, esters and ketones (and likely other organic solvents) over-predicted Chemistries that induce delayed effects (reactive chemistries, oxidizers, etc.) may be under-predicted Used in non-regulatory eye irritation safety assessments to predict a full range of responses Proposed for use in regulatory testing to identify GHS category 1, EPA category I severe eye irritant substances

Table 7.1 (Continued)

Method	Test system description	Dosing, exposure, and endpoints	Applicability and limitations
In Vitro Three-dimensional Reconstructed Tissue Systems			
Reconstructed Human Corneal Epithelium (HCE) (MatTek Corp. EpiOcular and SkinEthic HCE)	human, *in vitro*, commercially-available, reconstructed from primary, or extended life-span human epithelial cells Cultured to form 3D corneal epithelium, complete with proliferative basal cells, and squamous epithelial barrier Machine-read spectrophotometric determination of changes in vital dye (MTT) reduction	Neat (undiluted) application of liquids and solids at full formulation strength Exposure Protocols: 1) single fixed exposure time for specific threshold responses (i.e., to resolve between GHS 2 or non-classified) 2) multiple exposure times to determine the exposure time expected to reduce viability by 50% (ET_{50}). Evaluates a continuum of potential irritation responses Endpoint: relative viabilities at specified exposure times are used for irritation classifications	Small chain alcohols, esters and ketones (and likely other organic solvents) over-predicted, in protocols applying infinite dose volumes Chemicals that directly reduce the MTT in the absence of viable cells may be under-predicted if appropriate non-viable tissue controls are not included Currently two major commercially-available HCE models are undergoing validation for use in regulatory testing to discriminate among GHS 1/2 irritants and non-classified (NC) substances ET_{50} protocol using MatTek EpiOcular tissues accepted by EPA OPP Antimicrobial Cleaning Products Program for EPA I and III/IV labeling
In Vitro Non-cellular Eye Models			
Ocular Irritection Assay (InVitro International)	A substrate solution of proteins, glycoproteins, lipids and low molecular weight components that self-associate to form an optically clear matrix, modeling the optically clear cornea Machine-read measurement of changes in optical turbidity	Non-surfactant liquids, powders, waxes and gels are applied to the test system without dilution Surfactants are diluted in water to 20 mg/mL Substances are added to test system for 24 hours (25°C) prior to measurement of optical turbidity	Used in non-regulatory eye irritation safety assessments to identify eye irritants Currently undergoing validation for use in regulatory testing to discriminate among GHS 1/2 irritants and non-classified (NC) substances

US EPA OPP Antimicrobial Cleaning Products - Non-animal testing approach to EPA labeling for eye irritation of Antimicrobial Products with Certain Cleaning Claims.

Table 7.2 Characteristics and applications of *in vitro* skin irritation assays.

Method	Test system description	Dosing, exposure, and endpoints	Applicability and limitations
In Vitro Monolayer Cell Systems			
Neutral Red Uptake in Normal Human Epidermal Keratinocytes (NHEK NRU Assay)	NHEK cells, *in vitro*, monoculture in 96-well plate format Cell-based model to evaluate substances for dermal irritation potential, based upon cytotoxic effects Machine-read spectrophotometric determination of uptake and retention of vital dye (neutral red)	Aqueous dilution series of substances are prepared and dosed in assay medium Doses are applied for 48 hours at 37°C, followed by a viability assessment (by NRU) Endpoint: Dose dependent cytotoxicity response curves are used to calculate the concentration inhibiting viability by 50% (IC_{50}).	Limited to aqueous-soluble substances, and substances that can be solubilized in organic solvents and retain homogeneity upon transfer into the aqueous vehicle Generally not compatible with complex formulations with lipophilic ingredients Assay does not model skin barrier function Used in non-regulatory assessment to compare relative cytotoxic effects to predict relative dermal irritation potential
3T3 Neutral Red Uptake Acute Phototoxicity Test (3T3 NRU PT)	Balb/c 3T3 cells, *in vitro*, monoculture in 96-well plate format Cell-based model to evaluate substances for potential phototoxic effects after UVA exposure Machine-read spectrophotometric determination of uptake and retention of vital dye (neutral red)	Aqueous dilution series of substances are prepared and dosed in buffered saline Doses are applied for 60 minutes at 37°C, followed by a 50-minute exposure in the presence or absence of UVA exposure. Doses are rinsed off and incubated for 24 hours prior to a viability assessment (by NRU) Endpoint: A comparison of the dose responses in the presence and absence of UVA exposure is used for the phototoxic hazard prediction	Limited to aqueous-soluble substances, and substances that can be solubilized in organic solvents and retain homogeneity upon transfer into the aqueous vehicle Generally not compatible with complex formulations with lipophilic ingredients Assay does not model skin barrier function, and thus assumes substances predicted to be phototoxic have dermal relevance Used in non-regulatory assessment of phototoxic hazard potential Used in regulatory arena to identify substances with phototoxic hazard potential
Ex Vivo Tissue or Organ Systems			
Rat Skin Trans Epithelial Resistance Corrosivity Test (TER)	Freshly excised rat dorsal skin buttons (0.8 cm^2), with minimum trans epithelial resistance (TER) of 10 kΩ	Neat (undiluted) application of liquids and solids are applied at full formulation strength for 2 and 24 hours	Residues of substances on the skin may affect the TER, potentially resulting in under-prediction Used for non-regulatory evaluation of corrosive potential

Table 7.2 (Continued)

Method	Test system description	Dosing, exposure, and endpoints	Applicability and limitations
	A reduction in trans epithelial resistance (TER) in treated tissues is measured	Endpoint: TER readings of $<5\,k\Omega$ are predictive of corrosive potential	Used for regulatory classification and labeling of corrosive substances

In Vitro Three-dimensional Reconstructed Tissue Systems

Method	Test system description	Dosing, exposure, and endpoints	Applicability and limitations
In Vitro Skin Corrosion Test: Reconstructed Human Epidermis (RhE) (MatTek Corp. EpiDerm, SkinEthic EPISKIN, SkinEthic RHE, CellSystems epiCS)	Human, *in vitro*, commercially-available, reconstructed from primary, or extended life-span human epidermal cells Cultured to form 3D epidermis, complete with proliferative basal cells, spinous and granular cells and functional stratum corneum Machine-read spectrophotometric determination of changes in vital dye (MTT) reduction	Neat (undiluted) application of liquids and solids at full strength Fixed exposure times for corrosive prediction Endpoint: relative viabilities at specified exposure times are used for corrosion predictions	Chemicals that directly reduce the MTT in the absence of viable cells may be under-predicted if appropriate non-viable tissue controls are not included Used for non-regulatory evaluation of corrosive potential Used for regulatory classification and labeling of corrosive substances SkinEthic EPISKIN validated for identification of GHS corrosive substances and sub classification of GHS 1A *vs.* 1B/1C
In Vitro Skin Irritation Test: Reconstructed Human Epidermis (RhE) (MatTek Corp. EpiDerm, SkinEthic EPISKIN, SkinEthic RHE)	Human, *in vitro*, commercially-available, reconstructed from primary, or extended life-span human epidermal cells Cultured to form 3D epidermis, complete with proliferative basal cells, spinous and granular cells and functional stratum corneum Machine-read spectrophotometric determination of changes in vital dye (MTT) reduction	Neat (undiluted) application of liquids and solids at full strength Exposure Protocols: 1) single fixed exposure time to resolve between GHS 2 or non-classified) 2) multiple exposure times to determine the exposure time expected to reduce viability by 50% (ET_{50}). Evaluates a continuum of potential skin irritation responses Endpoint: relative viabilities at specified exposure times are used for irritation classifications	Chemicals that directly reduce the MTT in the absence of viable cells may be under-predicted if appropriate non-viable tissue controls are not included ET_{50} protocol used for non-regulatory evaluation of a wide range of skin irritation responses from non-irritant to corrosives Used for regulatory classification and labeling of GHS 2 *vs.* non-classified substances

In Vitro Non-cellular Skin Models

Model	Description	Protocol	Applicability
Dermal Irritection Assay (*In Vitro* International)	A membrane substrate containing covalently cross-linked mixture of keratin and collagen to model the stratum corneum, and a matrix solution of highly organized globulin proteins. Machine-read measurement of optical density	Non-surfactant liquids, powders, waxes and gels are applied to the test system without dilution. Substances are added to test system for 24 hours (25°C) prior to measurement of optical density	Used in non-regulatory skin irritation safety assessments to identify skin irritants and non-irritants
In Vitro Membrane Barrier Test for Corrosion: CORROSITEX Continuous Time Monitor Assay (*In Vitro* International)	A biobarrier membrane substrate composed of a hydrated collagen matrix on a supporting membrane to model the epidermis, and an aqueous chemical detection system (CDS) with pH indicator dyes. Operator-evaluated color changes indicate endpoint has been achieved	500 µL liquid or 500 mg solid substances are applied onto the biobarrier matrix and monitored for up to 4 hours to determine the time it takes for a substance to penetrate the biobarrier and produce a color change in the CDS. Endpoint: The time that the operator initially observes a color change in the CDS is indicative of the time of chemical penetration, and is used to classify corrosivity	Generally applicable for testing acids, acid derivatives, acyl halides, alkylamines and polyalkylamines, bases, chlorosilanes, metal halides, and oxyhalides (US DOT-SP 10904). Substances that do not cause a color change or bleach the CDS are limited from use. Used for non-regulatory evaluation of corrosive potential. Used for regulatory classification and labeling of corrosive subcategories according to the US DOT and UN Transport of Dangerous Goods regulations

US Department of Transportation Special Permit (DOT-SP) 10904 Special Permit Authorization granted to InVitro International's Application for Revision and Renewal of DOT-E 10904, September 13, 1994 and Revision letter to the Application (May 13, 1994), March 6, 1995; DOT-SP 10904, February 17, 2010 [expires January 31, 2014].

Table 7.3 Regulatory acceptance status of in vitro ocular irritation assays.

Method	USA	European Union (EU)	OECD[a]
In Vitro Monolayer Cell Systems			
Fluorescein Leakage Assay (FL)	Accepted via TG 460 for identifying severe or corrosive eye irritants in a top-down approach	Accepted via TG 460 for identifying severe or corrosive eye irritants in a Top-Down approach	OECD approved, TG 460 for identifying severe or corrosive eye irritants in a Top-Down approach
Cytosensor Microphysiometer (CM)	Accepted by EPA OPP Antimicrobial Cleaning Products Program for EPA III/IV labeling	ECVAM-approved to identify severe irritants of aqueous-soluble substances in a Top-Down approach ECVAM-approved for identifying non-irritant surfactant-containing materials in a Bottom-Up approach	Used for supporting information only Draft OECD Test Guideline, submitted in 2010 to identify severe irritants of aqueous-soluble substances in a Top-Down approach and for identifying non-irritant surfactant-containing materials in a Bottom-Up approach
Short Time Exposure Assay in SIRC cells (STE)	No current acceptance	Used for supporting information only	Standard Project Submission Form (SPSF) submitted 2012 for Draft Test Guideline
Ex Vivo Tissue or Organ Systems			
Bovine Corneal Opacity and Permeability Assay (BCOP)	ICCVAM approved for identifying severe or corrosive eye irritants Accepted by EPA OPP Antimicrobial Cleaning Products Program for EPA I/II labeling	ECVAM-approved for Globally Harmonized System Labeling for severe or corrosive categories; EU Test Method B.47	OECD-approved, TG 437 Revised TG 437 for identifying GHS non-classified eye irritants and GHS 1 eye irritants, pending September 2013 publication
Isolated Chicken Eye Assay (ICE)	ICCVAM-approved for identifying severe or corrosive eye irritants	ECVAM-approved for Globally Harmonized System Labeling for severe or corrosive categories; EU Test Method B.48	OECD-approved, TG 438 Revised TG 438 for identifying GHS non-classified eye irritants and GHS 1 eye irritants, pending September 2013 publication

Method			
Hen's Egg Test on the Chorioallantoic Membrane (HET-CAM)	Used for supporting information only		Used for supporting information only
Isolated Rabbit Eye Assay (IRE)	Used for supporting information only		Used for supporting information only
In Vitro Three-dimensional Reconstructed Tissue Systems			
Reconstructed Human Corneal Epithelium (HCE) (MatTek Corp. EpiOcular and SkinEthic HCE)	MatTek EpiOcular Model accepted by EPA OPP Antimicrobial Cleaning Products Program for EPA III/IV labeling	Used for supporting information only / Currently undergoing ECVAM / Cosmetics Europe validation for use in regulatory testing to discriminate among GHS 1/2 irritants and non-classified (NC) substances	Used for supporting information only
In Vitro Non-cellular Eye Models			
Ocular Irritection Assay	Used for supporting information only	Used for supporting information only / Currently undergoing ECVAM validation for use in regulatory testing to discriminate among GHS 1/2 irritants and non-classified (NC) substances	Used for supporting information only

[a]OECD – Organisation for Economic Co-operation and Development.
OECD TG 437 – (2009a). OECD GUIDELINE FOR THE TESTING OF CHEMICALS (OECD 437). Bovine Corneal Opacity and Permeability Test Method for Identifying Ocular Corrosives and Severe Irritants.
OECD TG 438 – (2009b). OECD GUIDELINE FOR THE TESTING OF CHEMICALS (OECD 438). Isolated Chicken Eye Test Method for Identifying Ocular Corrosives and Severe Irritants.

Table 7.4 Regulatory acceptance status of in vitro skin irritation assays.

Method	USA	European Union (EU)	OECD[a]
In Vitro Monolayer Cell Systems			
3T3 Neutral Red Uptake Acute Phototoxicity Test (3T3 NRU PT)	Accepted via TG 432 (for hazard identification relative to skin photo-irritation)	EU Test Method B.41	OECD-approved, TG 432
Ex Vivo Tissue or Organ Systems			
Rat Skin Trans Epithelial Resistance Corrosivity Test (TER)	Accepted via TG 430	Approved for Globally Harmonized System Labeling as Test Method B.40	OECD-approved, TG 430
In Vitro Three-dimensional Reconstructed Tissue Systems			
In Vitro Skin Corrosion Test: Reconstructed Human Epidermis (RhE) (MatTek Corp. EpiDerm, SkinEthic EPISKIN, SkinEthic RHE, CellSystems epiCS)	ICCVAM approved. Accepted by US agencies; 49 CFR 173.137 (2011)	Approved for Globally Harmonized System Labeling as Test Method B.40 Bis of Annex to 440/2008/EC	OECD-approved, TG 431
In Vitro Skin Irritation Test: Reconstructed Human Epidermis (RhE) (MatTek Corp. EpiDerm, SkinEthic EPISKIN, SkinEthic RHE)	Used for supporting information only	Approved for Globally Harmonized System Labeling as Test Method B.46 of Annex to 440/2008/EC	OECD-approved, TG 439
In Vitro Non-cellular Skin Models			
In Vitro Membrane Barrier Test for Corrosion: CORROSITEX Continuous Time Monitor Assay	ICCVAM approved. Accepted by US agencies; 49 CFR 173.137 (2011) Acceptable for US Department of Transportation Corrosive Packing Group subcategorization	Accepted via TG 435	OECD-approved, TG 435 UN Corrosive Packing Groups

[a]OECD – Organisation for Economic Co-operation and Development.
OECD TG 430 – (2004b). OECD GUIDELINE FOR THE TESTING OF CHEMICALS (OECD 430). Draft proposal for an update of test guideline 430. *In Vitro* Skin Corrosion: Transcutaneous Electrical Resistance Test (TER).
OECD TG 431 – (2004a). OECD GUIDELINE FOR THE TESTING OF CHEMICALS (OECD 431). Draft proposal for an update of test guideline 431. *In Vitro* Skin Corrosion: Reconstructed Human Epidermis (RhE) Test Method.
OECD TG 432 – (2004c). OECD GUIDELINE FOR TESTING OF CHEMICALS (OECD 432). *In Vitro* 3T3 NRU phototoxicity test.
OECD TG 435 – (2006). OECD GUIDELINE FOR THE TESTING OF CHEMICALS (OECD 435). *In Vitro* Membrane Barrier Test Method for Skin Corrosion.
OECD TG 439 – (2010). OECD GUIDELINE FOR THE TESTING OF CHEMICALS (OECD 439). *In Vitro* Skin Irritation: Reconstructed Human Epidermis Test Method.

These regulations either directly prohibit the use of animals for safety eye and skin irritation testing (as is the case of the 7th Amendment to the Cosmetics Directive) or provide strong animal welfare-focused language regarding the prioritization of testing using non-animal methods. Although these regulations were adopted by the European Commission, they have global impact as they affect chemicals and products manufactured in, transported through, or marketed in Europe, and must be heeded by European and non-European companies alike.

These validated *in vitro* assays may currently be acceptable to numerous international regulatory agencies for stand-alone predictions for hazard classification and labeling. However, some regulatory agencies may require confirmatory testing using additional valid test systems for non-corrosive/non-severe skin and eye irritants. Furthermore, the OECD Test Guideline predictions are increasingly designed for compliance with the GHS for classification, and thus may not be directly correlated to specific regulatory agency classification systems outside of the GHS.

7.6   Benchmarks and Reference Materials

There are notable differences among the various *in vitro* eye and skin irritation test methods, and selecting the appropriate assay to obtain the most accurate predictions can be difficult. Having confidence in the test results becomes more challenging when testing novel ingredients or chemistries, or when unique chemistries have only been tested *in vivo*. Sometimes the challenges occur near thresholds of irritant classifications, where small differences in assay behaviors drive over- or under-predictions. Fortunately the use of benchmark or reference materials can help improve the interpretation of test results by comparing the responses of the unknown materials to data available for the relevant benchmarks. The concept of using reference materials to improve assay interpretation applies to all test systems, *in vitro* and *in vivo* alike (and in fact has historically been in use in EPA regulatory testing programs), but they are especially germane to non-animal test systems, since these systems require greater data interpretation for accurate toxicological predictions. Ideally, reference materials should have the following properties for use in eye and skin irritation studies; they should:

1 be of similar chemical or product class as the candidate test material;
2 have the same expected mechanism of irritation as the candidate test material;
3 be applied to the test system in a similar vehicle, and dose exposure protocol, and most importantly
4 have reliable *in vivo* (preferably human) clinical or animal data to gauge interpretations

Industry uses for eye and skin irritation studies include final product testing such that complex mixtures, formulations containing novel active ingredients, or reformulation of existing product lines are typical. Accordingly, the ideal benchmark should also be of similar composition, since the entire formulation (active and inactive components) has a significant effect upon the exposure and penetration kinetics of the actives into the eye or skin models. Lastly, although not a regulatory-accepted approach, the use of benchmarks in existing assays can provide very useful irritation predictions outside of the ideal, validated assay response ranges. For example, although the IRE assay may not be accepted for discriminating between low and high moderate eye irritants (for example, GHS 2B vs. 2A), the inclusion of the appropriate benchmark(s) at the response threshold provides the criterion for confidently evaluating the candidate test material.

7.7   Limitations of Current Technologies and a Vision for Future Development

Whereas the validated *in vitro* and *ex vivo* assays provide eye or skin irritation/corrosion predictions after a single acute exposure, to date none of the assays provide adequate demonstrations of recovery (i.e., reversibility of effect), which from a regulatory perspective is critical to the full replacement of the Draize test. The ability to demonstrate ocular recovery is a particular challenge for the replacement of the *in vivo* eye irritation test, since recovery is dependent upon re-establishing meaningful visual acuity and function. Most corneal injuries, even those that result in minimal opacities *in vivo*, will affect visual acuity and function. Depending upon the degree of the initial injury to the cornea as well as any potential damaging effects of the consequent inflammatory responses, time is required for recovery of normal vision. To this end, various *ex vivo* corneal and whole globe eye models have been investigated for their use in evaluating corneal recovery after acute chemical insult. Some of these models even provide reasonable demonstration of the re-establishment of the corneal epithelium,[45–47] albeit these models are generally limited mechanistically to demonstrating recovery from mild eye irritants. A model for demonstrating meaningful recovery of moderate to severe irritants should also include a competent corneal stroma, where repair and recovery mechanisms would be demonstrated. But since all of these models utilize *ex vivo* tissues, interactions with the whole animal have been eliminated such that no circulatory-based inflammatory cells can be recruited to the sites of cell damage. It is suggested that the depth of the chemical insults into the stroma and the degree of the injuries to the keratocytes determines the degree of the inflammatory response and the potential for lasting fibrotic changes.[9–11] Consequently, the current *in vitro*/*ex vivo* ocular models lack this inflammatory response mechanism, undermining the ability to discriminate between moderate (recoverable) and severe/corrosive (non-recoverable) eye irritants. In contrast to the complexity of predicting recovery in the eye, the issues for predicting recovery in skin are simpler, since the fundamental difference between skin corrosives and skin irritants is defined by whether the skin injury penetrates into the dermis and is irreversible, or whether the injury is limited to the epidermis and thus likely to heal without notable scar formation. Fortunately, these initial injury events can readily be predicted using the available *in vitro* methods for skin corrosion and irritation.

7.8 Conclusions

Test system complexity may vary from simple monoculture cellular models, to more complex 3D tissue models, the latter of which more closely model the penetration kinetics and mode of action of chemicals in the native tissue environment, and would be expected to improve upon the relevance of the prediction. *Ex vivo* tissues present the current state of model complexity provided that the differences from model species to humans don't undermine their relevance. One must be careful not to over-interpret data from an *ex vivo* model simply because of its apparent similarities to the human tissues. By focusing on the various physiological events occurring in the target tissue, individual *in vitro* assays and endpoints can be selected to model these events, providing vastly more information about mechanisms of action, predictions for duration of irritant pathology, and potential for recovery.

Whereas individual *in vitro* test methods have been able to meet the specific eye and skin irritation assessment goals for individual companies, especially where a few representative chemistries typified the range of actives in the product lineup, it is recognized that for broad industry appeal across chemical and product classes, no single *in vitro* method is likely to generate all of the required safety information needed. Thus, a combination of methods providing distinct endpoints may provide a better understanding of how toxic substances reach their cellular targets and affect critical biological pathways.[48] To this end, a fundamental understanding of the applicability domain and limitations of each of the *in vitro* methods is paramount before selecting one or several methods for corporate safety testing programs. Depending upon the complexity of the test system and the selection of individual endpoints, the researcher can build upon the evidence obtained from individual assays to enhance the reliability of the predictions.

References

1. Cosmetics and Skin, Lash Lure. http://www.cosmeticsandskin.com/bcb/lash-lure.php/ Accessed 28 May 2013.
2. J. H. Draize, G. Woodard and H. O. Calvery, Methods for the study of irritation and toxicity of substances applied topically to the skin and mucous membrane, *J. Pharmacol. Exp. Therap.*, 1944, **82**, 377.
3. Appendix B, ICCVAM Summary Review Document: The Low Volume Eye Test. http://iccvam.niehs.nih.gov/docs/ocutox_docs/LVET/AppB-SRD.pdf/ Accessed 10 May 2013.
4. C. S. Weil and R. A. Scala, Study of intra- and interlaboratory variability in the results of rabbit eye and skin irritation tests, *Toxicol. Appl. Pharmacol.*, 1971, **19**, 276.

5. F. E. Freeberg, G. A. Nixon, P. J. Reer, J. E. Weaver, R. D. Bruce, J. F. Griffith and L. W. Sanders, 3rd, Human and rabbit eye responses to chemical insult, *Fundam. Appl. Toxicol.*, 1986, **7**, 626.

6. G. P. Daston and F. E. Freeberg, Ocular irritation testing, in *Dermal and Ocular Toxicology: Fundamentals and Methods*, ed. D. W. Hobson, CRC Press, Ann Arbor, MI, 1991, pp. 509–540.

7. L. H. Brunner, G. J. Carr, M. Chamberlain and R. D. Curren, Validation of alternative methods for toxicity testing, *Toxicol. In Vitro*, 1996, **10**, 479.

8. K. R. Robinson, R. Osborn and M. A. Perkins, in *Handbook of Cosmetic Science and Technology, In Vitro Tests for Skin Irritation*, ed. A. O. Barel, M. Paye and H. I. Maibach, Marcel Dekker Inc., New York, NY, 2005, chapter 11, pp. 95–106.

9. J. V. Jester, H. F. Li, W. M. Petroll, R. D. Parker, H. D. Cavanaugh, G. J. Carr, B. Smith and J. K. Maurer, Area and depth of surfactant-induced corneal injury predicts extent of subsequent ocular responses, *Investig. Ophthalmol. Visual Sci.*, 1998, **39**, 922.

10. J. K. Maurer and R. D. Parker, Light microscopic comparison of surfactant-induced eye irritation in rabbits and rats at three hours and recovery/day 35, *Toxicol. Pathol.*, 1996, **24**, 403.

11. J. K. Maurer, R. D. Parker and J. V. Jester, Extent of initial corneal injury as the mechanistic basis for ocular irritation: key findings and recommendations for the development of alternative assays, *Regulat. Toxicol. Pharmacol.*, 2002, **36**, 106.

12. P. Gautheron, M. Dukic, D. Alix and J. F. Sina, Bovine corneal opacity and permeability test: an in vitro assay of ocular irritancy, *Fundam. Appl. Toxicol.*, 1992, **18**, 442.

13. OECD, 2009, OECD Guideline for the Testing of Chemicals (OECD 437), Bovine Corneal Opacity and Permeability Test Method for Identifying Ocular Corrosives and Severe Irritants. Available at http://iccvam.niehs.nih.gov/SuppDocs/FedDocs/OECD/OECD-TG437.pdf/ Accessed August 2, 2011.

14. OECD, 2006, OECD Guideline for the Testing of Chemicals (OECD 435), *In Vitro* Membrane Barrier Test Method for Skin Corrosion. Available at http://iccvam.niehs.nih.gov/SuppDocs/FedDocs/OECD/OECDtg435.pdf/ Accessed October 24, 2011.

15. L. K. Earl, P. A. Jones, M. B. Dixit and K. A. O'Brien, Comparison of five potential methods for assessing ocular irritation *in vitro*, *Toxicol. In Vitro*, 1995, **9**, 245.

16. OECD, 2012, OECD Guideline for the Testing of Chemicals (OECD 460), Fluorescein Leakage Test Method for Identifying Ocular Corrosives and Severe Irritants. http://www.oecd-ilibrary.org/docserver/download/9712241e.pdf?expires = 1369327428&id = id&accname = guest&checksum = 693F25113F1521A0C3EB8491BCBEECF8/ Accessed 23 May 2013.

17. H. Sakaguchi, N. Ota, T. Omori, H. Kuwahara, T. Sozu, Y. Takagi, Y. Takahashi, K. Tanigawa, M. Nakanishi, T. Nakamura, T. Morimoto, S. Wakuri, Y. Okamoto, M. Sakaguchi, T. Hayashi, T. Hanji and

S. Watanabe, Validation study of the Short Time Exposure (STE) test to assess the eye irritation potential of chemicals, *Toxicol. In Vitro*, 2011, **25**, 796.

18. R. Osborne and M. A. Perkins, An approach for development of alternative test methods based on mechanisms of skin irritation, *Food Chem. Toxicol.*, 1994, **32**, 133.

19. V. Zuang, The neutral red release assay: a review, *Alt. Lab. Anim.*, 2001, **29**, 575.

20. R. L. Bronaugh, R. F. Stewart, E. F. Congdon and A. L. Giles, Jr., Methods for *in vitro* percutaneous absorption studies. I. Comparison with *in vivo* results, *Toxicol. Appl. Pharmacol.*, 1982, **62**, 474.

21. M. M. Riege, Factors Affecting Sorption of Topically Applied Substances, in *Skin Permeation: Fundamentals and Application*, ed. J. L. Zatz, Allured Publishing Corporation, Wheaton, IL, 1993, chapter 2, pp. 33–72.

22. OECD, 2004, OECD Guideline for the Testing of Chemicals (OECD 432), In Vitro 3T3 NRU phototoxicity test. http://iccvam.niehs.nih.gov/SuppDocs/FedDocs/OECD/OECDtg432.pdf/ Accessed 23 May 2013.

23. A. Natsch, The Nrf2-Keap1-ARE toxicity pathway as a cellular sensor for skin sensitizers—functional relevance and a hypothesis on innate reactions to skin sensitizers, *Toxicol. Sci.*, 2010, **113**, 284.

24. R. L. Bronaugh and S. W. Collier, *In Vitro* Methods for Measuring Skin Permeation, in *Skin Permeation: Fundamentals and Application*, ed. J. L. Zatz, Allured Publishing Corporation, Wheaton, IL, 1993, chapter 4, pp. 93–111.

25. A. B. G. Burton, M. York and R. S. Lawrence, The in vitro assessment of severe eye irritants, *Food Cosmet. Toxicol.*, 1981, **19**, 471.

26. The Rabbit Enucleated Eye Test, Invittox Protocol: 85, 2007. http://ecvam-dbalm.jrc.ec.europa.eu/view_doc.cfm?iddoc = 673&tdoc = prot/ Accessed 22 May 2013.

27. M. K. Prinsen and H. B. W. M. Koeter, Justification of the enucleated eye test with eyes of slaughterhouse animals as an alternative to the Draize eye irritation test with rabbits, *Food Chem. Toxicol.*, 1993, **31**, 69.

28. D. M. Bagley, L. H. Bruner, O. de Silve, M. Cottin and K. A. O'Brien, An evaluation of five potential alternatives *in vitro* to the rabbit eye irritation test *in vivo*, *Toxicol. in Vitro*, 1992, **6**, 275.

29. OECD, 2009, OECD Guideline for the Testing of Chemicals (OECD 438), Isolated Chicken Eye Test Method for Identifying Ocular Corrosives and Severe Irritants. http://ecvam.jrc.it/ft_doc/OECD%20TG%20438.pdf/ Accessed 23 May 2013.

30. K. J. Cooper, L. K. Earl, J. Harbell and H. Raabe, Prediction of ocular irritancy of prototype shampoo formulations by the isolated rabbit eye (IRE) test and bovine corneal opacity and permeability (BCOP) assay, *Toxicol. in Vitro*, 2001, **15**, 95.

31. OECD, 2004, OECD Guideline for the Testing of Chemicals (OECD 431), Draft proposal for an update of test guideline 431, *In Vitro* Skin Corrosion: Reconstructed Human Epidermis (RhE) Test Method.

http://www.oecd.org/chemicalsafety/testingofchemicals/43302385.pdf/ Accessed 23 May 2013.

32. OECD, 2010, OECD Guideline for the Testing of Chemicals (OECD 439), *In Vitro* Skin Irritation: Reconstructed Human Epidermis Test Method. http://iccvam.niehs.nih.gov/SuppDocs/FedDocs/OECD/OECD-TG439. pdf/ Accessed 23 May 2013.

33. Y. Kaluzhny, H. Kandarova, P. Hayden, J. Kubilus, L. d'Argembeau-Thornton and M. Klausner, Development of the EpiOcular(™) eye irritation test for hazard identification and labelling of eye irritating chemicals in response to the requirements of the EU cosmetics directive and REACH legislation, *Alt. Lab. Anim.*, 2011, **39**, 339.

34. M. V. Berridge, A. S. Tan, K. D. McCoy and R. Wang, The Biochemical and Cellular Basis of Cell Proliferation Assays That Use Tetrazolium Salts, *Biochemica*, 1996, **4**, 14.

35. J. Cotovio, M. H. Grandidier, P. Portes, R. Roguet and G. Rubinstenn, The *in vitro* skin irritation of chemicals: optimisation of the EPISKIN prediction model within the framework of the ECVAM validation process, *Alt. Lab. Anim.*, 2005, **33**, 329.

36. S. Gibbs, J. Vicanova, J. Bouwstra, D. Valstar, J. Kempenaar and M. Ponec, Culture of reconstructed epidermis in a defined medium at 33 degrees C shows a delayed epidermal maturation, prolonged lifespan and improved stratum corneum, *Arch. Dermatol. Res.*, 1997, **289**, 585.

37. S. Gibbs, H. Vietsch, U. Meier and M. Ponec, Effect of skin barrier competence on SLS and water induced IL-1alpha expression, *Exp. Dermatol.*, 2002, **11**, 217.

38. F. Netzlaff, C. M. Lehr, P. W. Wertz and U. F. Schaefer, The human epidermis models EpiSkin, SkinEthic and EpiDerm: an evaluation of morphology and their suitability for testing phototoxicity, irritancy, corrosivity, and substance transport, *Eur. J. Pharm. Biopharm.*, 2005, **60**, 167.

39. J. A. Bouwstra, H. W. Groenink, J. A. Kempenaar, S. G. Romeijn and M. Ponec, Water distribution and natural moisturizer factor content in human skin equivalents are regulated by environmental relative humidity, *J. Investig. Dermatol.*, 2008, **128**, 378.

40. In Vitro International Irritection Assay Overview. http://www.invitrointl. com/products/irritect.htm/ Accessed 24 May 2013.

41. OECD, 2006, OECD Guideline for the Testing of Chemicals (OECD 435), *In Vitro* Membrane Barrier Test Method for Skin Corrosion. http://iccvam. niehs.nih.gov/SuppDocs/FedDocs/OECD/OECDtg435.pdf/ Accessed 23 May 2013.

42. Official Journal of the European Union, 2006, Corrigendum to Regulation (EC) No 1907/2006 of the European Parliament and of the Council of 18 December 2006 concerning the Registration, Evaluation, Authorisation and Restriction of Chemicals (REACH), establishing a European Chemicals Agency, amending Directive 1999/45/EC and repealing Council Regulation (EEC) No 793/93 and Commission Regulation (EC) No 1488/ 94 as well as Council Directive 76/769/EEC and Commission Directives

91/155/EEC, 93/67/EEC, 93/105/EC and 2000/21/EC. L136/3, 1-278. http://eur-lex.europa.eu/LexUriServ/LexUriServ.do?uri = OJ:L:2007:136: 0003:0280:EN:PDF, Accessed 23 May 2013.

43. Official Journal of the European Union, 2008, Regulation (EC) No 1272/ 2008 of the European Parliament and of the Council of 16 December 2008 on classification, labelling and packaging of substances and mixtures, amending and repealing Directives 67/548/EEC and 1999/45/EC, and amending Regulation (EC) No 1907/2006. http://eur-lex.europa.eu/ LexUriServ/LexUriServ.do?uri = OJ:L:2008:353:0001:1355:en:PDF/ Accessed 23 May 2013.

44. Official Journal of the European Union, 2003, Directive 2003/15/EC of the European Parliament and of the Council of 27 February 2003 amending Council Directive 76/768/EEC on the approximation of the laws of the Member States relating to cosmetic products. http://eur-lex.europa.eu/ LexUriServ/LexUriServ.do?uri = OJ:L:2003:066:0026:0035:en:PDF/ Accessed 23 May 2013.

45. D. M. Foreman, S. Pancholi, J. Jarvis-Evans, D. McLeod and M. E. Boulton, A simple organ culture model for assessing the effects of growth factors on corneal re-epithelialization, *Exp. Eye Res.*, 1996, **62**, 555.

46. M. Piehl, M. Carathers, R. Soda, D. Cerven and G. DeGeorge, Porcine Corneal Ocular Reversibility Assay (PorCORA) predicts ocular damage and recovery for global regulatory agency hazard categories, *Toxicol. In Vitro*, 2011, **25**, 1912.

47. H. Raabe, L. Bruner, T. Snyder, N. Wilt and J. Harbell, Optimization of an in vitro long term corneal culture assay, Abstract In, *The Toxicologist*, 2005, **84**, 332.

48. L. Scott, C. Eskes, S. Hoffmann, E. Adriaens, N. Alepée, M. Bufo, R. Clothier, D. Facchini, C. Faller, R. Guest, J. Harbell, T. Hartung, H. Kamp, B. Le Varlet, M. Mclonim, P. McNamee, R. Osborne, W. Pape, U. Pfannenbecker, M. Prinsen, C. Seaman, H. Spielmann, W. Stokes, K. Trouba, C. Van den Berghe, F. Van Goethem, M. Vassallo, P. Vinardell and V. Zuang, A proposed eye irritation testing strategy to reduce and replace in vivo studies using Bottom-Up and Top-Down approaches, *Toxicol. In Vitro*, 2010, **24**, 1.

CHAPTER 8

Skin Sensitization Testing

DAVID BASKETTER

DABMEB Consultancy Ltd., 2 Normans Road, Sharnbrook, Bedfordshire
MK44 1PR, UK
Email: david.basketter@ukonline.co.uk

8.1 Introduction to Allergic Contact Dermatitis (ACD)

Skin sensitizers are chemicals that have the capacity to induce a state of
contact allergy in exposed humans. Contact allergy can be detected by a
diagnostic patch test, which confirms that an individual is sensitized. Such
individuals are then at risk of experiencing episodes of eczema, ACD, as a
consequence of further cutaneous exposure. The reality is that ACD is the
most frequent manifestation of immunotoxicity in humans and may even be
the most prevalent adverse health effect associated with chemical exposure.
Certainly, it is a very common occupational and environmental health issue
and many hundreds of chemicals have been shown to cause skin
sensitization.[1,2]

What is the immunobiological mechanism which underlies ACD? This has
been the subject of much review,[3–5] so the repetition of detail will be avoided
here. In brief, chemicals coming into contact with the skin react with skin protein, thereby producing non-self structures. With the involvement of dendritic
cells, these are brought to the attention of T lymphocytes in the draining
lymph nodes. When a T cell clone possesses a matching receptor, a response
occurs in which there is clonal expansion and systemic recirculation of these
T cells, resulting in the sensitized state. The basic mechanism is outlined in

Issues in Toxicology No. 19
Reducing, Refining and Replacing the Use of Animals in Toxicity Testing
Edited by David G. Allen and Michael D. Waters
© The Royal Society of Chemistry 2014
Published by the Royal Society of Chemistry, www.rsc.org

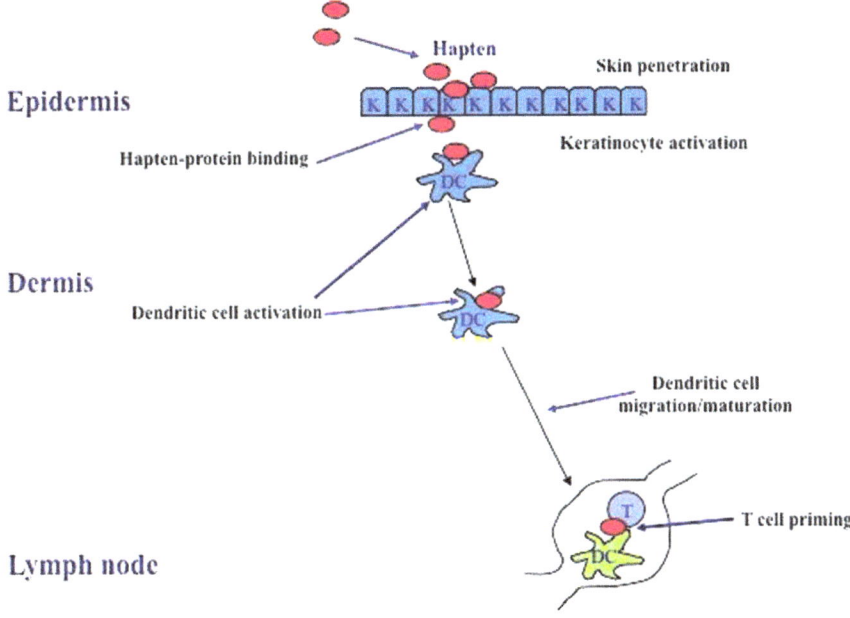

Figure 8.1 Basics of the induction of skin sensitization.

Figure 8.1. There are several generally agreed prerequisites for this sequence of events:

There must be sufficient skin contact with the chemical
The chemical must be reactive (or converted to a reactive species)
The chemical must derivatize protein(s) in a way that makes them immunogenic
The chemical must also trigger the release of danger signals

Failure to achieve any one of these requirements means that either there is a null event, or potentially that tolerance to the chemical is induced. This last matter is a subject of recent speculative debate.[6] The final stage in the expression of the disease, ACD, is elicitation, which requires a sufficient degree of cutaneous exposure to the original substance which induced the state of sensitization.

There are many variables which can impact upon the induction and elicitation of ACD: the primary aspect is the dose per unit area of the chemical (reviewed in[7]). Building upon that are matters such as the number, frequency and duration of exposures,[8] the vehicle in which those exposures occur,[9] the skin site and skin condition[10] and individual susceptibility, which of course is itself multifactorial and likely to be associated with proclivity for the expression of danger signals/skin irritation.[10–13] In contrast, gender, ethnic differences and age appear to be of relatively little importance and the underlying genetic factors (which undoubtedly exist) are masked by the single most dominant factor driving ACD – exposure.[14,15]

In the material which follows, the primary *in vivo* methods for the predictive identification of chemicals which have the capacity to cause ACD are reviewed, together with a discussion of the potential *in vitro* alternatives, some of which are in the final stages of validation (as this review is being prepared). Mention is made also of both *in silico* and human testing. Finally, it is essential to recognize that the identification of skin sensitization hazards is only the first step in a much more complex process of hazard characterisation, risk assessment and risk management. It is only when these latter processes are undertaken to a sufficient standard that human health can be protected.[16]

8.2 *In Vivo* Predictive Assays

8.2.1 Guinea Pig Test Methods

There is a long history of predictive assays for the identification of skin sensitising chemicals.[17] However, in these days of *in vitro* alternatives, primary attention here is devoted only to those assays which have the benefit of general regulatory acceptance – which in practice means two guinea pig methods and the local lymph node assay (LLNA). A general historical overview of *in vivo* test method development is given in Table 8.1, which includes also information on the development of test guidelines at the Organisation for Economic Co-operation and Development (OECD).

The two major methods for many years both involved the use of the guinea pig as the surrogate species of choice. A diagram of the basic aspects of the methods is presented as Figure 8.2. The earliest of these procedures was the Buehler test, which employed a series of 6 hour occluded patches for induction, with the extent of any induced skin sensitization being assessed two weeks later by topical challenge on the flank.[18] An overview of assay sensitivity was presented some years later.[19] This confirmed that, when well conducted, the Buehler test could be an acceptably sensitive assay. However, it had long been suspected that technical aspects of assay conduct could significantly affect the outcome and this was shown convincingly in a later paper.[20]

An early response to questions about the sensitivity and general reliability of the Buehler (and other) guinea pig tests led to the development of the guinea pig

Table 8.1 An overview of *in vivo* sensitization tests and the OECD test guidelines.

1944	Draize test (the original test) (BT)
1965	Buehler (occluded patch) test
1970	Magnusson and Kligman guinea pig maximization test (GPMT)
1981	OECD 406 (detailed 7 guinea pigs, including those above; included the mouse ear swelling test as a screen)
1993	OECD update (restricted tests to just BT and GPMT; LLNA accepted as a screen)
1999	LLNA validated by ICCVAM
2002	OECD 429 LLNA
2012	OECD 429 updated
2012	OECD 442a/442b (non-radioactive variants of the LLNA)

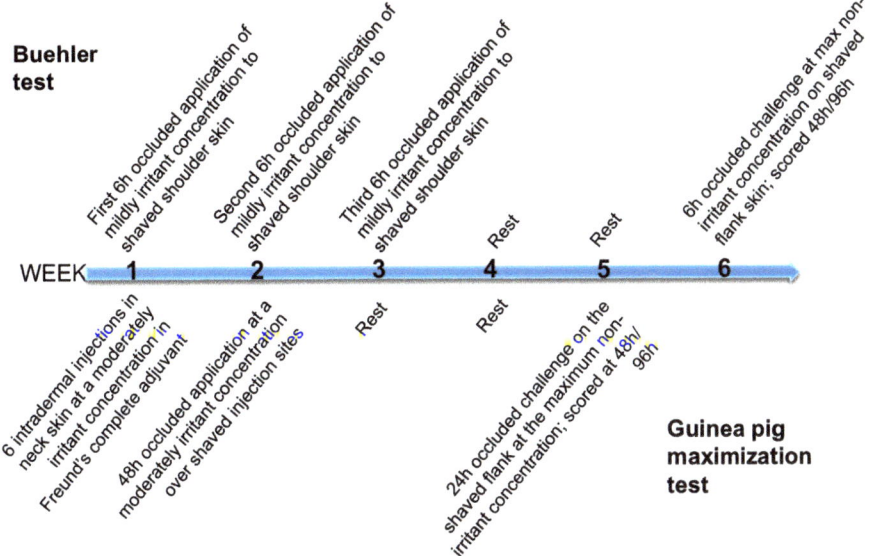

Figure 8.2 Summary of guinea pig sensitization tests used for regulatory purposes.

maximization test (GPMT).[21] Here, the first two topical applications of the Buehler test were replaced by a series of intradermal injections of test substance in combination with the powerful Freund's complete adjuvant. One week later, induction of sensitization was boosted by a 48 hour occluded patch applied over the injection sites. Some two weeks later, the extent of induction of sensitization was determined via a 24 hour occluded patch challenge applied to the flank.

For both the Buehler test and the GPMT, skin reactions are read on a simple scale, typically 0–3, reflecting negative, weak, moderate and strong individual responses. However, regulatory judgment on the response refers only to the frequency of reactions among test animals:

Buehler test: ≥15% positive, the substance is a skin sensitizer
GPMT: ≥30% positive, the substance is a skin sensitizer

Of course, these levels of response are judged by comparison with any skin reactions recorded in a set of concurrent sham treated controls (i.e., animals that have been through the induction process, but without the test substance, but which have then been treated with the test substance at challenge in an identical manner to the test guinea pigs).

Over many years there have been commentaries of various kinds on these two guinea pig assays.[22–24] However, the most interesting aspects, indeed potentialities, of guinea pig testing are those least undertaken in terms of regulatory toxicology assessment – rechallenge and cross challenge. The first of these, rechallenge, is a valuable tool for the discrimination of true weak responses from the background noise of minor irritation and non-immune-specific reactions.[23,25] A simple example is shown in Table 8.2, where

Table 8.2   Rechallenge in guinea pig skin sensitization testing.

Guinea pig[a]	Challenge 1 48 h[b]	Challenge 1 96h	Challenge 2 48 h[c]	Challenge 2 96 h	Result[d]
T1	2	1	0	0	−
T2	0	0	0	0	−
T3	0	0	0	0	−
T4	1	1	0	0	−
T5	0	0	0	0	−
T6	0	0	0	0	−
T7	0	0	0	0	−
T8	0	1	1	2	+
T9	0	0	0	0	−
T10	0	0	1	0	−
C1	0	0	0	0	NA
C2	0	0	0	0	NA
C3	1	0	0	0	NA
C4	0	0	0	0	NA
C5	0	0	0	1	NA[e]

[a]T = test; C = control.
[b]Scoring completed on a 0–3 scale.
[c]Conducted 2 weeks after challenge 1 with the same concentration and vehicle.
[d]Overall conclusion.
[e]NA = not applicable.

an uncertain degree of reaction is shown not to be a true positive result. The rationale is based on the premise that sensitization responses should generally be reproducible and, if anything, tend to be stronger at the later scoring time, reflecting their nature as delayed hypersensitivity. In this particular case, only a single guinea pig shows reproducible reactions and which cannot be negated by the evidence of minor reactions in the control animals.

The results for more than three hundred substances which have been tested in the guinea pig maximization test have been published.[26,27] There is only one published dataset of any size (37 substances) for the Buehler test.[19]

8.2.2   Murine Test Methods

In the mouse, two methods have gained some measure of regulatory acceptance. The mouse ear swelling test (MEST)[28] has been recorded in OECD Guideline 406 as a means of screening potential positives, but beyond this, the assay has not progressed towards formal validation and more general acceptance and usage. Accordingly, it will not be described further. In contrast, the murine local lymph node assay (LLNA) became the first alternative method to be formally validated and progressed into regulatory guidelines and accordingly, it is this method that will be the focus of the material that follows.

The LLNA arose from a desire to refine and reduce animal usage compared to the guinea pig methods, as well as to improve upon the subjective and non-quantitative endpoint of those older techniques. A simple overview of the LLNA is presented in Figure 8.3. A detailed description of the LLNA protocol

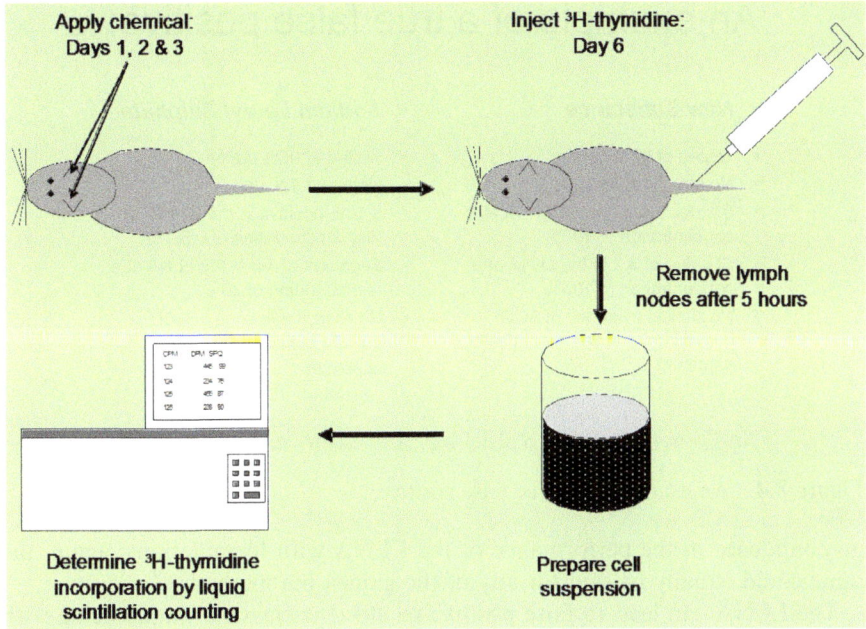

Apply chemical:
Days 1, 2 & 3

Inject ³H-thymidine:
Day 6

Remove lymph
nodes after 5 hours

Determine ³H-thymidine
incorporation by liquid
scintillation counting

Prepare cell
suspension

Figure 8.3 Outline of the LLNA protocol.

and associated aspects have already been published.[29–32] In brief, groups of mice receive daily topical applications of test substance on the dorsal skin of the ear; two days later, tritiated thymidine is injected via the tail vein. Five hours later, the draining lymph nodes are removed, processed, and the incorporated radioactivity quantified by scintillation counting. The activity in test groups is compared to concurrent vehicle controls. Where the response is at least 3-fold greater than control, the test substance is regarded as a skin sensitizer. It was this protocol and prediction model which underwent formal validation.[33,34] Since that time, compilations of the results of several hundred LLNA results with a wide range of chemicals have been reported.[35,36]

The fact that the LLNA offered animal welfare benefits and an objective endpoint made it an attractive prospect to toxicologists, including many in the regulatory community. Validation ensured that the method could be progressed to an OECD Test Guideline[29], which in turn led to adoption of the LLNA as the first choice regulatory assay in preference to the guinea pig methods.[37] However, with that regulatory acceptance came also some distinct challenges. Perhaps the most notable amongst these has been the question of false positives (from the industrial perspective) and the potential for false negatives (particularly for regulators reviewing tests on formulations). Of course, the reality is that no toxicology test is perfect.[38] The LLNA, like the guinea pig methods (and like the *in vitro* tests to come) suffers from a degree of false positive and false negative results. Furthermore, it was assessed and validated, at least originally, only with chemicals and therefore there could be

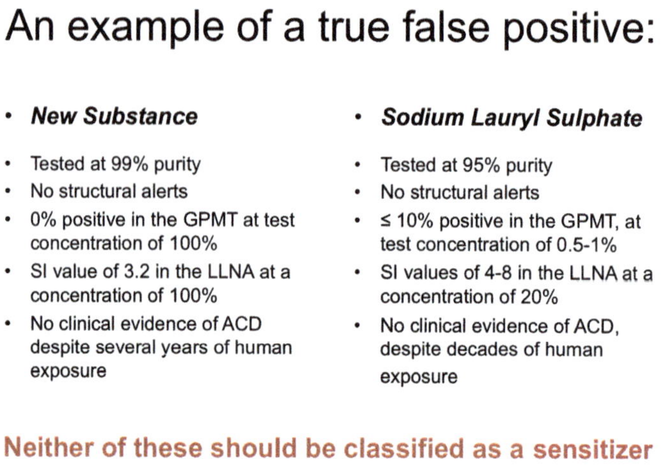

An example of a true false positive:

- **New Substance**

  - Tested at 99% purity
  - No structural alerts
  - 0% positive in the GPMT at test concentration of 100%
  - SI value of 3.2 in the LLNA at a concentration of 100%
  - No clinical evidence of ACD despite several years of human exposure

- **Sodium Lauryl Sulphate**

  - Tested at 95% purity
  - No structural alerts
  - ≤ 10% positive in the GPMT, at test concentration of 0.5-1%
  - SI values of 4-8 in the LLNA at a concentration of 20%
  - No clinical evidence of ACD, despite decades of human exposure

Neither of these should be classified as a sensitizer

Figure 8.4   An example of a true false positive.

no confidence in the performance of the LLNA with formulations; again, the same could equally be said for any of the guinea pig methods.[39]

The LLNA can lead to false positive results, the classic example being with the well know irritant, sodium dodecyl (lauryl) sulphate (SLS). This topic has been the subject of extensive debate and review, not least since it is important to distinguish true false positive results from those that are merely un-palatable.[40–42] Key to any decision is to balance the overall weight of evidence.[43–46] In this respect, it has also to be borne in mind that false positives and negatives occur with the guinea pig methods[47] and that predictions from chemistry are also prone to a degree of error.[48] An example of the application of this weight of evidence decision making is presented in Figure 8.4, which shows a true false positive result.

Finally in this section it is worth mentioning that two significant developments have occurred recently with the LLNA. One of these has been the adoption by the OECD of two variants of the assay which replace the radio-active endpoint with a non-radioactive alternative.[49,50] These methods offer a similar sensitivity and specificity to the original LLNA, but their use in the determination of relative sensitizing potency is less well established. The other development has been the acceptance of a version of the LLNA which uses only the top dose group, thus saving up to 50% of the animals. This protocol is referred to as the reduced LLNA (rLLNA).[51,52]

8.3   *In Vitro* Predictive Assays

Across the entire spectrum of toxicology there has always been a desire to avoid the use of experimental animals and this was brought into sharp focus some years ago by the decision in Europe to ban such testing in relation to cosmetic products.[53] The end date for the impact of this legislation of March 11, 2013 meant that within a decade it was expected that *in vitro* alternatives would be in place. Of course, toxicological scientists warned that this was highly improbable, and such was the clear conclusion contained within an extensive review of progress.[24] Nevertheless, at least in the area of skin sensitization, there have been some very substantial advances in that time period, such that at least for hazard identification, *in vitro* alternatives may well be just around the corner. In this section, it is those prospects which will be the focus, since they represent the most probable expression of skin sensitization testing in the next decade as *in vitro* finally replaces *in vivo*.

As already mentioned, skin sensitization is a consequence of reactive chemistry and accordingly, one of the most advanced *in vitro* alternatives is actually in effect no more than a simple reactivity test. This is the direct peptide reactivity assay (DPRA) reported by Gerberick and colleagues.[54,55] In its most mature form, synthetic oligopeptides containing either cysteine or lysine are exposed to a potential skin sensitizer and the loss of the native peptide is monitored by HPLC. With a panel of 82 selected chemicals, the predictive accuracy was 89%.[55] Predictive accuracy fell only slightly, to 85%, with a dataset of almost double the size (Leslie Foertsch/Frank Gerberick, personal communication). The experience with the assay was such that it was accepted by the European Centre for the Validation of Alternative Methods (ECVAM) for formal independent validation trial.[56] At the time of writing, the DPRA appears to have been successful in a blinded trial involving three laboratories, which tested within and between interlaboratory reproducibility, and thus it is in the final stages of independent peer review. As with any assay, there are some limitations, notably that the DPRA lacks any capacity for metabolic activation or inactivation of chemicals, nor does it pretend to measure the rate of reactivity.

A second assay, the human cell line activation test (h-CLAT) also has participated in the ECVAM trial. This method, developed in Japan, is based on the responses of a dendritic like cell line THP-1[57–61]. The method is based on a 24 hour exposure of THP-1 cells to chemicals at somewhat cytotoxic concentrations. Skin sensitizers are identified by a 150% increase in CD86 and/or a 200% increase in CD54 expression, measured by flow cytometry. Most interesting, albeit with a relatively modest dataset of some 53 chemicals, the predictive accuracy of the h-CLAT judged against human data was reported to be 83%.[59] As with the DPRA, the quality of the evidence in support of h-CLAT prompted ECVAM to accept the assay into the formal validation trial. Being a little more demanding than the DPRA, the h-CLAT work is a few months behind, but initial results look promising. Key limitations of the method appear to be the challenge associated with highly hydrophobic compounds (a difficulty common to many *in vitro* methods) and a reduced likelihood of the detection of skin sensitizing chemicals which possess very low cytotoxic potential.

A third assay which is in the final stages of formal validation combines what is, in essence, a reactivity chemistry sensor system into the HaCaT keratinocyte cell line and is called Keratinosens.[62–64] To develop the Keratinosens assay, a luciferase reporter gene under control of a single copy of the ARE-element of the human AKR1C2 gene was stably inserted into HaCaT keratinocytes. Evaluation of 67 selected chemicals showed a predictive accuracy of approximately 80%. This level of accuracy was maintained as the dataset was expanded to over 200 substances. Keratinosens also has been the subject of an independent blinded evaluation which has assessed interlaboratory transferability and reproducibility as well as intralaboratory reproducibility with a total of 28 chemicals.[65] The outcome was very positive in all respects and the information is currently with ECVAM, where it is in the final stages of peer review.[56]

So, three in vitro assays are now progressing well and seem likely to receive favorable comment from independent validation review. However, it is only fair to say that there are several other methods at various stages of development. Most of these are detailed in recent reviews.[66–69] Perhaps of particular note is an adaptation of the DPRA which aims to add elements to mimic the metabolic and/or air oxidation processes known to activate some otherwise unreactive substances.[70,71] These changes effectively enhance the applicability domain of the DPRA and with a wider chemical set whilst maintaining the overall predictive accuracy. Other methods include a suite of papers deriving from the EU project "Sens-it-iv"[72–74] and a technique called "SensCeeTox", which combines elements of reactivity and altered gene expression.[75,76] Yet more possibilities included the detection of skin sensitizers via oxidative stress[77] as well as via variations on the theme of peptide/protein reactivity.[78] Perhaps the overriding characteristics of all of these approaches is that generally they have been tested only in a single laboratory and on a relatively small set of chemicals. However, what it does serve to demonstrate (and the list of references is by no means exhaustive) is that there is much translation of research into potential assays and, at least in some cases, that work is continuing towards interlaboratory trials (James McKim, personal communication). Consequently, if for whatever reason the three lead assays mentioned above fall out of the running, there is a whole suite of hazard identification methods that are ready to take their place.

The material above indicates that the successful replacement of in vivo predictive assays can now be viewed as a probable short term prospect, but a key question remains: how is the data from multiple assays to be combined? A practical evaluation of this matter has been published recently.[79] This group assessed a number of in vitro methods (six in total) with 54 substances. The optimal predictive approach was found to be a combination of peptide reactivity (such as DPRA), activation of the Keap-1/Nrf2 signaling pathway (such as done with Keratinosens) and dendritic cell activation (such as h-CLAT). The specific test indications are those of the author, some variants were discussed in the original paper, but it serves to show how the thinking for hazard identification has coalesced. The basic strategy that arises in the publication is that to conclude a substance should be classified as a skin sensitizer, it must be positive in at least two of the three in vitro assays. This led to a predictive accuracy of >90%, which if proven with a somewhat larger dataset and supported by validated assays, would exceed the accuracy of the LLNA (and the guinea pig methods) as reported in the ICCVAM validation.[34]

8.4 Of Chemistry, Computers and Man

Since the very early days of work on skin sensitizing chemicals it has been recognized that reactive chemistry is at the heart of this area of toxicology[80] Indeed, already more than three decades ago, a mathematical basis was advanced on which quantitative structure activity relationships (QSARs) could be developed.[81] From this were derived a range of QSARs; later came the definition of structural alerts for skin sensitization, which were embedded into a computer based expert system.[82,83] However, from this work, to date it has not proven possible to develop globally applicable QSARs.[48] What has been derived are a number of computer based predictive systems, all of which have imperfections and are perhaps best used in parallel with the human brain, rather than as its replacement.[52] This remains the case, even with the sophisticated OECD QSAR toolbox, which has now reached it third version.[52]

Where chemistry and computers fail to provide meaningful insights, many expect that human data can fill the gap. However, clinical and human experimental data can also be hard to interpret. In predictive sense, reliance is often placed on the human repeated insult patch test (HRIPT), but whereas this was clearly developed for the testing of chemicals[84,85], it has commonly been usurped for the misplaced testing of formulations.[86] Typically, the test is portrayed as generating an expected negative result to confirm that a formulation is safe to place on the consumer market, but it has long been recognized that the test does not have sufficient predictive power for this purpose.[87] In reality, by far the most valuable human data is derived from diagnostic patch testing, this being the information which demonstrates the success or failure of risk assessment/management procedures.[88] Unfortunately, the most common use of diagnostic patch test data has been to illustrate risk assessment and/or management failures.

8.4.1 Categorisation of Human Potency of Skin Sensitizers

Clinical data can provide compelling evidence of situations where, despite successful predictive identification of skin sensitizers, the risks to human health have not been adequately assessed or managed. One obvious omission from hazard identification methods is the characterization of potency. However, with the advent of the LLNA, a new approach to potency prediction was developed – the EC3 value – the concentration required to cause a 3-fold stimulation in test animals compared to the concurrent vehicle control.[89] This approach can be expressed mathematically as:

$$EC3 = c + [(3 - d)/(b - d)] \times (a - c)$$

where (a,b) and (c,d) are the coordinates, respectively, of data points lying immediately above and immediately below the SI value of 3 in the LLNA dose response curve. Sometimes, where the data points do not meet the criteria for interpolation, a cautious approach to extrapolation has also been defined.[51] Importantly, EC3 values were shown to be robust, reproducible figures.[90] With this tool, skin sensitizing chemicals were shown to span a range of approximately 5 orders of magnitude in terms of their relative potency.[35,36] However, the true value of LLNA EC3 values is that they do correlate reasonably well with what is known of relative human potency of skin sensitizers that has been derived from experimental data.[91–94] As a consequence, the use of the LLNA EC3 value has been encouraged by an EU expert group[93] and by the WHO[95], such that it has been incorporated into the latest Globally Harmonized System (GHS) for potency subcategorization of skin sensitizing chemicals.[96]

Ultimately, what is missing for a more rigorous evaluation of the predictive power of the LLNA EC3 value (or the related strategies proposed for the guinea pig methods – see [93]) is a more substantial list of substances categorised according to their relative human potency solely on the basis of human data. However, initial work on this has been published[94] and a list of approximately 130 substances is being prepared for publication.[97] In the meantime, for selected materials, the correlation between the human potency category and the LLNA value is reproduced as Figure 8.5. For this purpose, true non-sensitizers have been excluded. The figure shows 35 substances, 7 from each category (extreme, strong, moderate, weak and very weak). The LLNA EC3 value is quite predictive for the two highest categories (i.e. 1 and 2), but is less accurate for the lowest categories, where it of course is least sensitive and the substances span the regulatory classification boundary: those in category 5 are definite human sensitizers, but whose potency is sufficiently low that ordinarily they would not be classified as skin sensitizers.[96]

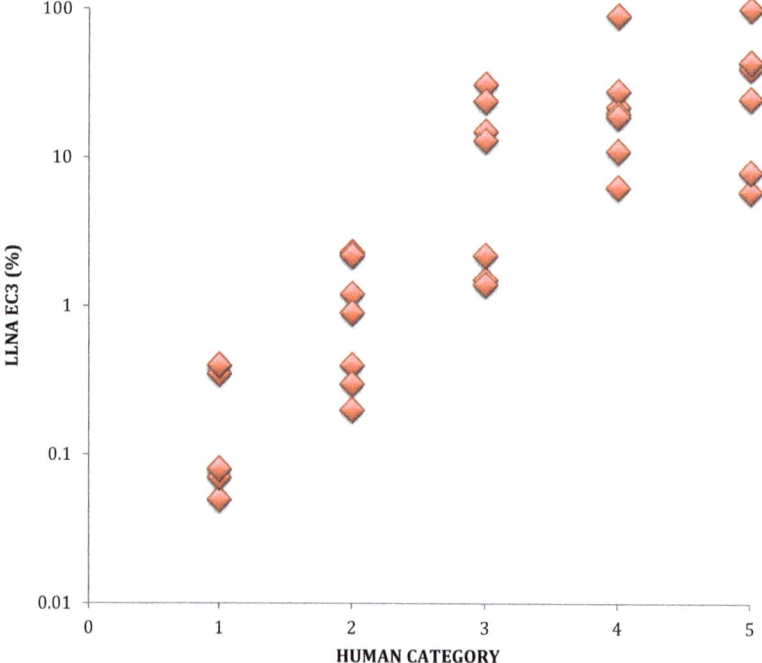

Figure 8.5   Prediction of human potency category by the LLNA EC3 value.

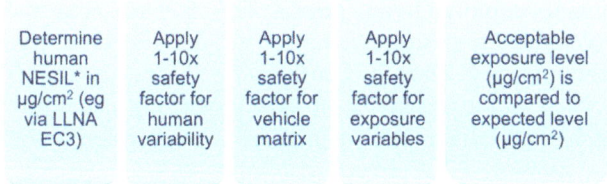

*NESIL = no expected sensitization
induction level in a human repeated
insult patch test predicted via the
LLNA EC3 value (Basketter et al., 2005)

Figure 8.6 Outline of skin sensitization quantitative risk assessment.

Armed with knowledge on skin sensitizer potency, it is possible to progress to a quantitative risk assessment (QRA) (fully detailed in [98]). An overview of the QRA approach is presented in Figure 8.6. In essence, the LLNA EC3 value is used to predict the threshold in a nominal HRIPT (the human study does not actually need to be conducted). This value is then reduced to take account of several uncertainty factors, including human variability, the vehicle matrix and exposure parameters. Practical examples of the use of QRA have been published.[98–101] Adaptations have also been proposed to take account of special circumstances.[102]

8.4.2   Non-Animal Methods for the Characterisation of Potency

The key expectation for toxicology in the 21st century is not for better hazard identification or risk assessment, but rather that its exponents conduct it without the use of animal testing. Nowhere is this more clearly evident than in the European Union, driven by the politically motivated demand that cosmetics and their ingredients must not be a cause of animal testing.[53] The current state of play in the alternatives area was reviewed recently[24,52], but such is outside the scope of this chapter. More specifically, the key question is what progress has there been towards the replacement of the *in vivo* methods discussed earlier, not in terms of hazard identification, but in the context of whether they deliver outputs which help to determine the potency of an identified skin sensitizer?

A theoretical basis for the use of *in vitro* test inputs into potency determination has already been expounded.[103,104] A first attempt to populate the theory was made, based largely on the *in vitro* methods close to validation.[105] However, although the performance was reasonable, it focused on prediction of potency in LLNA terms. Arguably, more interesting was the recent paper from Japan which focused on the capacity of the h-CLAT to predict human skin sensitizing effect.[59] Here, the *in vitro* assay was tested against 66 substances placed into one of five human potency categories, but although overall accuracy was acceptable (83%), the details of use of the assay for potency prediction were not explored to any degree. This may in fact be a wise decision, since many commentators suggest that it will require data combination as per[104] to predict potency. Thus, it is perhaps for this reason that in the DPRA (and similar assays), the reactivity of skin sensitising chemicals is typically referred to as high, medium, low or minimal, avoiding terms such as strong, weak, etc.[55]

The practical reality is that *in vitro* methods for the characterisation of potency, at least in an ideal world, will be based on the methods being validated for hazard identification. Thus, in effect, they would be used in the same manner as the LLNA, originally validated for hazard identification, but later extended in usefulness by interpolation of dose response data to derive EC3 values. For the *in vitro* alternatives which are now close, this means waiting for that final validation step and also for that body of gold standard human potency data that will be critical in deciding how to use/combine assay outputs in a meaningful and accurate manner.

8.5 Summary

Predictive testing for skin sensitization has come a long way since the initial methods dating back to the first half of the last century. In the second half of that century, guinea pig methods began to give way to more quantitative and objective mouse method, the LLNA.[33] From this change has evolved a way to measure the relative skin sensitizing potency of substances[90] and thence a new approach to skin sensitization risk assessment.[47,98] Whilst none of these changes represent perfection,[44] they have demonstrated the ability of toxicology, and even regulatory toxicology, to make positive progress in the direction of improving the protection of human health. As a consequence, it is a significant challenge for *in vitro* toxicology that it seeks to replace not only the *in vivo* methods, but to do so in a way that maintains the advances made during the last two decades. It is the case that a suite of *in vitro* methods for hazard identification are close to formal validation, but whether these methods, or others further back in development, can also deliver the quality of information for potency prediction remains an open question.

References

1. De Groot AC. 2008. Patch Testing, 3rd edition. acdegroot publishing, Wapserveen.
2. H. Johansson, A. S. Albrekt, C. A. Borrebaeck, M. Lindstedt The GARD assay for assessment of chemical skin sensitizers. *Toxicol In Vitro*. 2012, in press.

3. T. Rustemeyer, I. M. W. Van Hoogstraten, B. M. A. Von Blomberg, S. Gibbs and R. G. Scheper, 2011. Mechanisms of irritant and allergic contact dermatitis. In: J. D. Johansen, P. J. Frosch and J.-P. Lepoittevin, (Eds.), *Textbook of Contact Dermatitis,* 5th Ed., Springer, Berlin, pp. 43–90.
4. S. F. Martin, Contact dermatitis: from pathomechanisms to immunotoxicology, *Exp Dermatol.*, 2012, **21**, 382–389.
5. OECD. 2012a. The Adverse Outcome Pathway for Skin Sensitization. Organisation for Economic Cooperation and Development, Paris.
6. J. P. McFadden, J. M. L. White, D. A. Basketter, R. J. Dearman and I. Kimber, The hapten-atopy hypothesis II: the cutaneous hapten paradox, *Clin Exp Allergy*, 2011, **41**, 327–337.
7. I. Kimber, R. J. Dearman, D. A. Basketter, C. A. Ryan, G. F. Gerberick, J. Lalko and A. M. Api, Dose metrics in the acquisition of skin sensitization: thresholds and importance of dose per unit area, *Regulatory Toxicology and Pharmacology*, 2008, **52**, 39–45.
8. D. A. Basketter, D. Jefferies, R. J. Safford, N. J. Gilmour, J. McFadden, W. Chansinghakul, I. Duangdeeden and P. Kullavanijaya, The impact of exposure variables on the induction of skin sensitisation, *Contact Dermatitis*, 2006, **55**, 178–185.
9. I. R. Jowsey, C. J. Clapp, B. Safford, B. T. Gibbons and D. A. Basketter, The impact of vehicle on the relative potency of skin sensitising chemicals in the local lymph node assay, *Food and Chemical Toxicology*, 2008, **27**, 67–75.
10. D. A. Basketter, Methyldibromo glutaronitrile, skin sensitisation and quantitative risk assessment, *Cut. Ocul. Toxicol.*, 2010, **29**, 4–9.
11. J. P. McFadden and D. A. Basketter, Contact allergy, irritancy and danger, *Contact Dermatitis*, 2000, **42**, 123–127.
12. H. S. Smith, D. B. Holloway, D. K. B. Armstrong, D. A. Basketter and J. P. McFadden, Irritant thresholds in subjects with colophony allergy, *Contact Dermatitis*, 2000, **42**, 95–97.
13. M. H. Allen, S. H. Wakelin, D. Holloway, S. Lisby, O. Baadsgaard, J. N. Barker and J. P. McFadden, Association of TNFA gene polymorphism at position − 308 with susceptibility to irritant contact dermatitis, *Immunogenetics*, 2000, **51**, 201–205.
14. A. Schnuch, G. Westphal, R. Mössner, W. Uter and K. Reich, Genetic factors in contact allergy--review and future goals., *Contact Dermatitis, 2011 64*, 2011, 2–23.
15. A. Schnuch and B. C. Carlsen 2011. Genetics and individual predispositions in contact dermatitis. In: J. D. Johansen, P. J. Frosch and J.-P. Lepoittevin, (Eds.), *Textbook of Contact Dermatitis,* 5th Ed., Springer, Berlin, pp. 13–42.
16. D. A. Basketter, Skin sensitisation: strategies for risk assessment and risk management, *Brit J Dermatol.*, 2008, **159**, 267–273.
17. K. E. Andersen and H. I. Maibach 1985. Contact Allergy Predictive Tests in Guinea Pigs. *Karger*, Basel.
18. E. V. Buehler, Delayed contact hypersensitivity in the guinea pig, *Archives of Dermatology*, 1965, **91**, 171–177.

19. D. A. Basketter and G. F. Gerberick, Interlaboratory evaluation of the Buehler test, *Contact Dermatitis*, 1996, **35**, 146–151.
20. M. K. Robinson, T. L. Nusair, E. R. Fletcher and H. L. Ritz, A review of the Buehler guinea pig skin sensitization test and its use in a risk assessment process for human skin sensitization, *Toxicology*, 1990, **61**, 91–107.
21. B. Magnusson and A. M. Kligman Allergic contact dermatitis in the guinea pig. *Identification of contact allergens*. Springfield, Illinois, Charles C Thomas, 1970.
22. E. V. Buehler, A rationale for the selection of occlusion to induce and elicit delayed contact hypersensitivity in the guinea pig. A prospective test, *Current Problems in Dermatology*, 1985, **14**, 39–58.
23. A. M. Kligman and D. A. Basketter, A critical commentary and updating of the guinea pig maximization test, *Contact Dermatitis*, 1995, **32**, 129–34.
24. S. Adler, D. A. Basketter, S. Creton, O. Pelkonen, J. van Benthem and V. Zuang, *et al.*, Alternative (non-animal) methods for cosmetics testing: current status and future prospects, *Arch Toxicol.*, 2011, **85**, 367–485.
25. T. Ashikaga, M. Hoya, H. Itagaki, Y. Katsumura and S. Aiba, Evaluation of CD86 expression and MHC class II molecule internalization in THP-1 human monocytic cells as predictive endpoints for contact sensitizers, *Toxicol. In Vitro*, 2002, **16**, 711–716.
26. J. E. Wahlberg and A. Boman 1985. Guinea pig maximization test. In K. E. Andersen and H. I. Maibach, (Eds), *Contact Allergy: Predictive test in guinea pigs. Current Problems in Dermatology, Karger*, Basel, Switzerland, pp. 59–106.
27. M. T. D. Cronin and D. A. Basketter, Multivariate QSAR analysis of a skin sensitization database, *SAR and QSAR in Environmental Research*, 1994, **2**, 159–179.
28. S. C. Gad, B. J. Dunn, D. W. Dobbs, C. Reilly and R. D. Walsh, Development and validation of an alternative dermal sensitisation test: the mouse ear swelling test (MEST), *Toxicol Appl Pharmacol.*, 1986, **84**, 93–114.
29. OECD. 2002. Local Lymph Node Assay. Test Guideline no 429, Organisation for Economic Cooperation and Development, Paris.
30. OECD. 2012b. Local Lymph Node Assay. Update Test Guideline no 429, Organisation for Economic Cooperation and Development, Paris.
31. D. A. Basketter, I. Kimber, R. J. Dearman, C. A. Ryan and G. F. Gerberick 2012a. The local lymph node assay. *Dermatotoxicology*, 8th edition, Eds Wilhelm KP, Zhai H and Maibach HI, pp. 298–307.
32. C. Rovida, C. Ryan, S. Cinelli, D. Basketter, R. Dearman and I. Kimber, The local lymph node assay (LLNA)., *Current Protocols in Toxicology*, 2012, (Suppl. 51), Chapter 20, 20.7.1–20.7.14.
33. G. F. Gerberick, C. A. Ryan, I. Kimber, R. J. Dearman, L. J. Lea and D. A. Basketter, Local lymph node assay: validation assessment for regulatory purposes, *Am. J.Contact Derm.*, 2000, **11**, 3–18.
34. J. H. Dean, L. E. Twerdok, R. R. Tice, D. M. Sailstad, D. G. Hattan and W. S. Stokes, ICCVAM evaluation of the murine local lymph node assay.

II Conclusions and recommendations of an independent scientific peer review panel, *Reg. Toxicol. Pharmacol.*, 2001, **34**, 258–273.

35. G. F. Gerberick, C. A. Ryan, P. S. Kern, H. Schlatter, R. J. Dearman, I. Kimber, G. Y. Patlewicz and D. A. Basketter, Compilation of historical local lymph node data for evaluation of skin sensitization alternative methods, *Dermatitis*, 2005, **16**, 157–202.

36. P. S. Kern, G. F. Gerberick, C. A. Ryan, I. Kimber, A. Aptula and D. A. Basketter, Historical local lymph node data for the evaluation of skin sensitization alternatives: a second compilation, *Dermatitis*, 2010, **21**, 8–32.

37. ECHA, Guidance on information requirements and chemical safety assessment, *Chapter R.7a: Endpoint specific guidance, European Chemicals Agency, Helsinki*, 2008, 256–288.

38. D. A. Basketter, N. Ball, S. Cagen, J.-C. Carrillo, H. Certa, D. Eigler, H. Esch, C. Graham, D. Haux, R. Kreiling and A. Mehling, Application of a weight of evidence approach to analysing discordant sensitization datasets: implication for REACH, *Regulatory Toxicology and Pharmacology*, 2009, **55**, 90–96.

39. ICCVAM. 2010. Using the murine local lymph node assay for testing pesticide formulations, metals, substances in aqueous solutions, and other products. http://iccvam.niehs.nih.gov/docs/immunotox_docs/LLNA-AD/Front.pdf. Last accessed Oct 22 2012.

40. D. A. Basketter, G. F. Gerberick and I. Kimber, Strategies for identifying false positive responses in predictive sensitisation tests, *Food and Chemical Toxicology*, 1998, **36**, 327–333.

41. D. A. Basketter, J. McFadden, P. Evans, K. E. Andersen and I. Jowsey, Identification and classification of skin sensitisers: identifying false positives and false negatives, *Contact Dermatitis*, 2006, **55**, 268–273.

42. D. A. Basketter and I. Kimber, Skin irritation, false positives and the local lymph node assay: a guideline issue, *Regulatory Toxicology and Pharmacology*, 2011, **61**, 137–140.

43. R. Kreiling, H. M. Hollnagel, L. Hareng, D. Eigler, M. S. Lee, P. Griem, B. Dreesen, M. Kleber, A. Albrecht, C. Garcia and A. Wendel, Comparison of the skin sensitizing potential of unsaturated compounds and assessed by the murine local lymph node assay (LLNA) and the guinea pig maximization test (GPMT), *Food Chem. Toxicol.*, 2008, **46**, 1896–1904.

44. D. A. Basketter, J. F. McFadden, G. F. Gerberick, A. Cockshott and I. Kimber, Nothing is perfect, not even the local lymph node assay, *A commentary and the implications for REACH. Contact Dermatitis*, 2009, **60**, 65–69.

45. D. Basketter, J. Crozier, B. Hubesch, I. Manou, A. Mehling and J. Scheel, Optimized testing strategies for skin sensitization – the LLNA and beyond, *Regul Toxicol Pharmacol.*, 2012b, **64**, 9–16.

46. N. Balls, S. Cagen, J. C. Carrillo, H. Certa, D. Eigler, R. Emter, F. Faulhammer, C. Garcia, C. Graham, C. Haux, S. N. Kolle, R. Kreiling, A. Natsch and A. Mehling, Evaluating the sensitization

potential of surfactants: integrating data from the local lymph node assay, guinea pig maximization test, and in vitro methods in a weight-of-evidence approach, *Reg. Toxicol. Pharmacol.*, 2011, **60**, 389–400.

47. D. A. Basketter and I. Kimber, Skin sensitization, false positives and false negatives: experience with guinea pig assays, *J Appl Toxicol.*, 2010, **30**, 381–386.

48. G. Patlewicz, A. O. Aptula, D. W. Roberts, P. S. Kern, G. F. Gerberick, I. Kimber, R. J. Dearman, C. A. Ryan and D. A. Basketter, An evaluation of selected global (Q)SARs/expert systems for the prediction of skin sensitisation potential, *SAR and QSAR in Environmental Research*, 2007, **18**, 515–541.

49. OECD. 2010a. Organisation for Economic Cooperation and Development. Test Guideline 429: The Local Lymph Node Assay. Paris, France.

50. OECD. 2010b. Organisation for Economic Cooperation and Development. Guidelines for Test of Chemicals No 442a and 442b. Paris, France.

51. C. A. Ryan, J. G. Chaney, P. S. Kern, G. Y. Patlewicz, D. A. Basketter, C. J. Betts, R. J. Dearman, I. Kimber and G. F. Gerberick, The reduced local lymph node assay: the impact of group size, *J Applied Toxicol.*, 2008, **28**, 518–523.

52. OECD. 2012c. The OECD QSAR Toolbox. Version 3.0. http://www.oecd.org/chemicalsafety/assessmentofchemicals/theoecdqsartoolbox.htm. Last accessed 21st October 2012.

53. EC, Directive 2003/15/EC of the European Parliament and of the Council of 27 February 2003 amending Council Directive 76/768/EEC on the approximation of the laws of the Member States relating to cosmetic products, *Official Journal of the European Union*, 2003, **L66**, 26–35.

54. G. F. Gerberick, J. D. Vassallo, R. E. Bailey, J. G. Chaney, S. W. Morrall and J. P. Lepoittevin, Development of a peptide reactivity assay for screening contact allergens, *Toxicol Sci.*, 2004, **81**, 332–43.

55. G. F. Gerberick, J. D. Vassallo, L. M. Foertsch, B. B. Price, J. G. Chaney and J. P. Lepoittevin, Quantification of chemical peptide reactivity for screening contact allergens: a classification tree model approach, *Toxicol Sci.*, 2007, **97**, 417–27.

56. D. Basketter, A. Angers-Loustau, S. Casati and ECVAM, *ECVAM: Progressing skin sensitisation alternatives for hazard identification. Contact Dermatitis*, 2012c, **66**(Suppl 2), 24.

57. T. Ashikaga, Y. Yoshida, M. Hirota, K. Yoneyama, H. Itagaki, H. Sakaguchi, M. Miyazawa, Y. Ito, H. Suzuki and H. Toyoda, Development of an in vitro skin sensitization test using human cell lines: the human Cell Line Activation Test (h-CLAT). I. Optimization of the h-CLAT protocol, *Toxicol In Vitro*, 2006, **20**, 767–773.

58. T. Ashikaga, H. Sakaguchi, S. Sono, N. Kosaka, M. Ishikawa, Y. Nukada, M. Miyazawa, Y. Ito, M. Nishiyama and H. Itagaki, A comparative evaluation of in vitro skin sensitisation tests: the human cell-line activation test (h-CLAT) versus the local lymph node assay (LLNA), *ATLA*, 2010, **38**, 275–284.

59. Y. Nukada, T. Ashikaga, H. Sakaguchi, S. Sono, N. Mugita, M. Hirota, M. Miyazawa, Y. Ito, H. Sasa and N. Nishiyama, Predictive performance for human skin sensitizing potential of the human cell line activation test (h-CLAT), *Contact Dermatitis*, 2011, **65**, 343–353.

60. H. Sakaguchi, T. Ashikaga, M. Miyazawa, Y. Yoshida, Y. Ito, K. Yoneyame, M. Hirota, H. Itagaki, H. Toyoda and H. Suzuki, Development of an in vitro skin sensitization test using human cell lines; Human Cell Line Activation Test (h-CLAT). II. An international study of the h-CLAT, *Toxicol. In Vitro*, 2006, **20**, 774–784.

61. H. Sakaguchi, T. Ashikaga, M. Miyazawa, N. Kosaka, Y. Ito, K. Yoneyama, S. Sono, H. Itagaki, H. Toyoda and H. Suzuki, The relationship between CD86/CD54 expression and THP-1 cell viability in an in vitro skin sensitization test--human cell line activation test (h-CLAT), *Cell Biol Toxicol.*, 2009, **25**, 109–126.

62. A. Natsch and R. Emter, Skin sensitizers induce antioxidant response element dependent genes: application to the in vitro testing of the sensitization potential of chemicals, *Toxicol Sci.*, 2008, **102**, 110–119.

63. R. Emter, G. Ellis and A. Natsch, Performance of a novel keratinocyte-based reporter cell line to screen skin sensitizers in vitro, *Toxicol Appl Pharmacol.*, 2010, **245**, 281–290.

64. A. Natsch, R. Emter and G. Ellis, Filling the concept with data: integrating data from different in vitro and in silico assays on skin sensitizers to explore the battery approach for animal-free skin sensitization testing, *Toxicol Sci.*, 2009, **107**, 106–121.

65. A. Natsch, C. Bauch, L. Foertsch, F. Gerberick, K. Norman, A. Hilberer, H. Inglis, R. Landseidel, S. Onken, H. Reuter, A. Schepky and R. Emter, The intra- and inter-laboratory reproducibility and predictivity of the KeratinoSens assay to predict skin sensitizers in vitro: Results of a ring-study in five laboratories, *Toxicol In Vitro*, 2011, **25**, 733–744.

66. R. J. Vandebriel and H. van Loveren, Non-animal sensitization testing: state-of-the-art., *Crit Rev Toxicol.*, 2010, **40**, 389–404.

67. Kimber I, Basketter DA and Dearman RJ. 2012. Dendritic cells and the assessment in vitro of skin sensitizing potential. *Cut Ocul Toxicol.*, in press.

68. A. Mehling, T. Eriksson, T. Eltze, S. Kolle, T. Ramirez, W. Teubner, B. van Ravenzwaay and R. Landsiedel, Non-animal test methods for predicting skin sensitization potentials, *Arch Toxicol*, 2012, **86**, 1273–1295.

69. M. Vocanson, J.-F. Nicolas and D. A. Basketter, In vitro approaches to the identification and characterization of skin sensitisers., *Exp Rev Dermatol.*, 2013, in press.

70. G. F. Gerberick, J. A. Troutman, L. M. Foertsch, J. D. Vassallo, M. Quijano, R. L. Dobson, C. Goebel and J. P. Lepoittevin, Investigation of peptide reactivity of pro-hapten skin sensitizers using a peroxidase-peroxide oxidation system, *Toxicol Sci.*, 2009, **112**, 164–74.

71. J. A. Troutman, L. M. Foertsch, P. S. Kern, H. J. Dai, M. Quijano, R. L. Dobson, J. F. Lalko, J. P. Lepoittevin and G. F. Gerberick,

The incorporation of lysine into the peroxidase peptide reactivity assay for skin sensitization assessments, *Toxicol Sci.*, 2011, **122**, 422–436.

72. E. Corsini, V. Galbiati, M. Mitjans, C. L. Galli and M. Marinovich, 2012. NCTC 2544 and IL-18 production: A tool for the identification of contact allergens. *Toxicol In Vitro.*, in press.

73. L. Dietz, S. Kinzebach, S. Ohnesorge, B. Franke, I. Goette, D. Koenig-Gressel and H. J. Thierse, Proteomic allergen-peptide/protein interaction assay for the identification of human skin sensitizers, *Toxicol In Vitro*, 2012in press.

74. J. D. Johansen, P. J. Frosch, J.-P. Lepoittevin 2011. *Contact Dermatitis*, 5th edition. Springer, Berlin.

75. J. M. McKim, D. J. Keller and J. R. Gorski, A new in vitro method for identifying chemical sensitizers combining peptide binding with ARE/EpRE-mediated gene expression in human skin cells, *Cutan Ocul Toxicol.*, 2010, **29**, 171–192.

76. J. M. McKim, Jr., D. J. Keller 3rd and J. R. Gorski. 2012. An in vitro method for detecting chemical sensitization using human reconstructed skin models and its applicability to cosmetic, pharmaceutical, and medical device safety testing. *Cutan Ocul Toxicol.*, in press.

77. M. Miyazawa and A. Takashima, Development and validation of a new in vitro assay designed to measure contact allergen-triggered oxidative stress in dendritic cells, *J Dermatol Sci*, 2012, **68**, 73–81.

78. Y. H. Jeong, S. An, K. Shin and T. R. Lee. 2012. Peptide reactivity assay using spectrophotometric method for high-throughput screening of skin sensitization potential of chemical haptens. *Toxicol In Vitro.*, in press.

79. C. Bauch, S. N. Kolle, T. Ramirez, T. Eltze, E. Fabian, A. Mehling, W. Teubner, B. van Ravenzwaay and R. Landsiedel, Putting the parts together: combining in vitro methods to test for skin sensitizing potentials, *Regul Toxicol Pharmacol.*, 2012, **63**, 489–504.

80. J.-P. Lepoittevin, D. A. Basketter, A. Dooms-Goossens, A.-T. Karlberg 1997. *Allergic Contact Dermatitis; The Molecular Basis*. Springer-Verlag, Heidelberg.

81. D. W. Roberts and D. L. Williams, The derivation of quantitative correlations between skin sensitization and physicochemical parameters for alkylating agents, and their application to experimental data for sultones, *Journal of Theoretical Biology*, 1982, **99**, 807–825.

82. M. D. Barratt, D. A. Basketter, M. Chamberlain, G. Admans and J. Langowski, An expert system rulebase for identifying contact allergens, *Toxicol in Vitro*, 1994, **8**, 1053–1060.

83. K. Langton, G. Y. Patlewicz, A. Long, C. A. Marchant and D. A. Basketter, Structure activity-relationships for skin sensitisation: recent improvements to DEREK for windows, *Contact Dermatitis*, 2006, **55**, 342–347.

84. F. N. Marzulli and H. I. Maibach, Antimicrobials: experimental contact sensitization in man, *J. Soc. Cosmet. Chem.*, 1973, **24**, 399–421.

85. F. N. Marzulli and H. I. Maibach, The use of graded concentration in studying skin sensitizers: Experimental contact sensitization in man, *Food Cosmet Toxicol.*, 1974, **12**, 219–227.
86. D. A. Basketter, The human repeated insult patch test in the 21ˢᵗ century: a commentary on ethics and validity, *Cutaneous and Ocular Toxicology*, 2009, **28**, 49–53.
87. C. R. Henderson and E. C. Riley, Certain statistical considerations in patch testing, *Journal of Investigative Dermatology*, 1945, **6**, 227–229.
88. D. A. Basketter and I. R. White, Diagnostic patch testing – does it have a wider relevance?, *Contact Dermatitis*, 2012, **67**, 1–2.
89. D. A. Basketter, L. Lea, K. Cooper, A. Dickens, D. Briggs, I. Pate, R. J. Dearman and I. Kimber, *A comparison of statistical approaches to derivation of EC3 values from local lymph node assay dose responses. J. Appl. Toxicol.*, 1999, **19**, 261–266.
90. D. A. Basketter, G. F. Gerberick and I. Kimber, The local lymph node assay EC3 value: status of validation, *Contact Dermatitis*, 2007, **57**, 70–75.
91. P. Griem, C. Goebel and H. Scheffler, Proposal for a risk assessment methodology for skin sensitization based on sensitization potency data, *Regulatory Toxicology and Pharmacology*, 2003, **38**, 269–290.
92. K. Schneider and Z. Akkan, Quantitative relationship between the local lymph node assay and human skin sensitization assays, *Regulatory Toxicology and Pharmacology*, 2004, **39**, 245–255.
93. D. A. Basketter, C. Clapp, D. Jefferies, R. J. Safford, C. A. Ryan, G. F. Gerberick, R. J. Dearman and I. Kimber, Predictive identification of human skin sensitisation thresholds, *Contact Dermatitis*, 2005, **53**, 260–267.
94. D. A. Basketter and J. P. McFadden 2012. Cutaneous allergies. In *"Immunotoxicity, Immune Dysfunction and Chronic Disease*. Eds. R. R Dietert and R. W. Luebke, Humana Press, New York, pp. 103–126.
95. H. van Loveren, A. Cockshott, T. Gebel, U. Gundert-Remy, W. H. de Jong, J. Matheson, H. McGarry, L. Musset, M. K. Selgrade and C. Vickers, Skin sensitization in chemical risk assessment: report of a WHO/IPCS international workshop focusing on dose-response assessment., *Regulatory Toxicology and Pharmacology*, 2008, **50**, 155–199.
96. United Nations Nations. 2011. Globally Harmonized System of Classification and Labelling of Chemicals (GHS). Part 3: Health Hazards. http://www.unece.org/fileadmin/DAM/trans/danger/publi/ghs/ghs_rev04/English/03e_part3.pdf. Last accessed 22 October 2012.
97. Basketter DA, *et al.*, 2013. Categorisation of chemicals according to their relative human skin sensitising potency. MS in preparation.
98. A. M. Api and M. Vey, Special issue on QRA, *Regul Toxicol Pharmacol.*, 2008, **52**(1), 1–2.
99. G. F. Gerberick, M. K. Robinson, S. Felter, I. White and D. A. Basketter, Understanding fragrance allergy using an exposure-based risk assessment approach, *Contact Dermatitis*, 2001, **45**, 333–340.

100. D. A. Basketter, C. J. Clapp, B. J. Safford, I. R. Jowsey, P. M. McNamee, C. A. Ryan and G. F. Gerberick, Preservatives and skin sensitisation quantitative risk assessment: risk benefit considerations, *Dermatitis*, 2008, **19**, 20–27.
101. N. Corea, D. A. Basketter, A. van Asten, J.-P. Marty, A. Pons Guiraud and C. Laverdet, Fragrance allergy: assessing the risk from fabric washing products, *Contact Dermatitis*, 2006, **55**, 48–53.
102. M. A. Farage, D. L. Bjerke, C. Mahony, K. L. Blackburn and G. F. Gerberick, Quantitative risk assessment for the induction of allergic contact dermatitis: uncertainty factors for mucosal exposures, *Contact Dermatitis*, 2003, **49**, 140–147.
103. I. Jowsey, D. A. Basketter, C. Westmoreland and I. Kimber, A future approach to measuring relative skin sensitising potency, *J Appl Toxicology*, 2006, **26**, 341 –350.
104. D. A. Basketter and I. Kimber, Updating the skin sensitisation in vitro data assessment paradigm in 2009, *Journal of Applied Toxicology*, 2009, **29**, 545–550.
105. A. Natsch, The Nrf2-Keap1-ARE toxicity pathway as a cellular sensor for skin sensitizers – functional relevance and a hypothesis on innate reactions to skin sensitizers, *Toxicol. Sci.*, 2010, **113**, 284–292.

CHAPTER 9

Integrated Approaches to Safety Testing: General Principles and Skin Sensitization as Test Case

ANDREAS NATSCH

Givaudan Schweiz AG, Ueberlandstrasse 138, CH-8600 Duebendorf, Switzerland
Email: andreas.natsch@givaudan.com

9.1 Introduction

With the aim of replacing animal testing, an enormous body of research has been conducted over the last 20 years to develop *in vitro* toxicological tests and predictive computer algorithms. Despite significant progress, in relatively few cases is there an officially accepted one-to-one replacement, where one single *in vitro* test can fully replace the previous animal testing. Such cases are the epidermis equivalent assays to replace skin corrosion and skin irritation tests[1] or the 3T3-NRU assay to perform phototoxicity testing.[2,3] In other instances, alternative tests are used in tiered testing strategies to reduce animal testing (e.g. cytotoxicity tests to determine starting dose in the acute rat toxicity study)[4] or several *in vitro* assays are combined in a weight-of-evidence approach as has become a practice in genotoxicity testing.[5] Since one-to-one replacements are often difficult to achieve, the term ITS has become a buzzword recently– it stands for both 'Integrated Testing Strategies' and 'Intelligent Testing Strategies'. I prefer the former definition.

Issues in Toxicology No. 19
Reducing, Refining and Replacing the Use of Animals in Toxicity Testing
Edited by David G. Allen and Michael D. Waters
© The Royal Society of Chemistry 2014
Published by the Royal Society of Chemistry, www.rsc.org

Integrated testing strategies can be classified according to their structure or their goal. In a first distinction, an ITS can be either a *tiered* or a *battery* approach.

- In the *tiered ITS* different tests are performed sequentially, and the outcome of any test may lead either to a final call on the potential hazard or define the need for additional testing. Typically, the approach is summarized in a decision tree, but also Bayesian networks have been proposed to arrive at probabilities to decide on which next test may give more information.[6]
- In a *battery ITS*, a battery of tests are routinely performed in parallel, since all are considered to be necessary for the given endpoint – an algorithm integrating the different testing results then leads to an overall toxicological prediction.

  Another distinction for different ITS approaches relates to their goal, namely whether they aim at reducing or fully replacing animal testing.
- In a *reductive ITS*, alternative methods (*in vitro*, *in silico* or other) are used to arrive at a first conclusion which may be sufficient to optimize or waive the animal test, which is still considered as the last resort if the alternative data are not sufficiently conclusive. Frameworks for a reductive ITS have been proposed for different toxicological endpoints.[7–10] The use of this approach is in principle favoured by the REACH legislation – although it has been questioned based on analysis of submitted REACH dossiers whether it is really being applied in practice.[11–13] A reductive ITS, by definition, is a tiered approach, with the animal test defined as the last potential step. It is a very pragmatic approach which can reduce animal test numbers significantly in cases where a test (or algorithm) is available with either a very high sensitivity or a very high specificity to rapidly classify clearly non-toxic or clearly toxic agents upfront. An example is the proposal, for the acute toxicity endpoint, to use the cytotoxicity parameter not only for initial dose determination for the animal test, but also to identify the chemicals with very low toxicity and exclude them from further testing in the animal, an approach which is currently under ECVAM validation (http://reach.setac.eu/embed/presentations_reach/05_BREMER_In_vitro.pdf).
- In a *replacement ITS*, a set of alternative data should fully replace the animal test. This is the key target in the cosmetic field, as the cosmetic legislation is not about how many animal tests are done but envisages a complete ban.

Finally an ITS can be either qualitative, leading to a yes/no answer or it may aim at a quantitative (or at least semi-quantitative) classification of potency, which is crucial if it was to be applied for risk assessment.

This chapter will primarily address general underlying principles of integrating data from different alternative non-animal sources (i.e. for a replacement ITS) and then exemplify these principles in the case of skin sensitization, an endpoint for which these ideas have been widely discussed in recent years and for which data to prove the concept are accumulating.

9.2 General Underlying Principles for Data Integration

9.2.1 Why Do We Need Data Integration?

The animal test (or the human test sometimes conducted for few specific endpoints like skin sensitization and irritation) is a holistic approach. After chemical exposure, the response of the whole organism can be evaluated. Thus without a detailed mechanistic understanding of the pathway leading to the toxic action, the response of the whole organism can be observed. More importantly, in most cases the endpoint evaluated is *the actual endpoint* one wants to predict (mortality in acute toxicity, red and inflamed skin in skin irritation tests or acquisition of delayed hypersensitivity reactions in the guinea pig tests for skin sensitization) and *not a proxy measurement* with varying relevance for the predicted endpoint. However, this is not true for all animal tests: in the local lymph node assay (LLNA) for skin sensitization, to take an example, lymph node proliferation is determined as a proxy measurement for the acquisition of specific immunity.

As compared to the holistic full-organism approach with the direct observation of the toxic action, alternative testing information is always reductionist and in many cases it mirrors only a specific part of a more complex process. Moreover, in almost all cases a proxy or indirect measurement is recorded (cellular viability, cellular proliferation, inflammatory mediator release, gene expression changes, receptor binding, etc.) which can be related to the endpoint of interest based on experience and correlation or, preferably, based on a mechanistic understanding.

It is this reductionist and indirect nature which leads to the perceived limitations of the alternative tests. There are at least five specific reasons why single tests do have limitations, and it is the understanding of these limitations which may drive the definition of an ITS.

1 The more complex *in vivo* process could be a chain of events leading to toxic action, but individual tests or predictions represent *single steps in a chain of events* – in this case data integration should bring data together on the individual steps and the integrated approach tries to recapitulate the full chain of events. Such chains of events are currently formulated into so called adverse outcome pathways (AOP).[14] Data integration in this case recapitulates serial events which all are needed to lead to the toxic effect and/or events at different levels of biological organization (molecular events, cellular events, effects at organ and finally organism level).

2 The same toxic outcome may be caused by *different modes of action*; while individual *in vitro* tests can only cover a specific mode of action (e.g. genotoxicity caused by mutagenicity or chromosomal damage; acute toxicity due to narcotic toxicity or reactive toxicity or pharmacological

effects). In such cases a battery of tests covering the different modes of action is needed.

3 Incomplete prediction, due to *limited applicability domain of each test* (i.e. not all chemical classes are predicted) – in this case a complementary set of tests is also needed. This can be due to intrinsic limitations of the individual test systems (physicochemical parameters, solubility, etc.).

4 *Lack of ADME information* (absorption, distribution, metabolism and excretion). Data integration should help to model metabolism and extrapolate from the *in vitro* setup to the organ/organism level.

5 Incomplete prediction since a single *in vitro* test is only suitable for a *certain potency range*. Data integration then helps to predict chemicals/ preparations in the full potency range. Such an example has been described for eye irritation of antimicrobial cleaning products.[15]

9.2.2 Sources of Complementary Information

Depending on the toxicological endpoint, there are different relevant sources of alternative non-animal data which can be considered in an ITS.

9.2.2.1 *Biological Assays Based on Cell Culture Systems*

A wide number of endpoints, which can be evaluated in cell based systems, have been described. They range from simple cytotoxicity measurements to specific tests on apoptosis, changes in gene expression, activation of receptors and signalling pathways, expression of specific protein markers, formation of specific metabolic enzymes, expression of cell surface markers, changes or damage to the chromatin structure, to name only a few.

In recent years, with the publication of the 21th century vision for toxicology[16] and the advances in genomics, there has been an increasing focus on toxicogenomics and the potential to study changes at the level of gene expression; however, not for all toxic modes of action specific and predictive changes in transcription patterns can be expected. Furthermore, these changes in transcription patterns are by definition proxy measurements, although they can be very informative and mechanistically valuable. It is often difficult to define whether the observed response belongs: (i) to an adverse response (pathway of toxicity, PoT) (ii) or an adaptive response (pathway of defence, PoD).[17] In some cases more classical approaches will model the actual toxic endpoint more closely (e.g. toxicity to a gill-cell line may be a more direct indication of acute toxicity as compared to the proxy measurement of gene expression changes in a fish embryo), and, just to note another example, damage to DNA as noted in a Comet[18] assay may be a more direct indication as compared to GADD45 induction in a reporter cell line.[19] However, whether the measured endpoint is more directly or indirectly related to the toxic action itself does not define its predictive value. Activation of a pathway of defense may have a high predictivity due to the fact that organisms have evolved to specifically sense and avoid a particular toxic insult, even if this is part of the adaptive and not the toxic response.

Finally, most of these cellular endpoints can be measured either in primary cells or in cell lines and either in classical cell culture or 3D cultures, which multiplies the potential test system.

9.2.2.2 In Chemico and In Silico Evaluation of Specific Chemical and Physicochemical Parameters

Physicochemical parameters will largely influence the bioavailability of a compound and affect penetration of specific biological barriers. This can be modelled with *in silico* models, or directly measured in penetration assays using different biological membranes or surrogates thereof.

Besides these physicochemical parameters, chemical reactivity is an important parameter which is considered a modifying factor for acute toxicity (referred to as 'excess toxicity' in aquatic toxicology)[20] and which is often a determining factor in mutagenicity and skin sensitization.[21] Therefore *in chemico* assays and *in silico* models describing intrinsic reactivity of chemicals are an important part of an ITS for several toxic endpoints.

9.2.2.3 Metabolism Assays

Finally, many test systems lack intrinsic metabolic capacity. Assays or algorithms to model the metabolism of tests compounds may thus be another important component of an ITS. Otherwise metabolic activation systems can directly be added to several *in vitro* assays, the most classical example being the liver S9 fractions added to many biological tests[22] for mutagenicity or liver microsomes added to glutathione trapping assays[23] in pharmacological research.

9.2.3 Different Weight of Information: Essential Versus Modifying Factors

The different sources of complementary information listed above will be used based on a mechanistic understanding of the toxic action. This basic understanding will also inform on the potential weight the information may have in an ITS. Some parameters will only be modifying factors (e.g. bioavailability may affect the potency of a compound, but it is rather rare that a low bioavailability will make an inherently toxic compound completely non-toxic). On the other hand, some parameters are essential prerequisites (e.g. a certain reactivity is essential for skin sensitization).[24] Finally, the closer the endpoint measured *in vitro* is to the real toxic action (i.e. the less it is a proxy measurement) the more weight it may have. For skin sensitization, the *in vitro* proliferation of allergen-specific T-cell clones would be an observation with a high weight, as it is a very close *in vitro* reflection of the *in vivo* toxic action.

9.2.4 How to Integrate Data from Different Sources: General Possibilities

Which of the five principle reasons leads to the need for an ITS in the first instance may also partly determine the way data are integrated (Figure 9.1).

If a multi-step *in vivo* process is rebuilt with alternative information, a serial approach to data evaluation may appear logical. The rationale is that a chemical must negotiate a series of hurdles to finally lead to the toxic action and the positive outcomes of the tests are connected with the logical AND operator to arrive at a positive conclusion: The toxic action can only take place if two (or more) tests modelling individual links in the chain are positive or above a given threshold.

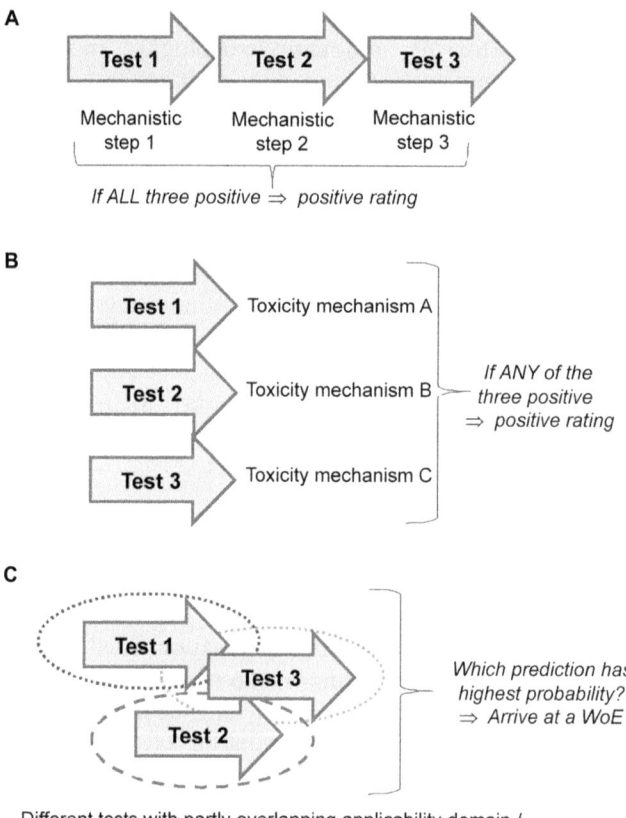

Figure 9.1 Schematic representation for three potential reasons for an ITS and the resulting data integration paradigm. (A) the tests mimic essential sequential steps in a toxicity pathway; (B) the tests cover parallel pathways of toxicity and (C) tests with overlapping information content/applicability domain requiring a WoE integration.

If, on the other hand, the tests mimic different modes of action all leading to the same adverse outcome, then the results will be connected with the logical OR operator, positivity in one of the tests is sufficient. If a chemical leads to DNA breaks OR DNA point mutations OR chromosomal damage (all tested in different assays), it will in all cases be genotoxic, and the data are combined in a parallel manner.

Similarly, if different tests recognize only a portion of the chemicals due to limitations in the applicability domain, data may need to be combined with the OR operator – if any test able to recognize toxic chemicals from only a part of the chemical space is positive, the final call will be a positive rating.

However – this is also where the crystal-clear distinctions stop – is it AND or OR? In many cases this is unclear and a weight-of-evidence (WoE) assessment may be needed. Several tests with partly overlapping information content are performed to reduce uncertainty (see Figure 9.1); the data from all tests are then needed to finally arrive at a 'best guess' estimation. Most *in vitro* tests are first developed with their individual prediction model for a yes/no rating of a hazard. In the simplest approach, integration of data would then be to just count the pluses and minuses in different tests to arrive at the WoE. However, ITS may be used not only for hazard identification but to arrive at potency information. Many *in vitro* tests yield a quantitative output, which may contain partial information on the quantitative risk. Integrating these quantitative data rather than just the yes/no answers from the individual prediction models may thus be a more informative way forward in an ITS. Integration of these quantitative data from different tests can be performed with different mathematical approaches, which are not further discussed here. A Bayesian net approach is just one possibility lately discussed in depth.[6] What is critical here is that the data integration to arrive at the best WoE algorithm will be data driven and cannot be made from a purely theoretical understanding of different steps in a complex process.

Finally an important point: most of the discussion on ITS is to overcome the limitations of the *in vitro* assays by combining data from several assays – but there is currently another movement away from combining data from individual tests towards combining the tests in a single, more complex *in vitro* model. With the creation of 3D-tissues comprising several cell types, micro-tissues and organs on a chip, a lot of current research tries to overcome the limitations of the simplistic *in vitro* assays by building *integrated in vitro systems*. This idea is based on the paradigm that cells need to have proper cell-to-cell interactions and communication between relevant cell types to generate reasonable phenotypic responses and proper metabolic competencies. Whether these integrated multicellular systems will become a reality in routine testing and finally will make integration at the data level obsolete is a fascinating question.

The principles outlined above describe the concept of ITS in general. Below, these principles are discussed in the same logical order for the case of skin sensitization.

9.3   ITS for Skin Sensitization as a Test Case

9.3.1   Background: Skin Sensitization as a 'Chain of Events'

As detailed in Chapter 8, skin sensitization is a T-cell mediated immunological reaction to exogenous molecules. Small, protein-reactive molecules can modify endogenous skin proteins to render them immunogenic.[24] Upon repeated contact with the chemical, a T-cell mediated immune reaction against the chemically modified skin proteins may then occur.

In the *sensitization phase* of skin sensitization, the naïve immune system is primed by the novel epitopes formed by the reactive chemical. This process is a chain of events involving the following steps.

- The molecule applied topically must reach the viable epidermis, and bioavailability in the skin is therefore thought to be a prerequisite for the skin sensitization reaction.[25]
- Certain molecules are considered prohaptens, indicating that they themselves are non-sensitizers and need to be metabolically converted by skin enzymes to form reactive metabolites.[24]
- A key step in the process is the formation of a covalent adduct between the skin sensitizer and endogenous proteins and/or peptides in the skin.
- The modified peptides then must be presented by the Langerhans cells (i.e. the dendritic cells in the skin) on their MHC molecules.
- At the same time, a 'danger signal' needs to be induced. It appears to be part of the innate immune reaction[26] and is formed in absence of specificity conveyed by specific T-cells. Based on the work of Kimber *et al.*[27,28] secretion of cytokines appears to be a central element of this danger signal.
- Finally, stimulated by the triggers/danger signals coming from the innate reactions, the dendritic cells mature and emigrate from the skin into the local lymph node where they finally present the modified peptides to the T-cells and stimulate the proliferation of specific T-cell clones.

Upon repeated contact with the sensitizing agent (which may occur at much lower concentration) the primed T-cells residing in the skin will react to their specific epitopes, thus mounting an inflammatory response, which is the actual disease state known as contact allergy or contact dermatitis. This second phase is called the elicitation phase. The elicitation phase had been the endpoint of the older animal tests evaluating inflamed skin upon repeated application of test chemicals (e.g. Guinea pig maximization test[29] and Buehler test),[30] and it is also the endpoint evaluated when performing human tests, be it the diagnostic patch tests conducted by dermatologists on sensitized individuals or the predictive human repeated insult patch test (HRIPT) occasionally used to assess the sensitization potential of individual chemicals in humans.[31]

Currently, the skin sensitization risk of novel chemicals is predicted by the local lymph node assay in mice which measures the proliferative response in the draining lymph nodes of the ears of mice topically treated thrice on consecutive days with the test agent.[32] This assay thus only includes the sensitization phase, and lacks the elicitation step. Similarly, most *in vitro* assays in advanced development only mimic a particular aspect or step of the sensitization phase.

9.3.2 Why Do We Need Data Integration for Skin Sensitization

Data integration for skin sensitization is perceived necessary for four of the five theoretical reasons outlined above.

1 The sensitization phase of skin sensitization is clearly composed of a serial *chain of events*, all necessary to arrive at the sensitized state of the organisms. It was thus proposed that an ITS must reflect these serial events to arrive at a reasonable conclusion.[33]

2 *Different modes of action*: sensitizers fall in different classes based on their reactivity: a majority has a strong reactivity with cysteine residues in proteins and a minority is able to (rather specifically) react with lysine residues.[34,35] An even lower fraction (mainly corticosteroids) has also been reported to be specifically reactive to arginine residues. This different reactivity is also reflected in different patterns of biological responses at the cellular level, with many cell-based *in vitro* assays not detecting specifically amine-reactive compounds.[36,37]

3 *Limited applicability domains* have been reported for several assays. A key limitation to all submerged cell culture assays is the limited solubility of chemicals in cell culture media, which makes testing of hydrophobic chemicals with a c Log $P > 5$ difficult.

4 Finally a lack of ADME information in the *in vitro* tests is a further perceived limitation.

9.3.3 Sources of Complementary Information for Skin Sensitization Assessment

9.3.3.1 Biological Assays Based on Cell Culture Systems

A large number of biological assays and molecular endpoints have been described and they have recently been reviewed elsewhere,[38] and a summary is given in Table 9.1 – reproduced from the OECD adverse outcome pathway document.[39] They include activation of surface markers in dendritic cells and cell lines,[40,41] activation of gene expression signatures[42,43] or reporter genes,[36] secretion of specific interleukins,[37] triggering formation of reactive oxygen species,[44] induction of migration of dendritic cells[45] and proliferation of blood derived T-cells by haptenized dendritic cells from autologous donors.[46]

The majority of these assays is related to dendritic cell maturation and activation, and to some degree they may mimic the danger signal. However, especially the assays based on gene expression and reporter gene activation may also reflect activation of pathways of defence triggered by sensitizers.

Table 9.1   The key events in skin sensitization according to the OECD adverse outcome pathway document.

Events	Experimental support	Strength of evidence
Key event 1 (**initial event**)	Site of action proteins; covalent binding at cysteine and/or lysine	**Strong**; well-accepted mode of toxic action associated with skin sensitisation
Key event 2	a) Keratinocyte inflammatory responses	a) **Adequate**; well-accepted cytokine IL-18 associated with skin sensitisation
	b) Gene expression of antioxidant response element in keratinocytes	b) **Strong**; well-accepted cell signalling pathway antioxidant/ electrophile response element ARE/EpRE-dependent pathways
Key event 3	Activation of dendritic cells	**Adequate**; well-accepted expressions of cell adhesion and co-stimulatory molecules, and cytokines associated with skin sensitization; various endpoints
Key event 4	T-cell proliferation	**Strong**; two decades of development and testing with the Local Lymph Node Assay (LLNA)
Adverse outcome	Allergic contact dermatitis in humans or its rodent equivalent contact hypersensitivity	**Well-established**; test guidelines and data for guinea pig, as well as data for human

Reproduced in abbreviated form from ref. 39.

It is interesting to note that, with the exception of IL-18 release,[47] the endpoints measured are not directly linked to the recently emerging picture of danger signal formation, which appears to involve sensitizer-triggered ATP release from cells and hyaluronic acid degradation.[48,49] Whether, based on this recent mechanistic understanding, further predictive tests on danger signal formation can be developed remains an open question.

9.3.3.2   In Chemico and In Silico Evaluation of Specific Chemical and Physicochemical Parameters

Since the molecular initiating event in skin sensitization is well understood and involves the formation of a hapten–protein conjugate, methods to mimic peptide or protein binding were found to have a high predictivity and to model a key event. The reactivity with peptides has been studied in greatest detail[50] but also reactions with small nucleophiles have been proposed as surrogate systems.[51]

The reactivity can also be modelled in silico. The structural alerts for reactivity are relatively well understood, and systems which use this information to predict reactivity (and based thereon skin sensitization) of unknown molecules were developed.[52,53]

Finally, also the step of skin penetration can be modelled *in silico*. A system specifically aimed at predicting finite dose exposure (as is the case in animal tests for by skin sensitization) was recently developed.[54] On the other hand penetration can also be measured with classical diffusion cells, and extensive work was done to evaluate epidermal disposition of 10 key sensitizers applied in the vehicle of the animal test.[55]

9.3.3.3 Metabolism Assays

There is still a very limited knowledge on the metabolic capacity of skin. At the mRNA level, expression of several P450 enzymes in skin was shown, and based on these data a 'skin-like' cocktail of recombinant P450 enzymes was developed.[56] On the other hand, recent work at the proteomic level showed that P450 enzymes in skin explants, keratinocytes and an epidermal model are below detection level.[57] Also at the activity level, there is partly contradicting evidence for P450/phase I enzyme levels in the skin,[58] whereas both at the proteomic and at the activity level the skin appears to have a high detoxification/phase II potential.[59] These results stand in a certain conflict with the empirical observation that for a number of skin sensitizers a metabolic activation appears a prerequisite to form a reactive hapten. However, it should be kept in mind that all the experiments on skin metabolism in skin homogenates or keratinocyte models may overlook the possibility that metabolism might take place in microlocations such as hair follicles or in minor cell types, which were reported to have particular metabolic capacities, such as neutrophils[60] or Langerhans cells.[61]

9.3.4 Different Weight of Information: Essential vs. Modifying Factors in the Sensitization Phase

Reactivity is clearly an essential event, hence reactivity information will always have a high predictive weighting (see Table 9.1).

What is currently unknown is whether *danger signal formation* is an independent parameter, strongly modifying sensitizer potency (i.e. whether two chemicals with similar reactivity can have widely differing potencies by triggering danger signal formation with differing efficacy). It was postulated that the activation of innate immune and stress responses might be due to contact allergen-dependent chemical modification of specific regulatory proteins.[48] In this case reactivity, next to the formation of the epitopes, is the key factor providing the danger signal and a reactive molecule will always provide the danger signal AND protein haptenization.

Bioavailability is certainly a prerequisite for sensitization to occur; however, it is very rare that a reactive chemical was found to be non-sensitizing due to the fact that it does not reach the target site in the skin. Interestingly, extremely reactive chemicals that are highly unstable in contact with water or which react with peptides within seconds[62] can in most cases enter the skin sufficiently well

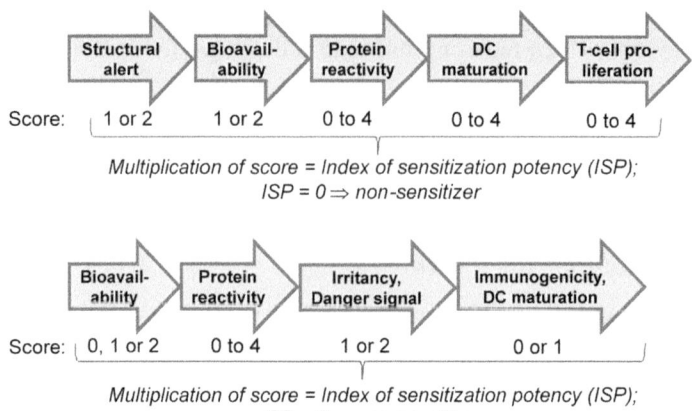

Figure 9.2 The proposed sequential paradigms for data integration for skin sensitization.[33,65]

to trigger sensitization, although one would expect them to be rapidly consumed by reactions with proteins at the skin surface (for more discussion on this issue see references 63 and 64). Based on the limited empirical evidence for bioavailability as a limiting factor, its importance was questioned,[21] yet it was also considered both as a modifying[25,33] or even an essential factor (see below and Figure 9.2).[65]

Much less is known about *antagonizing mechanisms*. For example, it was found that some reactive molecules have, linked to their reactivity, anti-inflammatory potential, and that such anti-inflammatory events may dampen the sensitization reaction.[62,66,67] While such anti-inflammatory effects start to be documented, antioxidant effects may also have a dampening action. Thus the antioxidant luteolin[68] was found to inhibit sensitization reactions despite the fact that it contains structural motifs of strongly sensitizing chemicals (α,β-unsaturated ketone and a catechol unit, like many other plant-derived phenolic antioxidants). Such reactive chemicals with antioxidant and/or anti-inflammatory activity may be over-predicted in an ITS, as long as most assays are linked to reactivity and danger signal formation and do not take possible dampening factors into account. Considering dampening effects as modifying factors may thus become important in a future quantitative ITS for skin sensitization.

9.3.5 How to Integrate Data: General Possibilities and How They Have Been Explored in the Skin Sensitization Field

9.3.5.1 Data Integration by a Serial Combination

The need for an ITS in skin sensitization has been widely discussed, but experimental and data driven approaches are only slowly emerging.

The first conceptual paradigm[33] was based on the 'serial chain of events'. It proposed that data from skin penetration and structural alerts should be combined with data on protein reactivity, dendritic cell activation and T-cell proliferation (see Figure 9.2). In this proposal, the latter three factors would be linked by an AND operator: all these three tests must be positive (all three hurdles overcome by the test chemical) in order that it would be rated as a sensitizer, whereas skin penetration and structural alerts would only obtain the status of a modifying factor.

In an update of this paradigm,[65] skin penetration, protein reactivity and dendritic cell activation are considered essential factors, whereas danger signal formation/irritancy was added as a modifying factor (Figure 9.2). Structural alert and T-cell proliferation were omitted in this updated paradigm, the latter mainly for the practical reason that T-cell tests currently are not sufficiently developed.

A serial recapitulation of events is also exemplified in the recently published adverse outcome pathway (AOP) developed by the OECD.[39]

If all the factors considered add to potency, combining them in a serial manner (the paradigms proposed multiplication of '*in vitro* scores') would indeed be a logical approach to model potency.

On the other hand, one may criticize this approach for a number of reasons. From a purely practical viewpoint, it would indicate that very sensitive assays are needed at each stage. If a false-negative result is generated at any of the stages considered essential, this would lead to a false-negative final rating and a very low overall sensitivity. We have presented a data integration approach based on data from the DPRA and an Nrf2-activation assay on 116 chemicals.[69] If we applied a serial combination of data, a specificity of 97% and a sensitivity of only 66% was obtained indicating this limitation. But there is a second issue: the steps proposed are indeed sequential events, but some *in vitro* assays may already mimic several of them. Especially a T-cell proliferation assay measures by definition the binding of the chemical to proteins in the dendritic cells and activation of dendritic cells and T cell proliferation. Thus this assay itself would already represent several steps, and if it was sufficiently robust and predictive, it might make protein binding and dendritic cell activation assays obsolete.

9.3.5.2 Data Integration by a Parallel Combination of Data

Other authors have explored the possibility of combining data from different tests by adding positive evidence from different tests (positive in A *OR* B). Data from the h-CLAT and the DPRA assay for a set of 101 chemicals were combined, taking positive evidence from either prediction model as sufficient for a positive rating. This led to a very high sensitivity of 96.1%, but this came at the cost of a low specificity (56%).[70] We have proposed a very simple combination of positive evidence from the KeratinoSens™ assay and an LC-MS based peptide binding assay.[36] If only observation of direct adduct formation from the peptide reactivity is taken as positive evidence, the latter assay has a limited sensitivity and a high specificity. Since the peptide binding assay can recognize specifically lysine-reactive chemicals (next to those directly reactive to cysteine), a minimalistic ITS combining positive evidence from these two assays enhances sensitivity[35] without compromising specificity. Adding tests with a high specificity in a parallel approach of data integration is crucial to avoid an ITS with very low overall specificity as was the case in genotoxicity, where a parallel integration of tests led to an extremely high false-positive rate.[71]

9.3.5.3   Data Integration by WoE Assessment

Since the tests which are available often do not fit nicely in a particular box of the serial events, a combination by a WoE may be pragmatic way forward, and a couple of data-driven examples to generate the WoE have been published in the last three years.

Data on 54 chemicals[72] tested in parallel in dendritic cell activation assays,[73,74] the direct peptide reactivity assay[75] and Nrf2-reporter assays (KeratinoSens™,[36] and LuSens) were evaluated for their use in an ITS for hazard identification. The assays related to reactivity (DPRA and Nrf2-reporters) had a high sensitivity, whereas the dendritic cell assay with U937 cells had a high specificity and the following integration was proposed: a positive rating in the U937 test leads to a sensitizer classification, while a negative result in both DPRA and the Nrf-2 reporter assay triggers a negative rating. If the results are contradicting, the WoE would be simply to take a 'majority voting' – the prediction obtained in two of the three tests would be the final classification. With this approach an amazingly good predictivity was achieved, with 94% accuracy to predict human data and 83% to predict the LLNA result, and this combination could predict the human data better than the LLNA did. However, no attempts to predict potency were made. This majority voting was recently also applied on a larger dataset of 145 chemicals.[76]

We had presented earlier a dataset on 116 chemicals tested in the DPRA (Cys-peptide only), in an Nrf2-reporter assay[77] and *in silico* with the TIMES-SS software.[52] As described above, a serial combination did not appear to be the most appropriate approach for this dataset, and different WoE approaches were discussed. Basically, as proposed in the Jowsey *et al.* paradigm,[33] the *in vitro* results were transformed to discrete scores/classes. Average scores (instead of products of scores) were then used (taking the average or sum of scores is making the simplest possible arithmetic WoE). With this approach a better accuracy was obtained, and a relationship of the average scores to potency was observed, although the prediction of potency at the level of the individual chemical was not yet satisfactory.

A very similar evaluation was later also presented for the dataset on 101 chemicals tested in the h-CLAT and the DPRA assay[70] and *in silico* with DEREK. Sum of scores from the three approaches were taken, and used for a WoE hazard and potency prediction.

The data integration approaches discussed above are, mathematically speaking, trivial. Also they are heavily influenced by the prediction model of the individual tests – class 0 is normally attributed based on the negative prediction from the individual prediction model. So far only one attempt was made to take the numerical data from individual tests and use them in a more sophisticated mathematical modelling approach: Jaworska *et al.*[78] developed a Bayesian net based on a larger dataset from the DPRA, the Nrf2-reporter assay, TIMES-SS, a bioavailability model[54] and (for a smaller number of chemicals) the U937 dendritic cell activation assay. This model tries to recapitulate the known chain of events, and then used the quantitative *in vitro* data to assign for each chemical probabilities of whether it falls into a particular LLNA potency class. A Bayesian net was constructed where the measured variables are linked to mechanistically relevant latent variables. The net then summarized the mutual information between the measured and latent variables. Next to assigning probabilities based on a full set of data for a given chemical, it was shown that this approach can also be used to predict, in cases of chemicals with partial datasets, whether additional testing will significantly improve the prediction. This model was recently updated with a more comprehensive dataset on 145 chemicals tested in the DPRA, KeratinoSens™ assay, U937 activation assay, TIMES-SS and the bioavailability model.[79]

9.3.5.4 Data Integration by a Tiered Strategy

Tiered strategies have not been widely explored to use exclusively non-animal data. They were mainly described for a reductive ITS or an ITS taking also human/historical data into account. An overall scheme on how *in vitro* data might be used in a more global assessment was recently presented.[80]

9.3.5.5 Reduction to the Rate-limiting Step

While the battery approach and ITS have a broad support in the scientific community, a logically valid criticism had been put forward.[21] As stated in Section 9.3.4, it is not clear how important bioavailability parameters are, and it is equally unclear whether danger signal formation is an event not mainly governed by the reactivity of chemicals. Thus it was proposed that the assessment can be made based on the rate-limiting or key-initiating event, which is the reactivity of chemicals towards relevant nucleophiles, as all other events can be taken for granted or do directly follow from the reactivity. In this paradigm it is proposed that such assessments are always made within specific mechanistically defined applicability domains, and thus this proposal contains the idea of read across. If we know a number of chemicals with a similar reaction mechanism, a new chemical in the same class can be predicted purely based on its kinetic reaction rate. This may be perceived as an over-simplification, yet it may hold true at least for groups of chemicals for which (i) reaction mechanism is clear, (ii) rate constants can be measured and (iii) which contain a sufficiently large number of chemicals with animal/human data. We have presented such simple kinetic models in a kinetic modification of the DPRA for the groups of Michael acceptors[62,81] and chemicals reacting by addition-elimination reactions.[62] However, there are two important notes of caution: (a) only some chemicals will fall in such clear groups and (b) reaction

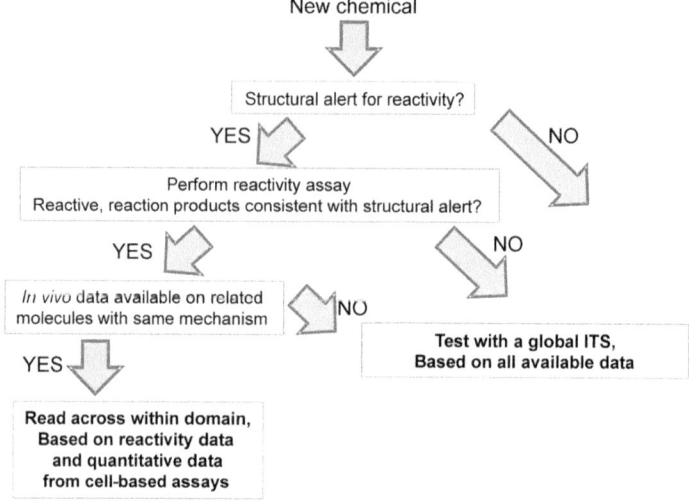

Figure 9.3 A potential paradigm how a holistic ITS on all available data can be combined with predictions in specific applicability domains.

mechanisms were often predicted based on structural alerts, but these *in cerebro* predictions do not always stand empirical tests. Thus for both the extreme sensitizers oxazolone[82] and chloratranol[83] we found a reaction mechanism with the DPRA-peptides differing from the published theoretical predictions.

9.3.5.6   Data Integration Within Mechanistic Applicability Domains and Read-across

The battery approaches presented above treat all chemicals equal, and do not specifically take into account *a priori* knowledge on the chemical structure of the test chemical. However, different classes of chemicals may have an intrinsically differing behaviour in *in vitro* tests (e.g. due to differing dampening factors). Thus instead of comparing everything with everything (as is done in an ITS based on a database of all available LLNA data), it may be a smarter way to compare apples to apples and pears to pears. *In vitro* data may be used for a read-across to related chemicals only, for which both *in vitro* and *in vivo* data are available. Instead of discussing whether a global ITS is correct or whether all can be solved within applicability domains, these two concepts could be used in parallel – a possible way to do this is shown in Figure 9.3 as final food for thought.

References

1. H. Spielmann, S. Hoffmann, M. Liebsch, P. Botham, J. H. Fentem, C. Eskes, R. Roguet, J. Cotovio, T. Cole, A. Worth, J. Heylings, P. Jones, C. Robles, H. Kandarova, A. Gamer, M. Remmele, R. Curren, H. Raabe,

A. Cockshott, I. Gerner and V. Zuang, The ECVAM international valid-ation study on in vitro tests for acute skin irritation: report on the validity of the EPISKIN and EpiDerm assays and on the Skin Integrity Function Test, *Altern. Lab. Anim.*, 2007, **35**, 559–601.

2. M. Ceridono, P. Tellner, D. Bauer, J. Barroso, N. Alepee, R. Corvi, A. De Smedt, M. D. Fellows, N. K. Gibbs, E. Heisler, A. Jacobs, D. Jirova, D. Jones, H. Kandarova, P. Kasper, J. K. Akunda, C. Krul, D. Learn, M. Liebsch, A. M. Lynch, W. Muster, K. Nakamura, J. F. Nash, U. Pfannenbecker, G. Phillips, C. Robles, V. Rogiers, F. Van De Water, U. W. Liminga, H. W. Vohr, O. Wattrelos, J. Woods, V. Zuang, J. Kreysa and P. Wilcox, The 3T3 neutral red uptake phototoxicity test: practical ex-perience and implications for phototoxicity testing--the report of an ECVAM-EFPIA workshop, *Regul. Toxicol. Pharmacol.*, 2012, **63**, 480–488.

3. K. Maier, R. Schmitt-Landgraf and B. Siegemund, Development of an in vitro test system with human skin cells for evaluation of phototoxicity, *Toxicol. In Vitro*, 1991, **5**, 457–461.

4. M. Bouhifd, G. Bories, J. Casado, S. Coecke, H. Norlen, N. Parissis, R. M. Rodrigues and M. P. Whelan, Automation of an in vitro cytotoxicity assay used to estimate starting doses in acute oral systemic toxicity tests, *Food. Chem. Toxicol*, 2012, **50**, 2084–2096.

5. D. Kirkland, L. Reeve, D. Gatehouse and P. Vanparys, A core in vitro genotoxicity battery comprising the Ames test plus the in vitro micro-nucleus test is sufficient to detect rodent carcinogens and in vivo geno-toxins, *Mutat Res.*, 2011, **721**, 27–73.

6. J. Jaworska, S. Gabbert and T. Aldenberg, Towards optimization of chemical testing under REACH: a Bayesian network approach to Inte-grated Testing Strategies, *Regul. Toxicol. Pharmacol.*, 2010, **57**, 157–167.

7. R. Combes, C. Grindon, M. T. Cronin, D. W. Roberts and J. F. Garrod, Integrated decision-tree testing strategies for acute systemic toxicity and toxicokinetics with respect to the requirements of the EU REACH legis-lation, *Altern. Lab Anim.*, 2008, **36**, 45–63.

8. C. Grindon, R. Combes, M. T. Cronin, D. W. Roberts and J. F. Garrod, An integrated decision-tree testing strategy for skin sensitisation with re-spect to the requirements of the EU REACH legislation, *Altern. Lab Anim.*, 2008, **36**(Suppl 1), 75–89.

9. C. Grindon, R. Combes, M. T. Cronin, D. W. Roberts and J. F. Garrod, An integrated decision-tree testing strategy for repeat dose toxicity with respect to the requirements of the EU REACH legislation, *Altern. Lab Anim.*, 2008, **36**, 93–101.

10. C. Grindon, R. Combes, M. T. Cronin, D. W. Roberts and J. F. Garrod, An integrated decision-tree testing strategy for eye irritation with respect to the requirements of the EU REACH legislation, *Altern. Lab Anim.*, 2008, **36**, 81–92.

11. C. Rovida, Food for thought ... why no new in vitro tests will be done for REACH by registrants, *ALTE*, 2010, **27**, 175–183.

12. C. Rovida and F. Longo, R. R. Rabbit, How are reproductive toxicity and developmental toxicity addressed in REACH dossiers?, *ALTEX*, 2011, **28**, 273–294.

13. C. Rovida, Local lymph node assay: how testing laboratories apply OECD TG 429 for REACH purposes, *ALTEX*, 2011, **28**, 117–129.

14. G. T. Ankley, R. S. Bennett, R. J. Erickson, D. J. Hoff, M. W. Hornung, R. D. Johnson, D. R. Mount, J. W. Nichols, C. L. Russom, P. K. Schmieder, J. A. Serrrano, J. E. Tietge and D. L. Villeneuve, Adverse outcome pathways: a conceptual framework to support ecotoxicology research and risk assessment, *Environ. Toxicol. Chem.*, 2010, **29**, 730–741.

15. B. De Wever, H. W. Fuchs, M. Gaca, C. Krul, S. Mikulowski, A. Poth, E. L. Roggen and M. R. Vila, Implementation challenges for designing integrated in vitro testing strategies (ITS) aiming at reducing and replacing animal experimentation, *Toxicol. In Vitro*, 2012, **26**, 526–534.

16. National-Research-Council, *Toxicity Testing in the 21st Century. A Vision and a Strategy*. National Academies Press: Washington, DC, 2007.

17. T. Hartung, E. van Vliet, J. Jaworska, L. Bonilla, N. Skinner and R. Thomas, Food for thought ... Systems toxicology, *ALTEX*, 2012, **29**, 119–128.

18. J. Dumont, R. Josse, C. Lambert, S. Antherieu, L. Le Hegarat, C. Aninat, M. A. Robin, C. Guguen-Guillouzo and A. Guillouzo, Differential toxicity of heterocyclic aromatic amines and their mixture in metabolically competent HepaRG cells, *Toxicol. Appl. Pharmacol.*, 2010, **245**, 256–263.

19. L. Birrell, P. Cahill, C. Hughes, M. Tate and R. M. Walmsley, GADD45a-GFP GreenScreen HC assay results for the ECVAM recommended lists of genotoxic and non-genotoxic chemicals for assessment of new genotoxicity tests, *Mutat. Res.*, 2010, **695**, 87–95.

20. P. C. von der Ohe, R. Kuhne, R. U. Ebert, R. Altenburger, M. Liess and G. Schuurmann, Structural alerts--a new classification model to discriminate excess toxicity from narcotic effect levels of organic compounds in the acute daphnid assay, *Chem. Res. Toxicol.*, 2005, **18**, 536–555.

21. D. W. Roberts and A. O. Aptula, Determinants of skin sensitisation potential, *Journal of Applied Toxicology*, 2008, **28**, 377–387.

22. C. Jagger, M. Tate, P. A. Cahill, C. Hughes, A. W. Knight, N. Billinton and R. M. Walmsley, Assessment of the genotoxicity of S9-generated metabolites using the GreenScreen HC GADD45a-GFP assay., *Mutagenesis*, 2009, **24**, 35–50.

23. J. Zheng, L. Ma, B. Xin, T. Olah, W. G. Humphreys and M. Zhu, Screening and identification of GSH-trapped reactive metabolites using hybrid triple quadruple linear ion trap mass spectrometry, *Chemical Research in Toxicology*, 2007, **20**, 757–766.

24. A. T. Karlberg, M. A. Bergstrom, A. Borje, K. Luthman and J. L. Nilsson, Allergic contact dermatitis--formation, structural requirements, and reactivity of skin sensitizers, *Chem. Res. Toxicol.*, 2008, **21**, 53–69.

25. D. Basketter, C. Pease, G. Kasting, I. Kimber, S. Casati, M. Cronin, W. Diembeck, F. Gerberick, J. Hadgraft, T. Hartung, J. P. Marty,

E. Nikolaidis, G. Patlewicz, D. Roberts, E. Roggen, C. Rovida and J. Van De Sandt, Skin sensitisation and epidermal disposition: The relevance of epidermal disposition for sensitisation hazard identification and risk assessment: The report and recommendations of ECVAM workshop 59a, *ATLA Alternatives to Laboratory Animals*, 2007, **35**, 137–154.

26. M. A. Freudenberg, P. R. Esser, T. Jakob, C. Galanos and S. F. Martin, Innate and adaptive immune responses in contact dermatitis: analogy with infections, *G. Ital. Dermatol. Venereol.*, 2009, **144**, 173–185.

27. R. J. Dearman, M. Cumberbatch and I. Kimber, Cutaneous cytokine expression: Induction by chemical allergen and paracrine regulation, *Journal of Toxicology - Cutaneous and Ocular Toxicology*, 2003, **22**, 69–86.

28. I. Kimber, M. Cumberbatch, R. J. Dearman, M. Bhushan and C. E. M. Griffiths, Cytokines and chemokines in the initiation and regulation of epidermal Langerhans cell mobilization, *British Journal of Dermatology*, 2000, **142**, 401–412.

29. J. E. Wahlberg and A. Boman, Guinea pig maximization test, *Current problems in dermatology*, 1985, **14**, 59–106.

30. D. A. Basketter and G. F. Gerberick, An interlaboratory evaluation of the Buehler test for the identification and classification of skin sensitizers, *Contact Dermatitis*, 1996, **35**, 146–151.

31. P. M. McNamee, A. M. Api, D. A. Basketter, G. Frank Gerberick, D. A. Gilpin, B. M. Hall, I. Jowsey and M. K. Robinson, A review of critical factors in the conduct and interpretation of the human repeat insult patch test, *Regul. Toxicol. Pharmacol.*, 2008, **52**, 24–34.

32. D. A. Basketter, P. Evans, R. J. Fielder, G. F. Gerberick, R. J. Dearman and I. Kimber, Local lymph node assay - Validation, conduct and use in practice, *Food Chem. Toxicol.*, 2002, **40**, 593–598.

33. I. R. Jowsey, D. A. Basketter, C. Westmoreland and I. Kimber, A future approach to measuring relative skin sensitising potency: A proposal, *Journal of Applied Toxicology*, 2006, **26**, 341–350.

34. M. Aleksic, E. Thain, D. Roger, O. Saib, M. Davies, J. Li, A. Aptula and R. Zazzeroni, Reactivity profiling: Covalent modification of single nucleophile peptides for skin sensitization risk assessment, *Toxicol. Sci.*, 2009, **108**, 401–411.

35. A. Natsch, The Nrf2-Keap1-ARE toxicity pathway as a cellular sensor for skin sensitizers--functional relevance and a hypothesis on innate reactions to skin sensitizers, *Toxicol. Sci.*, 2010, **113**, 284–292.

36. R. Emter, G. Ellis and A. Natsch, Performance of a novel keratinocyte-based reporter cell line to screen skin sensitizers in vitro, *Toxicol. Appl. Pharmacol.*, 2010, **245**, 281–290.

37. E. Corsini, M. Mitjans, V. Galbiati, L. Lucchi, C. L. Galli and M. Marinovich, Use of IL-18 production in a human keratinocyte cell line to discriminate contact sensitizers from irritants and low molecular weight respiratory allergens, *Toxicol. In Vitro*, 2009, **23**, 789–796.

38. A. Mehling, T. Eriksson, T. Eltze, S. Kolle, T. Ramirez, W. Teubner, B. van Ravenzwaay and R. Landsiedel, Non-animal test methods

for predicting skin sensitization potentials, *Arch Toxicol.*, 2012, **86**, 1273–1295.

39. OECD, The Adverse Outcome Pathway for Skin Sensitisation Initiated by Covalent Binding to Proteins, Part 1: Scientific Evidence. OECD ENVIRONMENT, HEALTH AND SAFETY PUBLICATIONS, SERIES ON TESTING AND ASSESSMENT, 2012, NO. 168.

40. H. Reuter, J. Spieker, S. Gerlach, U. Engels, W. Pape, L. Kolbe, R. Schmucker, H. Wenck, W. Diembeck, K. P. Wittern, K. Reisinger and A. G. Schepky, In vitro detection of contact allergens: development of an optimized protocol using human peripheral blood monocyte-derived dendritic cells, *Toxicol. In Vitro*, 2011, **25**, 315–323.

41. H. Sakaguchi, T. Ashikaga, N. Kosaka, S. Sono, N. Nishiyama and H. Itagaki, The in vitro skin sensitization test; human cell line activation test (h-CLAT) using THP-1 cells, *Abstracts of the 44th Congress of the European Societies of Toxicology. Toxicology Letters*, 2007, **172**(Supp 1), S93.

42. H. Johansson, M. Lindstedt, A. S. Albrekt and C. A. Borrebaeck, A genomic biomarker signature can predict skin sensitizers using a cell-based in vitro alternative to animal tests, *BMC Genomics*, 2011, **12**, 399.

43. J. W. van der Veen, T. E. Pronk, H. van Loveren and J. Ezendam, Applicability of a keratinocyte gene signature to predict skin sensitizing potential, *Toxicol. In Vitro*, 2013, **27**, 314–322.

44. M. Miyazawa and A. Takashima, Development and validation of a new in vitro assay designed to measure contact allergen-triggered oxidative stress in dendritic cells, *J. Dermatol. Sci.*, 2012, **68**, 73–81.

45. K. Ouwehand, S. W. Spiekstra, J. Reinders, R. J. Scheper, T. D. de Gruijl and S. Gibbs, Comparison of a novel CXCL12/CCL5 dependent migration assay with CXCL8 secretion and CD86 expression for distinguishing sensitizers from non-sensitizers using MUTZ-3 Langerhans cells, *Toxicol. In Vitro*, 2010, **24**, 578–585.

46. M. Vocanson, M. Cluzel-Tailhardat, G. Poyet, M. Valeyrie, C. Chavagnac, B. Levarlet, P. Courtellemont, A. Rozieres, A. Hennino and J. F. Nicolas, Depletion of human peripheral blood lymphocytes in CD25+ cells allows for the sensitive in vitro screening of contact allergens, *J. Invest. Dermatol.*, 2008, **128**, 2119–2122.

47. C. Antonopoulos, M. Cumberbatch, J. B. Mee, R. J. Dearman, X. Q. Wei, F. Y. Liew, I. Kimber and R. W. Groves, IL-18 is a key proximal mediator of contact hypersensitivity and allergen-induced Langerhans cell migration in murine epidermis, *Journal of Leukocyte Biology*, 2008, **83**, 361–367.

48. S. F. Martin, P. R. Esser, F. C. Weber, T. Jakob, M. A. Freudenberg, M. Schmidt and M. Goebeler, Mechanisms of chemical-induced innate immunity in allergic contact dermatitis, *Allergy*, 2011, **66**, 1152–1163.

49. P. R. Esser, U. Wolfle, C. Durr, F. D. von Loewenich, C. M. Schempp, M. A. Freudenberg, T. Jakob and S. F. Martin, Contact Sensitizers Induce Skin Inflammation via ROS Production and Hyaluronic Acid Degradation, *PLoS ONE*, 2012, **7**, e41340.

50. F. Gerberick, M. Aleksic, D. Basketter, S. Casati, A. T. Karlberg, P. Kern, I. Kimber, J. P. Lepoittevin, A. Natsch, J. M. Ovigne, C. Rovida, H. Sakaguchi and T. Schultz, Chemical reactivity measurement and the predicitve identification of skin sensitisers. The report and recommendations of ECVAM Workshop 64, *Altern. Lab. Anim.*, 2008, **36**, 215–242.

51. I. Chipinda, R. O. Ajibola, M. K. Morakinyo, T. B. Ruwona, R. H. Simoyi and P. D. Siegel, Rapid and simple kinetics screening assay for electrophilic dermal sensitizers using nitrobenzenethiol, *Chem. Res. Toxicol.*, 2010, **23**, 918–925.

52. G. Patlewicz, S. D. Dimitrov, L. K. Low, P. S. Kern, G. D. Dimitrova, M. I. Comber, A. O. Aptula, R. D. Phillips, J. Niemela, C. Madsen, E. B. Wedebye, D. W. Roberts, P. T. Bailey and O. G. Mekenyan, TIMES-SS--a promising tool for the assessment of skin sensitization hazard. A characterization with respect to the OECD validation principles for (Q)SARs and an external evaluation for predictivity, *Regul. Toxicol. Pharmacol.*, 2007, **48**, 225–239.

53. M. D. Barratt, D. A. Basketter, M. Chamberlain, G. D. Admans and J. J. Langowski, An expert system rulebase for identifying contact allergens, *Toxicology in Vitro*, 1994, **8**, 1053–1060.

54. Y. Dancik, M. A. Miller, J. Jaworska and G. B. Kasting, Design and performance of a spreadsheet-based model for estimating bioavailability of chemicals from dermal exposure, *Adv. Drug Deliv. Rev.*, 2013, **65**, 221–236.

55. M. Davies, R. U. Pendlington, L. Page, C. S. Roper, D. J. Sanders, C. Bourner, C. K. Pease and C. MacKay, Determining epidermal disposition kinetics for use in an integrated nonanimal approach to skin sensitization risk assessment, *Toxicol. Sci.*, 2011, **119**, 308–318.

56. M. A. Bergström, H. Ott, A. Carlsson, M. Neis, G. Zwadlo-Klarwasser, C. A. M. Jonsson, H. F. Merk, A. T. Karlberg and J. M. Baron, A skin-like cytochrome P450 cocktail activates prohaptens to contact allergenic metabolites, *Journal of Investigative Dermatology*, 2007, **127**, 1145–1153.

57. S. van Eijl, Z. Zhu, J. Cupitt, M. Gierula, C. Gotz, E. Fritsche and R. J. Edwards, Elucidation of xenobiotic metabolism pathways in human skin and human skin models by proteomic profiling, *PLoS ONE*, 2012, **7**, e41721.

58. C. Gotz, R. Pfeiffer, J. Tigges, V. Blatz, C. Jackh, E. M. Freytag, E. Fabian, R. Landsiedel, H. F. Merk, J. Krutmann, R. J. Edwards, C. Pease, C. Goebel, N. Hewitt and E. Fritsche, Xenobiotic metabolism capacities of human skin in comparison with a 3D epidermis model and keratinocyte-based cell culture as in vitro alternatives for chemical testing: activating enzymes (Phase I), *Exp. Dermatol.*, 2012, **21**, 358–363.

59. C. Gotz, R. Pfeiffer, J. Tigges, K. Ruwiedel, U. Hubenthal, H. F. Merk, J. Krutmann, R. J. Edwards, J. Abel, C. Pease, C. Goebel, N. Hewitt and E. Fritsche, Xenobiotic metabolism capacities of human skin in comparison with a 3D-epidermis model and keratinocyte-based cell culture as

in vitro alternatives for chemical testing: phase II enzymes, *Exp. Dermatol.*, 2012, **21**, 364–369.

60. E. Kiorpelidou, B. Foster, J. Farrell, M. O. Ogese, L. Faulkner, C. E. Goldring, B. K. Park and D. J. Naisbitt, IL-8 release from human neutrophils cultured with pro-haptenic chemical sensitizers, *Chem. Res. Toxicol.*, 2012, **25**, 2054–2056.

61. B. G. Modi, J. Neustadter, E. Binda and J. Lewis, R. B. Filler, S. J. Roberts, B. Y. Kwong, S. Reddy, J. D. Overton, A. Galan, R. Tigelaar, L. Cai, P. Fu, M. Shlomchik, D. H. Kaplan, A. Hayday, M. Girardi, Langerhans cells facilitate epithelial DNA damage and squamous cell carcinoma, *Science*, 2012, **335**, 104–108.

62. A. Natsch, T. Haupt and H. Laue, Relating skin sensitizing potency to chemical reactivity: reactive Michael acceptors inhibit NF-kappaB signaling and are less sensitizing than S(N)Ar- and S(N)2- reactive chemicals, *Chem. Res. Toxicol.*, 2011, **24**, 2018–2027.

63. C. Pickard, F. Louafi, C. McGuire, K. Lowings, P. Kumar, H. Cooper, R. J. Dearman, M. Cumberbatch, I. Kimber, E. Healy and P. S. Friedmann, The cutaneous biochemical redox barrier: a component of the innate immune defenses against sensitization by highly reactive environmental xenobiotics, *J. Immunol.*, 2009, **183**, 7576–7584.

64. K. Samuelsson, C. Simonsson, C. A. Jonsson, G. Westman, M. B. Ericson and A. T. Karlberg, Accumulation of FITC near stratum corneum-visualizing epidermal distribution of a strong sensitizer using two-photon microscopy, *Contact Dermatitis*, 2009, **61**, 91–100.

65. D. A. Basketter and I. Kimber, Updating the skin sensitization in vitro data assessment paradigm in 2009, *Journal of Applied Toxicology*, 2009, **29**, 545–550.

66. A. Del Bufalo, J. Bernad, C. Dardenne, D. Verda, J. R. Meunier, F. Rousset, S. Martinozzi-Teissier and B. Pipy, Contact sensitizers modulate the arachidonic acid metabolism of PMA-differentiated U-937 monocytic cells activated by LPS, *Toxicol. Appl. Pharmacol.*, 2011, **256**, 35–43.

67. C. Lass, M. Vocanson, S. Wagner, C. M. Schempp, J. F. Nicolas, I. Merfort and S. F. Martin, Anti-inflammatory and immune-regulatory mechanisms prevent contact hypersensitivity to *Arnica montana* L, *Exp. Dermatol.*, 2008, **17**, 849–857.

68. U. Wolfle, P. R. Esser, B. Simon-Haarhaus, S. F. Martin, J. Lademann and C. M. Schempp, UVB-induced DNA damage, generation of reactive oxygen species, and inflammation are effectively attenuated by the flavonoid luteolin in vitro and in vivo, *Free Radic. Biol. Med.*, 2011, **50**, 1081–1093.

69. A. Natsch, R. Emter and G. Ellis, Filling the concept with data: Integrating data from different in vitro and in silico assays on skin sensitizers to explore the battery approach for animal-free skin sensitization testing, *Toxicol. Sci.*, 2009, **107**, 106–121.

70. M. Miyazawa, Y. Nukada, K. Saito, H. Sakaguchi, N. Nishiyama and G. Gerberick, A novel strategy for in vitro assay platform to predict skin

sensitizing potential of broad raw materials. The Toxicologist, Supplement to Toxicological Sciences, 2011, **120**(Supplement 2), 548.

71. D. Kirkland, M. Aardema, L. Henderson and L. Muller, Evaluation of the ability of a battery of three in vitro genotoxicity tests to discriminate rodent carcinogens and non-carcinogens I. Sensitivity, specificity and relative predictivity, *Mutat. Res.*, 2005, **584**, 1–256.

72. C. Bauch, S. N. Kolle, T. Ramirez, T. Eltze, E. Fabian, A. Mehling, W. Teubner, B. van Ravenzwaay and R. Landsiedel, Putting the parts together: Combining in vitro methods to test for skin sensitizing potentials, *Regul. Toxicol. Pharmacol.*, 2012, **63**, 489–504.

73. N. Ade, S. Martinozzi-Teissier, M. Pallardy and F. Rousset, Activation of U937 cells by contact sensitizers: CD86 expression is independent of apoptosis, *Journal of Immunotoxicology*, 2006, **3**, 189–197.

74. H. Sakaguchi, T. Ashikaga, M. Miyazawa, Y. Yoshida, Y. Ito, K. Yoneyama, M. Hirota, H. Itagaki, H. Toyoda and H. Suzuki, Development of an in vitro skin sensitization test using human cell lines; human Cell Line Activation Test (h-CLAT) II. An inter-laboratory study of the h-CLAT, *Toxicology in Vitro*, 2006, **20**, 774–784.

75. G. F. Gerberick, J. D. Vassallo, R. E. Bailey, J. G. Chaney, S. W. Morrall and J. P. Lepoittevin, Development of a peptide reactivity assay for screening contact allergens, *Toxicol. Sci.*, 2004, **81**, 332–343.

76. A. Natsch, C. A. Ryan, L. Foertsch, R. Emter, J. Jaworska, F. Gerberick, P. Kern, A dataset on 145 chemicals tested in alternative assays for skin sensitization undergoing prevalidation. *J. Appl. Toxicol.*, 2013, published online, 9 April 2013, DOI:10.1002/jat.2868.

77. A. Natsch and R. Emter, Skin sensitizers induce antioxidant response element dependent genes: Application to the in vitro testing of the sensitization potential of chemicals, *Toxicol. Sci.*, 2008, **102**, 110–119.

78. J. Jaworska, A. Harol, P. S. Kern and G. F. Gerberick, Integrating non animal test information into an adaptive testing strategy - skin sensitization proof of concept case, *ALTEX*, 2011, **28**, 211 225.

79. J. Jaworska, Y. Dancik, P. Kern, F. Gerberick, A. Natsch, Bayesian integrated testing strategy to assess skin sensitization potency: from theory to practice, *J. Appl. Toxicol.*, 2013, published online, 14 May 2013, DOI: 10.1002/jat.2869.

80. C. Goebel, P. Aeby, N. Ade, N. Alepee, A. Aptula, D. Araki, E. Dufour, N. Gilmour, J. Hibatallah, D. Keller, P. Kern, A. Kirst, M. Marrec-Fairley, G. Maxwell, J. Rowland, B. Safford, F. Schellauf, A. Schepky, C. Seaman, T. Teichert, N. Tessier, S. Teissier, H. U. Weltzien, P. Winkler and J. Scheel, Guiding principles for the implementation of non-animal safety assessment approaches for cosmetics: skin sensitisation, *Regul. Toxicol. Pharmacol.*, 2012, **63**, 40–52.

81. D. W. Roberts and A. Natsch, High throughput kinetic profiling approach for covalent binding to peptides: Application to skin sensitization potency of michael acceptor electrophiles, *Chem. Res. Toxicol.*, 2009, **22**, 592–603.

82. A. Natsch, H. Gfeller, F. Kuhn, T. Granier and D. W. Roberts, Chemical Basis for the Extreme Skin Sensitization Potency of (E)-4-(Ethoxymethylene)-2-phenyloxazol-5(4H)-one, *Chem. Res. Toxicol.*, 2010, **23**, 1913–1920.

83. A. Natsch, H. Gfeller, T. Haupt and G. Brunner, Chemical reactivity and skin sensitization potential for benzaldehydes: can schiff base formation explain everything?, *Chem. Res. Toxicol.*, 2012, **25**, 2203–2215.

CHAPTER 10

In Vitro Receptor Binding Assays

J. CHARLES ELDRIDGE

Department of Physiology & Pharmacology, Wake Forest School of
Medicine, Winston-Salem, NC 27157-1083, USA
Email: eldridge@wakehealth.edu

10.1 Hormones and Impersonators

10.1.1 Hormone Receptors

Although an increasing number of present-day studies are revealing important
actions of hormones that appear not to utilize a discrete receptor, the overwhelming body of endocrine physiology identifies a hormone–receptor interaction as the initial step of tissue response. Indeed, the central necessity of this
interaction is so crucial that whole generations of researchers have obtained
support by postulating that "because there's a receptor in cell A that binds
hormone B, then B must be doing something in A." Furthermore, regulatory
agencies around the world want to conclude, "If contaminant X binds to a
hormone receptor for B, it might be imitating or inhibiting B's activity."

The logic of this postulate is easy to rationalize because hormones are in fact
discrete chemicals. So are their receptors. Some hormones are fairly large
proteins of over 100 amino acids, some are small peptides, some are highly
complex organic molecules, and some are essentially hydrocarbons of very
low solubility in water. Despite diverse sizes and properties, each one has a
host partner built with a high affinity site to attract and retain a coupling event.

Issues in Toxicology No. 19
Reducing, Refining and Replacing the Use of Animals in Toxicity Testing
Edited by David G. Allen and Michael D. Waters
© The Royal Society of Chemistry 2014
Published by the Royal Society of Chemistry, www.rsc.org

The coupling then promotes a change in the target cell identified as a *hormone-mediated response*.

In the classic model, hormone chemicals are secreted by specialized tissues and distributed in blood throughout the body. Generally their role is to modulate activity of target responses as part of an overall scheme to support the organism's life and equilibrium (homeostasis). They are often viewed as a chemical signal or switch, to initiate, halt, or gently alter the speed of a process. Each cell's functions are uniquely programmed and potential modulators need a way to work with a cell's equipment. To this end a target cell expresses its own host proteins (receptors) with a domain that uniquely binds the proper hormone.

10.1.2   Hormone Deception

Unfortunately, our environment includes tens of thousands of chemical substances of which more than a few resemble animal hormones. Many occur in nature while others are synthetic contaminants. In fact, a number of important hormonally active pharmaceuticals are copies or modifications of similar materials found in plants, other animals, and microorganisms. Such is the conservation of nature. Nonetheless the potential exists for our environment, whether the origins are natural or human-induced, to influence endogenous homeostasis, and interest is growing to find ways to identify these substances, being now called "*endocrine disruptors* (ED)."

Suspicious observations reported decades ago led to accumulated evidence of specific hormonal properties of some chemicals. One recent catalog, called the Endocrine Disruption Exchange (TEDX), is a comprehensive listing of some 870 putative substances.[1] Many are identified as receptor binders or substances that express direct hormonal activity *in vivo*. Other citations appear to be reports of functional disruptions (e.g., of fertility or developmental anatomy) that may or may not have resulted from interaction with a specific hormone-mediated process. While no avenue to toxic damage should be overlooked, the point needs to be made that current technology permits more precise testing for true EDs than with whole-animal exposure studies of function. At present the preponderance of identified EDs in the TEDX listing are reported to have estrogen-related activity. Lesser but still important listings of androgen- and thyroid-related ED activities are also included.

Regulatory officials around the world have begun to mandate testing of manufactured goods, food, and water to identify the presence and especially the potency of EDs, as a step toward reducing their presence. For example, the United States EPA's Endocrine Disruptor Screening Program (EDSP)[2] presently requires a battery of 11 Tier 1 screens for registration of manufactured substances that might pervade the environment.[3,4] Two of the present battery are estrogen and androgen receptor binding assays. In other words, a substance with demonstrated ability to bind to the estrogen or androgen receptor may be considered for additional (Tier 2) evaluation as an endocrine disruptor.

This chapter describes the rationale, methodology, advantages, and limitations of using receptor binding assays, of the type included in the EDSP Tier 1 battery, to identify putative estrogens and androgens. Although totally *in vitro* technology does exist for determination of receptor binding properties, the present EDSP screen battery requires analysis of receptor binding using preparations of prostate (androgen) or uterus (estrogen) tissue from live rats.

10.2 Kinetics of Receptor Binding Assays

Each receptor molecule contains one hormone binding domain (HBD). Receptor binding is typically very strong as well as specific, and is known as *high-affinity* binding; once associated, the two molecules are slow to dissociate. However, the binding is not permanent. In incubation (or in living cells) hormone molecules are always dissociating and reassociating with their receptors. A computation of the strength of hormone–receptor attraction becomes possible, if one can measure all three populations:

$$H + R \rightleftharpoons HR$$

At equilibrium a stronger affinity for binding produces a higher proportion of HR. This relationship is often characterized mathematically by measuring the concentration at which *half* the receptor population is occupied (R and HR are equal), and is called the *dissociation constant* (K_d). The K_d values of estradiol and testosterone for their respective receptors are routinely determined to be 1 nM (10^{-9} M) or less. This means that an *in vivo* concentration of 1 nM estradiol or testosterone should occupy approximately 50% of the available receptors, and it also represents a useful estimate of the concentration of each hormone necessary to begin biologic expression. K_d measurements can become useful values when testing the *comparative potential* for a foreign substance to interact with a receptor. For example, if a substance is determined to have a K_d of 1 μM (10^{-6} M) for the estrogen receptor, it means the substance must exist at a 1000-fold *greater concentration* than estradiol to produce the same population of *occupied receptor*. As suggested above, proper risk assessment should include both qualitative and quantitative measures.

Many natural hormones are produced with radioactive labels and high activity sufficient to observed a rate of binding at equilibrium. However, few suspected synthetic ligands are available with highly radioactive moieties. Therefore, estimates of a putative ligand's affinity for a hormone receptor is more typically conducted by a *competitive incubation* approach. Since a population of receptors saturated with its natural hormone (e.g. ER and estradiol) will experience dissociation, introduction of a competitor will gradually replace some of the hormone molecules, in accordance with the *competitor's respective affinity* for the receptor. By arranging a series of incubation tubes with identical amounts of receptor and radiolabeled hormone, but with varying amounts of competitor, a sigmoid-shaped dose–response plot is produced.

Figure 10.1 illustrates an example of ER incubated to equilibrium with radiolabeled estradiol and increasing concentrations of unlabeled estradiol. The regression plot covers 2 orders of competitor concentration in accordance with the law of mass action: a range of 100-fold molar excess displaces tracer binding from about 90% to 10%. The sigmoid plot is continuously curving but changes direction near to the 50% concentration of inhibitor (where half of the receptors are occupied with each ligand). A plotting program can compute the precise point of curve change, known as the K_i, or inhibition constant (5.41 × 10^{-10} M, or 0.541 nM in this example) and also the IC$_{50}$ (7.415 × 10^{-10} M, or 0.7415 nM in this example). Because the competing ligands in this example are nearly identical ([^{3}H]-estradiol and unlabeled estradiol), the two computed values are also very close.

If the competing ligand is not the same as the natural or reference hormone, yet does bind to the receptor, the type of plot shown in Figure 10.2 can appear. The competing substance, *p*-cumyl phenol (CAS 599-644), is also known to express estrogenic action *in vivo*.[5,6] When incubated with ER and [^{3}H]-labeled estradiol, *p*-cumyl phenol inhibited tracer binding, but at higher concentrations than for estradiol. The plot IC$_{50}$ was observed at 9.16 µM, and the curve K_i was calculated to be 6.11 µM. The competitor binding was some 10 000-fold weaker than the natural hormone.

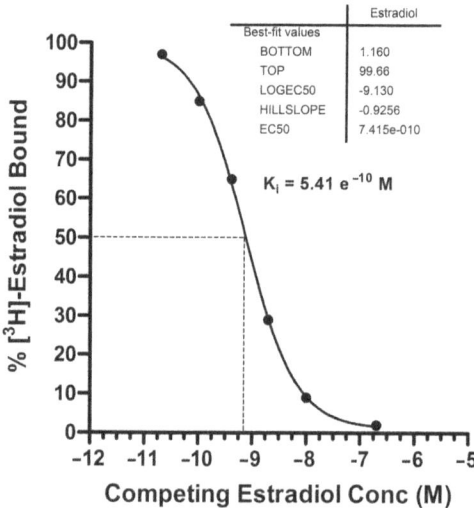

Figure 10.1 Displacement by increasing concentrations of 17β-estradiol against radiolabeled estradiol binding to rat estrogen receptor (ER). Concentrations of estradiol (20 pM–200 nM) were incubated with 1 nM ^{3}H-estradiol and rat uterine cytosol as described in Section 10.3.3. Plot of net radioactivity bound is sigmoid-shaped. The break point of this curve (K_i) was 0.541 nM, while the intercept at 50% binding (IC$_{50}$) was 0.7415 nM. The plot identifies the dissociation constant (K_d) for this ligand (17β-estradiol) with this receptor (ER).

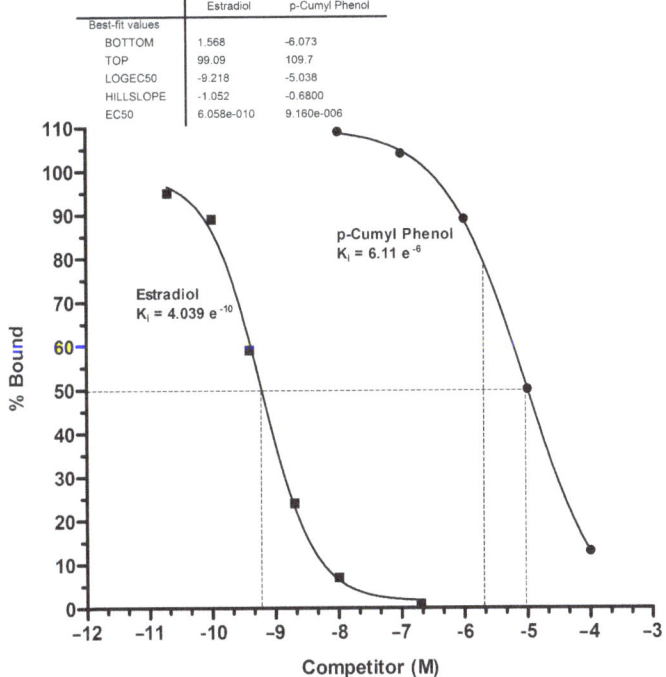

Figure 10.2 Displacement by increasing concentrations of *p*-cumyl phenol (CAS 599-64-4) against radiolabeled estradiol binding to rat estrogen receptor (ER). The displacement curve was shifted to the right, resulting in a computed K_i of 6.11 µM, approximately 3 orders of magnitude higher. The 50% intercept of the curve was shifted to 9.16 µM. In essence, 1000 molecules of this ligand must be present to displace each molecule of estradiol.

It is important to recognize that ligand binding to a hormone receptor does not necessarily lead to biologic expression. A substance may occupy binding sites (the LBD) but not transform the receptor to a bioactive conformation. A substance may also associate with sites distant from the hormone binding domain, yet impede natural hormone binding at the LBD (often referred to as *non-competitive inhibition*). The plot shown in Figure 10.2 does not indicate the *type* of receptor competition, and this requires additional analyses.

One type of follow-on analysis, now in use by the EDSP process, makes use of the Lineweaver–Burk approach, often referred to as *double-reciprocal* plots.[7] This is conducted with a series of re-incubations of hormone receptor with several concentrations of both tracer and competitor. Competitor concentrations are chosen to bracket closely the previously observed IC_{50}. The resulting binding data are plotted as shown in Figure 10.3. Note that both axes are reciprocals: 1/total ligand in each incubation versus 1/bound radioligand in each incubation. For each concentration of competitor, including zero, the data plot to straight lines that intersect near the origin.

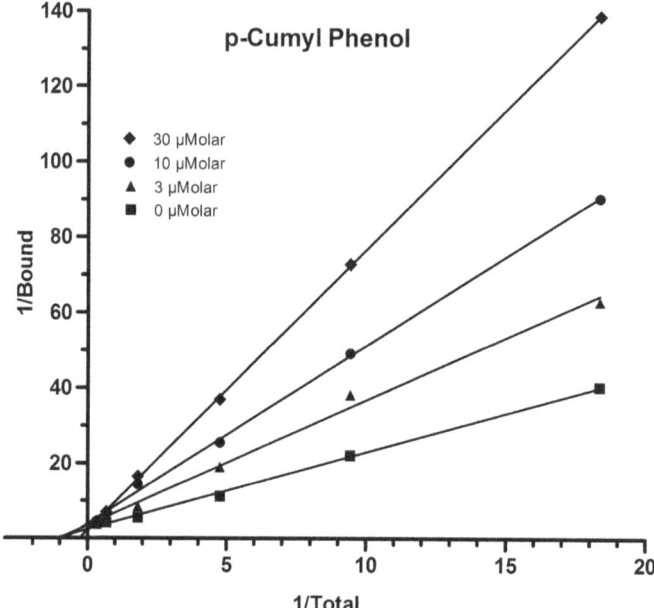

Figure 10.3 Lineweaver-burk plots of *p*-Cumyl phenol against radiolabeled estradiol binding to rat estrogen receptor (ER). Six concentrations of [^{3}H]-estradiol were incubated with a fixed amount of rat uterine cytosol. Four sets of incubations were prepared with *p*-cumyl phenol at $3\times$, $1\times$, $0.3\times$ and zero times the estimated EC_{50} from displacement plots (e.g., Figure 10.2). Binding data were plotted as 1/Bound vs. 1/Total (double reciprocals). Results show 4 linear plots intersecting near the origin.

Each plotted line has a slope. As shown in Figure 10.4, when the slopes of these four lines are replotted (ordinate) versus the competitor concentrations in the four sets of incubations (abscissa), another regression line is described. If the inhibitor is competing at the HBD, this line intersects the *x*-axis to the left of the origin, and identifies a K_i in molar concentration. Inhibitors that do not compete directly at the HBD produce non-linear, non-intersecting plots.[8]

An alternative method for expressing binding and competition results is called a *Scatchard* plot.[9] The principle underlying these plots is essentially the same as the methods illustrated by plots in Figures 10.1–10.4. Using a slightly different arrangement of ligand and competitor concentrations in the assay incubation, the binding data are plotted as a ratio of bound/unbound radioligand (ordinate) versus bound radioligand (abscissa). Scatchard analysis of specific binding produces a descending linear plot. Incubation with a true competitor produces another linear plot with a steeper slope than the reference ligand (altered K_d) but with the same *x*-intercept. Plots from incubations with substances that do not interact (compete) directly at the HBD are often parallel to the reference compound plot with a shifted *x*-intercept that suggests a reduction of apparent binding capacity (B_{max}). Our lab has used this approach

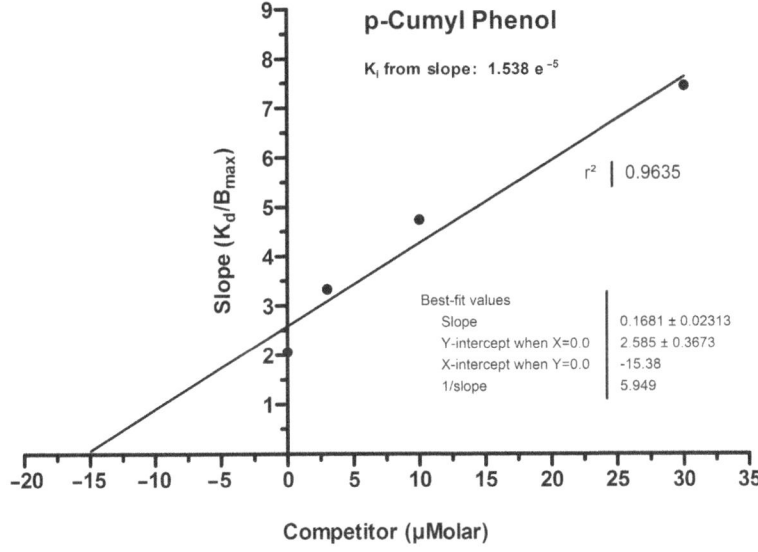

Figure 10.4 Plot of slopes of the 4 Lineweaver-Burk plots from Figure 10.3 vs. *p*-cumyl phenol concentration in the incubations. Points produce a linear plot that intersects the abscissa near the molar K_i for this ligand with ER.

over many years to study steroid receptor quantity and quality,[10–12] and also to characterize the nature of putative competitors.[13,14] The US EPA's EDSP process also permits submission of screen results that used Scatchard plotting.

10.3 Basics of Solubilized Assay Methodology

10.3.1 Tissue Preparation and Reagents

Protocols to assess ligand binding to natural steroid receptors of a given species begin with a fairly simple extraction from the tissue of interest. When whole tissue homogenate is centrifuged at very high speed (typically 105 000*g*), the resulting supernatant, termed *cytosol*, contains the intracellular receptor population. Use of an ice-cold hypotonic extraction buffer (generally about 10 mM salts) that includes glycerol releases the receptors from associated sites while preserving binding integrity. Typical receptor preparations are made from tissues with high density, e.g., female uterus for estrogen and male prostate for androgen.

Two different estrogen receptor (ER) subtypes have been identified, and while both bind estradiol very well, there are known differences in their propensity to bind other molecules.[15–17] However, receptor-dense animal tissues typically contain all subtypes that are not easily isolated in quantity from one another. Thus a solubilized receptor assay will include both ER subtypes. Assessment of binding of a putative ED to only one subtype requires a cloned receptor or at least the binding-domain fragment of one receptor.

Tissues of small mass or with lower receptor content can be examined with attention to maximizing sensitivity. Our lab has, for example, participated in studies of estrogen receptor binding in individual rat hypothalami and pituitary[10] and of glucocorticoid receptor binding in rat hippocampus.[11–14] In our hands a minimum of 0.5 mg/mL total protein in the final assay volume is needed for reliable binding sensitivity, but higher protein content is recommended if tissue and receptor density are plentiful. In a fairly recent study of binding characteristics to estrogen receptor in rat uterine cytosol, our preparations were 1.9–2.4 mg/mL protein in the final assay volume.[8]

It is reasonable to prepare large batches of cytosol for faster throughput of repeated assays. We have had no trouble homogenizing uteri of a dozen or more animals, centrifuging several full tubes in the same rotor, pooling the cytosols and storing very small aliquots of very high protein content. For one binding assay, several aliquots can be diluted in assay buffer to achieve the desired protein concentration. These high-protein cytosols (>20 mg/mL) can be stored for several months at –80 °C and used successfully when diluted just prior to assay.

There are numerous buffer formulations for receptor extraction but all are commonly hypotonic, active in the physiologic pH range, contain a strong antioxidant to inhibit proteases and help maintain protein structure, and include glycerol. Our version (called TEGD) is: 10 mM Tris, 1.5 mM EDTA, 1 mM dithiothreitol, 10% v/v Glycerol, pH 7.4. Some protocols also add molybdate to enhance receptor stability,[18] and we include this oxoanion in our androgen receptor systems. Reagents for the binding assays can be prepared in the same TEGD extraction buffer. However, for the final assay step, that separates unbound from receptor-bound hormone, we prepare a slurry of dextran-charcoal or hydroxyapetite in TED buffer without glycerol.

Radiolabeled hormone ligands for each respective steroid receptor assay are available with [^{3}H] or [^{125}I] tags. A higher specific activity (SA, recorded as curies per molar unit) provides greater sensitivity. Assay sensitivity is important to consider because reducing the molar concentration of radioligand improves the chance to observe competition by a weak competitor. Very few competitors have affinities to match the endogenous hormones (estradiol, testosterone), and most are much weaker.

[^{3}H]-labeled ligands possess some advantages over [^{125}I]-labeled materials. They have a relatively long half-life, so the SA of a stock batch will remain constant over several months. Large stocks are less expensive to buy, easier to store, and less is wasted due to decay. [^{125}I] activity declines at nearly 30% per month, so the tracer SA needs to be recalculated for each assay run. In addition, attachment of iodinated moiety to a steroid substantially alters the molecular shape, while [^{3}H] substitution effects minimal alteration. Although [^{125}I]-labeled hormone has been successfully used in receptor binding studies, the binding kinetics may be somewhat different than for the native hormone.[19]

On the other hand, [^{125}I]-ligands have much higher SA values than [^{3}H]-labeled compounds and confer greater assay sensitivity. In addition, [^{125}I] activity is measured with gamma counters that typically process multiple samples simultaneously, and with a shorter counting time. Measures of [³H] activity by liquid scintillation technology requires a distinct instrument, a solvent-based counting fluid for each sample, and plastic vials. Waste disposal of these scintillation vials is typically more expensive and bothersome than disposal of specimens containing [¹²⁵I], which can often be stored until legally decayed out.

10.3.2   Choice of Primary Stock Solvent

A critical point to understand is that steroid hormones are substituted hydrocarbons with very poor solubility in water; they bind to their receptors by inserting smoothly into a hydrophobic cleft (known as the *hormone binding domain*, HBD). However, the receptors themselves are complex proteins that must be used in *aqueous* buffers described above, to maintain their structure. Most tested ligands are similarly organic molecules with poor solubility in the assay systems and high concentrations will be difficult to achieve.

The traditional approach to preparing organic test materials for receptor binding analysis in an aqueous buffer system in to prepare primary stocks in ethanol and then dilute in assay buffer. Our lab typically prepares a fairly concentrated primary stock of the test ligand in ethanol, e.g. 1 mg/mL or 1 mM. The effect of ethanol is then diluted in assay buffer by preparing the most concentrated competitor at 10–50 µM. Recall that the K_d of estradiol and testosterone for their own receptors is less than 1 nM, so a 10 µM competitor is upwards of a 100 000-fold molar excess.

Nevertheless, some highly non-polar solutes remain poorly soluble in ethanol. We have frequently observed U-shaped displacement curves at high micromolar concentrations of competitor because the test substance had become insoluble.[8] Thus the test ligand was unable to gain access to the solublized receptor at these high concentrations. Furthermore, our preliminary studies of protocol methodology revealed ethanol itself in excess of 5% interfered with ER binding. In addition, many putative estrogen-related endocrine disruptors appear to exhibit very weak activity in bioassays or for ER binding, and require testing at high concentrations. We therefore evaluated the use of dimethyl sulfoxide (DMSO) in the primary stock solvent, with the premise that higher stock concentrations might be possible.

Initial tests showed that the inclusion of up to 24% DMSO in the ER assay buffer had no effect on assay performance. Figure 10.5 shows that estradiol in a primary stock of 50% DMSO in TEG buffer displayed the same displacement curve as ethanol-based stock in the RUC ER binding assay.

Figure 10.6 illustrates the importance of using DMSO for preparing the primary stock. Trimethylborate appeared to be a very weak binder in excess of 1 mM, but calculating the degree of ER binding displacement from an ethanol primary stock became compromised because the test material was no longer soluble, and the high ethanol concentration destroyed assay performance. Use of DMSO permitted higher solute concentrations without compromising assay performance. The complete sigmoid dose–response plot yielded more accurate kinetic measures when DMSO was used.

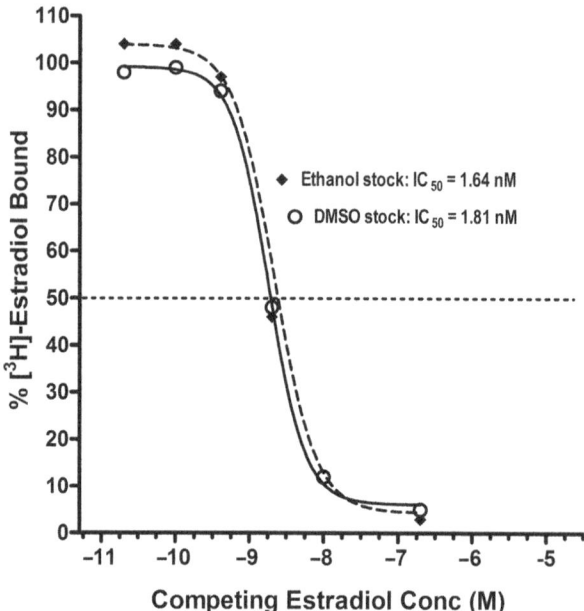

Figure 10.5 Effect of Dimethyl Sulfoxide (DMSO) stock solvent on 17β-estradiol displacement in the rat estrogen receptor (ER) binding assay. Primary stock concentrations of estradiol were prepared in 100% ethanol or 50% DMSO in TEG buffer, and binding assays were conducted in an identical manner.

We have recommended that 50% DMSO in TEG solvent can be safely adopted for preparation of primary solute stocks at up to 30 mM; this in contrast to a 1 mM maximum in ethanol. One might question whether testing mM concentrations of competitor is relevant to a hormone receptor system designed to operate with natural ligands at nM concentrations. Nonetheless, many putative competitors have very weak affinities for ER, and it remains important to remove as many artifacts as possible when evaluating these substances.

One final point: While it is tempting to evaluate competition at very high concentrations, it should also be understood that the mere presence of a foreign substance in the assay incubation can itself be non-specifically disruptive. Because the measured parameter is a reduction of radiolabeled tracer binding to receptor, a substance that alters the *incubation environment* in any way may similarly reduce binding. It is extremely important to use binding kinetic analysis to distinguish true from false competitors.

10.3.3 Assay Protocols

Each assay run must be a complete test that includes suitable controls. For example, for a dose–response plot of competition, a similar set of reference

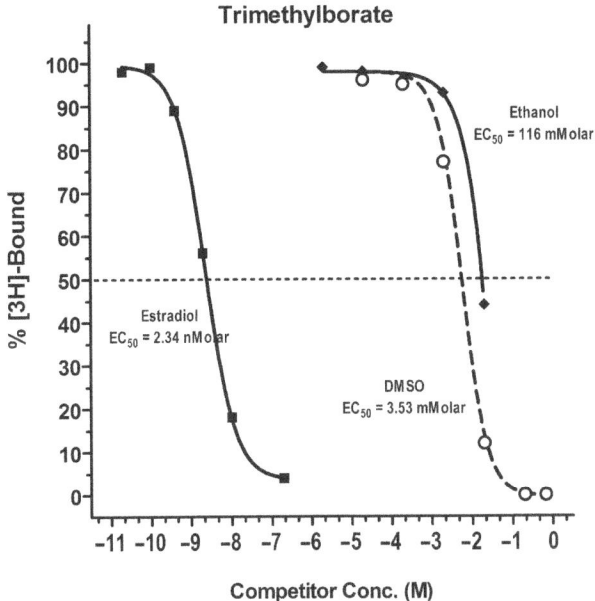

Figure 10.6 Comparison of displacement curves of trimethylborate in ethanol and DMSO primary stocks. Concentrations ranged from 0.67 M–20 μMolar (DMSO) and 20 mMolar – 2 μMolar (ethanol). Trimethylborate is a liquid at ambient temperatures.

hormone controls must be included, and relative comparison must be calculated against the control responses *within the same run*. The assay separation step is extremely time-dependent, so run sizes must be small, perhaps 50 70 tubes. Thus, in a screen to identify displacement at a range of concentrations (e.g., 10-fold dilutions from 100 μM to 10 nM), only a few unknowns should be tested at once. For a complete analysis of competition kinetics, following a suggestive reading in a screen, the assay set should ideally contain only one compound, along with control tubes of reference hormone. Include at least duplicate tubes of each analytical point, including controls. Once started, the assay set must be maintained at ice-cold temperatures; binding association dissipates rapidly as ambient temperature increases.

A basic assay method that uses very small volumes in glass tubes is best: 50 μL each of tracer, competitor, and cytosol. Appropriate concentrations of tracer and competitors are prepared and added to tubes already on ice. Cytosol is diluted just prior to addition to the assay, and tubes are incubated overnight at 4 °C. On the following day 150 μL of the ice-cold separation slurry is introduced, incubated for 15 min at 4 °C, and centrifuged to separate receptor-bound from unbound tracer. A precisely measured portion of the desired fraction is then counted. If dextran-charcoal is used, the receptor fraction is in the supernatant; if hydroxyapetite is being used, the receptor-bound portion is in the precipitate.

10.4 Solubilized Binding Assays in the US EPA's EDSP Tier 1 Screen Battery

The US EPA Endocrine Disruptor Screening Program has a set of 11 assays in its Tier 1 Screening Battery that must be satisfactorily performed on manufactured substances identified by the Agency for this testing.[3,4] Included are assays that assess binding to estrogen and androgen receptors.

10.4.1 Estrogen Receptor Binding

The detailed ER assay protocol is available at the EDSP website.[20] The basic assay is similar to that described here in Section 10.3. Notably, the ER preparation is extracted from live rat uterine tissue by ultracentrifugation, the analysis itself is a competition assay for ER binding between the test substance and radiolabeled [³H]-estradiol, and the kinetic analyses are conducted as described and illustrated above. The EPA protocol uses hydroxyapatite slurry instead of dextran-charcoal to separate receptor-bound and unbound radioactivity. Our lab has always preferred the charcoal method, but either is equally effective when conducted properly. An additional aspect of the EPA screen is the use of DMSO as the primary solvent for stock preparation of the test substance.

The tissue source is fresh uteri dissected from recently ovariectomized rats. The protocol does permit storage of aliquots of prepared cytosol at −80 °C for use in subsequent binding assays. The EPA protocol also permits kinetic analysis of binding properties by Scatchard plots.

10.4.2 Androgen Receptor Binding

The detailed AR binding assay is also found at the EPA's EDSP website.[21] The protocol is quite similar to the EPA's ER assay protocol, and differs essentially only in the tissue source and hormonal ligands. Ventral prostate tissue is dissected from orchidectomized young adult rats and AR is extracted as described for uterus. Frozen cytosols can be stored and used at a later time. The radiolabeled tracer and high affinity reference competitor are not testosterone (the prototype androgenic hormone) but R-1881, a synthetic AR ligand of higher affinity than testosterone. The EPA protocol also recommends Scatchard analysis of the competition results. However, it should be noted here that this author knows of instances when Lineweaver–Burk kinetic analyses, as shown earlier in this chapter, were successfully used for regulatory purposes in the EDSP.

10.5 Identification of ER Binders With Reporter Technology

10.5.1 Basic Principles

As previously described, steroid hormone receptors are large single-strand proteins that fold into complex, yet precise, shapes with a single hormone binding domain (HBD) cleft. The domain functions much as a lock into which only the appropriate hormone inserts as a key. Even steroids with very similar shapes do not fit and do not activate the particular receptor. For this reason alone, the principle of "receptor binding" has long served as a valid basis for identifying putative environmental contaminant disruptors. Recent analytical advances have moved beyond observations of receptor binding to measures of known hormone-stimulated biological responses *in vitro*. For example, Sumpter and Jobling identified a number of estrogenic pollutants by measuring production of vitellogenin from male trout heptaocyte cultures.[24] Another useful response is to assess survival of hormone-sensitive cells. Soto *et al.* developed an "E Screen" that measured growth of estrogen-dependent MCF-7 cells; the screen was then used to identify relative potency of many dozens of putative xenoestrogens.[25]

An increasingly common approach is to synthesize constructs of gene-linked reporter technology in cultured cells. In short, cells are made to express the proper receptor that, when activated by a ligand, links to an easily measured and sensitive dose-related response. Basic researchers have utilized many versions of this model over the past two decades.[26] The activated reporter can be growth regulators such as chloramphenicol acetyltransferase (CAT) that confers antibiotic resistance to cells when exposed to the proper hormone in question. However, a more simple and rapid reporter for screening purposes is one that produces light after expression of luciferase, or of a green fluorescent protein (GFP), in the presence of hormone.

Numerous transfected cellular models have been generated; among the more popular are MCF-7, HeLa, Chinese hamster oocytes, yeast cells, HepG2 (hepatoma cell), and BG-1, an ovarian carcinoma line. For a screening assay, cells are typically incubated for 24 h with the test substances, followed by reagents to promote the illuminated signal. Microtiter 96-well plates are easily adapted so that an entire system of controls, reference standards, and test samples can be incubated together; plate reader software analyzes the responses, constructs plots and computes results. Many hundreds of substances have been screened for estrogenic activity in this manner.[1,5,27–29]

10.5.2 Advantages and Limitations of Reporter Systems

A number of other reviews of this methodology have appeared over many years.[30–33] An instructive example of evaluating a cell-based reporter system that is intended for regulatory screening of environmental xenoestrogens is found in a recent National Toxicology Program report.[23] Any evaluation process must recognize and identify both limitations and advantages of this technology; there are significant examples of both. As discussed at the end of this chapter, *in vitro* systems cannot completely replace *in vivo* models for a complete appraisal of a potential disruptor.

10.5.2.1 Comparative Advantages of In Vitro Reporters

Perhaps the most favorable feature of a cell-based reporter system is that it provides some answers that are unavailable with *in vivo* testing, or with tissue preparations, notably the ability to create any specific hormone receptor construct. With estrogens, for example, two different binding receptors are known, called ERα and ERβ. Both receptors bind natural estrogens to promote biologic expression, but the HBDs differ, and binding characteristics of synthetic substances are not identical.[17,33] While rat uterine cytosol contains primarily ERα receptors,[33] no easily extracted tissue preparation contains exclusively ERβ. An *in vitro* reporter construct containing the ERβ binding domain can easily determine whether, and to what degree, a suspect ligand binds specifically to ERβ.[34] Indeed, cDNAs can be customized to assess binding of any natural or modified HBD, for any nuclear-active hormone (e.g., androgen, progestin, thyroid), while continuing to use virtually the same cell type, reporter, and laboratory protocols.

A second advantage is that a reporter system is actually an "*in vitro* bioassay", moving beyond receptor binding to measure transcribed and translated expression. Receptor binding alone does not necessarily correspond to an agonist response and indeed, some binders can be antagonists with little or no intrinsic activity. The RUC assay measures only strength of binding but a reporter assay, when including suitable agonist/antagonist controls, can measure the ability of a test substance to imitate *or inhibit* the hormone-related response.

Another advantage to reporter cells is the relative precision of methodology. RUC binding assay protocols can be quite different in various laboratories, and yield variable results. In a review of various ER binding studies of the herbicide atrazine, all conducted with rat uterine cytosol, this author counted four published studies showing a weak interaction but two other studies reporting no interaction.[27] For regulatory decisions, precise standardization of RUC protocols by performing laboratories is critically important.

Inter-laboratory qualitative inconsistencies seem fewer with reporter systems, even with varied cell models. In the review of atrazine's putative estrogen-like responses, this author cited 17 published reports using 8 different cell types; all yielded the same negative result.[27]

A more general, yet increasingly important, advantage of *in vitro* reporter technology is a reduced cost structure. A typical RUC binding assay (see Section 10.3) consumes several ovariectomized rats for one complete analysis of binding kinetics, encompassing cost of animal purchase, husbandry facility and staff, and technician time to prepare the cytosol. Although a single protocol using cultured cell responses also consumes a comparable amount of calendar time, one laboratory technician should be able to process more samples more efficiently with cell reporters. Furthermore, receptor binding assays require the purchase and maintenance of expensive specialized equipment (ultracentrifuge, radioisotope scintillation counters), and include the troublesome need to buy, store, and maintain inventory records on radioisotopes. The economics of lab testing using cell cultures versus live animals are moving in opposite directions.

10.5.2.2 Comparative Disadvantages of In Vitro Reporters

Receptor-linked reporter systems are also not without complications and inconsistencies. A major problem is that any response measured in a living system is limited by the agent's basic toxicity to the system. Indeed, because putative ED values are typically weak binders of hormone receptors, a toxicity threshold may often be reached before receptor binding occurs. This becomes a critical issue when interpreting results of an "inhibition assay." A test substance that non-specifically compromises living cell function or survival may show the same response as a specific inhibitor of a hormone-stimulated reporter. One artifact occurs when cell reporters that rely on growth or mitosis are affected by numerous non-specific factors, even by other hormones. For example, progestins and glucocorticoids antagonize estrogen stimulation of growth.[35]

In vivo tolerance may exceed a cell's tolerance. Metabolism of a tested substance may become a significant factor if the cell culture converts a tested substance to metabolites that are more or less active than the parent compound.[35] On the other hand, an animal model could also metabolize a tested substance to products that are inactive or more active with respect to a hormone-mediated response.

The point is that a cell-based initial screen may fail to pick up a disruptor that is active *in vivo*: this is once again the classic type II error that must never be permitted in a screen. Reporter assays need suitable and reliable controls to distinguish "toxicity" from "hormone-mediated effect."

Another issue, discussed previously with regard to binding assays, is that many test substances are highly non-polar, with very low solubility in the aqueous culture medium. Indeed the steroids themselves, bioactive at very low nanomolar levels, are delivered to cells *in vivo* by a set of specific plasma proteins. While media often include animal serum that should facilitate solubility, it is unlikely that bioavailable concentrations of non-polar test substances can be created *in vivo* at high micromolar levels. Said another way, an *in vitro* test result may never be possible to achieve *in vivo*, thus leading to a type I error (false positive) in a screen.

One significant and unique advantage to the RUC ER binding assay is the ability to describe the kinetics of a disruptor's interaction. It could be very useful to know whether a substance's interference with a biomarker in a living reporter system is due to specific, competitive binding at the receptor's HBD, to a non-competitive receptor interaction, or to a non-specific (non-receptor) interference. Only a binding assay can answer this question (i.e., is the disruptor truly hormonally active?). Thus, the RUC assay, and similar protocols for other steroid receptor systems, may retain important value as a follow-on test when a reporter system yields an inhibition response.

A different issue with reporter systems, rather unique to the regulatory process, is that these constructs and their uses are typically patented. Given the need for mandated testing protocols to use precise reagents and methods, it becomes difficult for regulatory officials to require use of a specific licensed product. Protocols using animals, or a long-used non-specific method such as measures of receptor binding in tissue cytosol, are not in conflict with policies about licensed products. This problem appears to have delayed inclusion of an ER-mediated reporter assay in screening batteries of many countries including the US.[23]

To summarize, hormone-mediated reporter systems add a very important piece of endocrine disruptor information, namely the ability to identify inhibitors as well as stimulators of a response. But the results remain incomplete in a way that a screen battery must still resolve by other means. Is an observed response caused by a true hormone-mediated mechanism, involving the receptor or post-receptor components and, very importantly, is an observed "inhibited" response due to a specific hormone-linked action, or to a non-specific inhibition of biological events in the test system?

10.6   Receptor Binding and the Regulatory Process

10.6.1   Value of the Results

As with any interpretive test or screen, the principal issues for receptor binding assays are accuracy and sensitivity. For the environmental regulatory process, the question of accuracy may be stated as: "Do the test results indicate a true interaction with a hormone receptor?" The sensitivity question can be "Do the test results occur at a level that may lead to endocrine disruption in living organisms?"

Considering the stakes, it is obviously most important to avoid type II, or "false negative" errors for both accuracy and sensitivity. Testing must reliably identify true disruptors, because lack of response in a screen sends an "all clear" signal. If one accepts that steroid hormones must interact with target site receptors to express bioactivity, then absence of binding by a tested substance should establish a reliable conclusion of no disruptive potential *in vivo* (by this mechanism).

It must be noted that interpretation of test results from an *in vitro* procedure are limited to only the actual substance evaluated. Materials introduced into a complete animal system are likely altered or metabolized to other substances whose properties may vary widely from the original agent. A receptor non-binder could be activated to a potent binder *in vivo*, or a substance showing receptor binding in test tubes might be easily inactivated and eliminated *in vivo*, with no functional consequence from exposure. Considering the precise substance being tested, the risk of a type II error in a binding assay is essentially zero, but for extrapolation to living systems, a binding assay could yield a small amount of false negative information.

While type I errors (false positive results) are of lower concern for risk analysis, the potential result has considerable consequence for the process of establishing regulatory status. As described previously, binding assays do not directly measure attachment of a test substance to a receptor's HBD; it observes *reduction* (elimination, displacement) of a radiolabeled reference hormone (e.g., estradiol or R-1881) binding to the respective receptor. Unfortunately, there are many indirect ways for this to occur, particularly when high concentrations of test substance are incubated relative to receptor-hormone concentration. Our lab has frequently observed disruptive effects of a test substance on the assay conditions themselves,[8] and these effects could lead to much follow-up kinetic testing, as described in Section 10.2. Diminished tracer binding may be falsely interpreted a sign of hormone-related activity but only further analysis can establish whether the initial sign was true or false. By the time a correct final conclusion is eventually reached, a type I error can lead to expensive and complicated additional testing. In the case of SBAs, it also means that additional animals will be consumed.

To summarize the concept, solubilized receptor binding assays appear to be highly accurate as a screen, especially for achieving the crucial identification and elimination of non-disruptors. However, the incidence of type I failures may result in an inefficient pathway to a non-disruptor conclusion. As for final accuracy of identifying true disruptors, binding assays are very reliable.

Regarding the second major question, of sensitivity, binding assays are also quite reliable. As described earlier in this chapter, natural bioactive hormones bind to their receptors with extremely high affinity, generally 1 nM or less. In other words, bioactivity is expressed with tissue concentrations of hormone at 1 nM or less. At present, the large share of environmental contaminants that have been reliably established as receptor binders interact with much lower affinities. Often the necessary concentrations are several orders of magnitude higher, and one would need to achieve extreme concentration levels (e.g. µM) in order to elicit a significant effect on hormone expression. As illustrated by Figures 10.2–10.4, the binding assay system provides a reliable estimate of comparative potency, that should help regulators and manufacturers establish a good margin of safety for exposure. Indeed, these high-concentration interactions may be inconsequential if the active substances are already regulated at lower exposure levels due to observed *in vivo* toxicity from other mechanisms.

That said, pharmaceutical research has identified a number of synthetic compounds with very high affinities for ER and AR, often referred to as *selective estrogen (or androgen) receptor modulators*. Some of these are in development or on the commercial market as antagonists (i.e. blockers of hormonal activity), so there is clearly the potential for discovery of contaminants that express high-affinity interaction with hormonal receptors. Because solubilized assay protocols were among the reference methods used by pharmaceutical researchers to identify true and sensitive ligands, the methods should continue to be reliable in EDSP screening.

In summary, the value of receptor binding analysis for identification of endocrine disruptors is high in most respects. Non-disruptors are confidently identified and true disruptors can be well characterized regarding the nature and sensitivity of interaction. Indeed, the library of receptor binding information, particularly for estrogen, is more comprehensive than for any other type of ED analysis.[1]

10.6.2 Value of the Technology

Evaluating the assay technology principle as currently used in the EDSP Tier 1 screen battery should be conducted on two levels. A binding assay identifies possible endocrine disruptors based on the concept that natural hormones nearly always interact with cellular receptors as an initial step in target action. Any substance that binds to a receptor as the natural hormone does can be regarded as a candidate for exerting (or blocking) the same action *in vivo*. Binding interactions between hormones and their receptors are very specific. Indeed the steroids estradiol, testosterone, progesterone, cortisol, aldosterone, and vitamin D have quite similar basic structures that are in fact all derived from cholesterol by just a few enzymatic alterations on the edges of the steroid nucleus. Yet each binds to a specific receptor that demonstrates extremely low proclivity to bind the other hormones. A very specific set of characteristic shapes would be necessary for any non-hormonal molecule to bind to one of the receptors as the natural hormone does. There have been some serious attempts to configure the necessary shapes and binding moieties using computer modeling, which could potentially replace all tests of binding and expression, but reliable success remains elusive.[22] Confidence in the principle remains high for reasons discussed earlier.

On the other hand, one must also consider whether the present SBAs continue to provide the best technology for the question at hand. There are some challenging limitations to this technology, foremost being a binding interaction does not prove the presence of bioactivity. A binding assay should be only a screen.

A second limitation of a simple competition assay, as shown in Figure 10.1, is that the nature of the competition is unclear without additional testing. True ligand interaction with the hormone binding domain (HBD) displays certain kinetic characteristics that are revealed only with follow-up binding tests. In our experience, many compounds that appear to displace radiolabeled hormone from the receptor acted at other portions of the receptor, or simply interfered with the conditions of the binding assay.[8] As discussed before, these kinds of interference typically occur only at very high concentrations and would be impossible to achieve in most cases by *in vivo* exposure.

A third issue with present binding assays is the considerable expense, in terms of money, required technical skill, equipment, and use of animals. The EPA's mandate could potentially lead to Tier 1 endocrine disruptor screening of up to 70 000 substances. Each binding assay (and each substance should be tested at least twice) consumes 2–4 rats for the initial screen, and several more animals if the screen must be followed by additional kinetic evaluation. The assays require experienced hands, performing numerous micropipetting steps, and (in the EPA's Tier 1 protocols), all data points must be set up in triplicate for quality control purposes. One test may require two whole days labor for one technician. The solubilized receptor assay is not a simple, efficient, low-cost screen.

A fourth drawback with whole-tissue extraction is, as noted earlier, that tissue preparations contain more than one subtype of each receptor, yet research has shown that subtypes may have dissimilar binding profiles. A tissue-based receptor population can't distinguish specific subtype interaction.

A completely different approach to screening is the use of computer modeling (see ref. 22). The 3D designs of receptor binding domains are now completely modeled and screening for "fit" is easily done. This has the obvious advantage of high throughput and low cost (except for the hardware and software), without the use of animals, living systems, radioisotopes, or legions of technicians. It still relies on the "key and lock" principle that active substances must interact with a receptor's HBD. As with the solubilized receptor method described here, a disadvantage is that one may not distinguish agonists from antagonists, or know if the tested substance is active *in vivo*.

10.6.3 Assessing the Future

As a widely used screen, the solubilized receptor binding assays presently included in the EDSP Tier 1 battery will result in an extraordinary consumption of animals. As discussed in the previous sections, the assays remain popular primarily because of their long history of confident results and a well-understood methodology. Nevertheless, this method, which has served an invaluable historical purpose, is likely to decline in value. In addition to the need to reduce, refine, and replace animal consumption, solubilized tissue assays will become progressively more expensive, slower, and less precise. Purely *in vitro* methods that address the same principle (i.e., receptor interaction) are gaining acceptance but total replacement as the new reference method remains in the future, again for reasons discussed earlier, and also discussed to a larger extent in the following chapter in this volume, as well as in the NICEATM-ICCVAM report.[23]

This is not to suggest *in vitro* testing will never become an effective first-line screen for hormone-mediated responses. Any reliable receptor binding methodology should be able to screen out non-interactive substances without further question, and it is expected that the vast majority of environmental substances will prove to be non-reactive. Hormone receptors are, after all, highly specific capture devices. Biological life could not survive if receptors were confused, so the chance for substantial or widespread artificial interaction with any steroid receptor is predictably low. The key challenges for reliably identifying hormone-related activities by screening will remain· (1) the incidence of the aforementioned type 1 errors, and (2) accurate identification of a test substance as a true hormonally active material. Follow-on testing will be needed in both cases, and use of live animals may be required in order to settle on a confident conclusion.

It is no longer a question whether substances in our environment can interact with hormone expression, production, or clearance. They can and do. A far more complex issue is to identify the type and potency of each disruptor's interaction. The latter point is often given minimal attention yet it would seem to be the more critical issue. It is of course impossible for us all to escape exposure to all EDs, or to thousands of other toxins in our midst. Indeed, many EDs occur naturally and are regularly consumed from food and water. Millions of other species survive and even thrive because the *level of exposure* to toxins is not harmful. Nonetheless, potential harm from contamination by *synthetic* substances is a serious, pressing matter. Species have not evolved or thrived in the presence of these additional contaminants. How many of these are sufficiently potent to cause significant disruption or harm? Identification methods need to be both qualitative and quantitative.

One can predict that a substance that interacts with a hormone receptor will do something with regard to that hormone's action. One can even gain a reasonable estimate of potency from kinetic analysis. But appreciating the complete nature of a biologic response will always require biology-related testing. *In vitro* test results may suggest a positive effect that turns out to be absent *in vivo*. No single *in vitro* test, or combination of tests, will replace the complete physiologic operation of a living animal. So long as the essential question is, "Does it affect human life (or wildlife)?" the complete positive answer will require some testing on living surrogates.

References

1. The Endocrine Disruption Exchange (TEDX) Home Page. http://www.endocrinedisruption.com/endocrine.TEDXList.overview.php/ Accessed 19 Oct 2012.
2. US EPA Endocrine Disruptor Screening Program (EDSP) Home Page. http://epa.gov/endo/ Accessed 4 Oct 2012.
3. J. C. Eldridge and S. C. Laws, The U.S. EPA's Tier 1 screening battery for endocrine disruptor compounds, in *Endocrine Toxicology*, ed. J. C. Eldridge and J. T. Stevens, Informa Healthcare Publishing, New York, NY, 3rd edition, 2010, chapter 1, pp. 1–26.
4. US EPA Endocrine Disruptor Screening Program (EDSP), Tier 1, Battery. http://www.epa.gov/ocspp/pubs/frs/publications/Test_Guidelines/series890.htm/ Accessed 15 Oct 2012.
5. K. Yamasaki, M. Takeyoshi, Y. Yakabe, M. Sawaki, N. Imatanaka and M. Takatsuki, Comparison of reporter gene assay and immature rat uterotrophic assay of twenty-three chemicals, *Toxicology*, 2002, **170**, 21.
6. K. Yamasaki, M. Takeyoshi, M. Sawaki, N. Imatanaka, K. Shinoda and M. Takatsuki, Immature rat uterotrophic assay of 18 chemicals and Hershberger assay of 30 chemicals, *Toxicology*, 2003, **183**, 93.
7. H. Lineweaver and D. Burk, The determination of enzyme dissociation constants, *J. Am. Chem. Soc.*, 1934, **56**, 658.
8. S. C. Laws, S. Yavanhxay, R. L. Cooper and J. C. Eldridge, Nature of the binding interaction for 50 structurally diverse chemicals with rat estrogen receptors, *Toxicol. Sci.*, 2006, **94**, 46.
9. G. Scatchard, The attractions of proteins for small molecules and ions, *Ann. NY Acad. Sci.*, 1949, **51**, 660.
10. J. C. Eldridge, J. A. Cidlowski and T. G. Muldoon, Correlation between LH and estrogen receptor turnover in pituitary and hypothalamus of castrate rats following estrogen agonists and antagonists, *J. Steroid Biochem.*, 1986, **24**, 623.
11. J. C. Eldridge, D. G. Fleenor, D. S. Kerr and P. W. Landfield, Impaired up-regulation of type II corticosteroid receptors in hippocampus of aged rats, *Brain Res.*, 1989, **478**, 248.
12. P. W. Landfield and J. C. Eldridge, Increased affinity of type II corticosteroid binding in aged rat hippocampus, *Exp. Neurol.*, 1989, **106**, 110.

13. J. C. Eldridge and P. W. Landfield, Cannabinoid interactions with gluco-corticoid receptors in rat hippocampus, *Brain Res.*, 1990, **534**, 135.

14. J. C. Eldridge, L. L. Murphy and P. W. Landfield, Cannabinoids and the hippocampal glucocorticoid receptor: recent findings and possible signifi-cance, *Steroids*, 1991, **56**, 226.

15. S. Mosselman, J. Polman and R. Dijkema, ER beta: identification and characterization of a novel human estrogen receptor, *FEBS Lett.*, 1996, **392**, 49.

16. G. G. J. M. Kuiper, B. Carlsson, K. Grandien, E. Enmark, J. Häggblad, S. Nilsson and J.-Å. Gustafsson, Comparison of the Ligand Binding Specificity and Transcript Tissue Distribution of Estrogen Receptors α and β, *Endocrinology*, 1997, **138**, 863.

17. M. Younes and N. Honma, Estrogen receptor β, *Arch. Pathol. Lab. Med.*, 1997, **135**, 63.

18. K. Noma, K. Nakao, B. Sato, Y. Nishizawa, K. Matsumoto and Y. Yamamura, Effect of molybdate on activation and stabilization of steroid receptors, *Endocrinology*, 1980, **107**, 205.

19. H. J. Grill, B. Manz, O. Belovsky, B. Krawielitzki and K. Pollow, Comparison of [³H]oestradiol and [¹²⁵I]oestradiol as ligands for oestrogen receptor determination, *J. Clin. Chem. Clin. Biochem.*, 1983, **21**, 175 I-125 v 3H.

20. EDSP Test Guidelines: OPPTS 890.1250: Estrogen Receptor Binding Assay Using Rat Uterine Cytosol (ER-RUC) [EPA 740-C-09-005]. http://www.regulations.gov/#!documentDetail;D = EPA-HQ-OPPT-2009-0576-0005/ Accessed 4 Oct 2012.

21. EDSP Test Guidelines: OPPTS 890.1150: Androgen Receptor Binding (Rat Prostate Cytosol) [EPA 640-C-09-003]. http://www.regulations.gov/#!documentDetail;D = EPA-HQ-OPPT-2009-0576-0003/ Accessed 4 Oct 2012.

22. J. Devillers, Structure-Activity Modeling of Endocrine Disruptors, in *Endocrine Toxicology*, ed. J. C. Eldridge and J. T. Stevens, Informa Healthcare Publishing, New York, NY, 3rd edition, 2010, chapter 7, pp. 175–188.

23. Interagency Coordinating Committee on the Validation of Alternative Methods, 2011, Test Method Evaluation Report. The LUMI-CELL® ER (BG1Luc ER TA) Test Method: An *In Vitro* Assay for Identifying Human Estrogen Receptor Agonist and Antagonist Activity of Chemicals. NIH Publication No. 11-7850. Research Triangle Park, NC: National Institute of Environmental Health Sciences. http://iccvam.niehs.nih.gov/docs/endo_docs/EDPRPRept2011.pdf/ Accessed 15 Oct 2012.

24. J. P. Sumpter and S. Jobling, Vitellogenesis as a biomarker for estrogenic contamination of the aquatic environment, *Environ. Health Perspect.*, 1995, **103**(Suppl. 7), 173.

25. A. M. Soto, C. Sonnenschein, K. L. Chung, M. F. Fernandez, N. Olea and F. O. Serrano, The E-SCREEN assay as a tool to identify estrogens: an update on estrogenic environmental pollutants, *Environ. Health Perspect.*, 1995, **103**(Suppl. 7), 113.

26. M. Pons, D. Gagne, J. C. Nicholas and M. Mehtali, A new cellular model of response to estrogens: a bioluminescent test to characterize (anti) estrogen molecules, *Biotechniques*, 1990, **9**, 450.

27. J. C. Eldridge, J. T. Stevens and C. B. Breckenridge, Atrazine interaction with estrogen expression systems, *Rev. Environ. Contam. Toxicol.*, 2008, **196**, 147.

28. V. A. Baker, Endocrine disrupters-testing strategies to assess human hazard, *Toxicol. In Vitro*, 2001, **15**, 413.

29. S. H. Safe, L. Pallaroni, K. Yoon, K. Gaido, S. Ross, B. Saville and D. McDonnelic, Toxicology of environmental estrogens, *Reprod. Fertil. Dev.*, 2001, **13**, 307.

30. J. R. Reel, J. C. Lamb IV and B. H. Neal, Survey and assessment of mammalian estrogen biological assays for hazard characterization, *Fundam. Appl. Toxicol.*, 1996, **34**, 288.

31. T. Zacharewski, Identification and assessment of endocrine disruptors: limitations of *in vivo* and *in vitro* assays, *Environ. Health Perspect.*, 1998, **106**(Suppl. 2), 577.

32. S. Scholz, P. Renner, S. E. Belanger, F. Busquet, R. Davi, B. A. Demeneix, J. S. Denny, M. Leonard, M. E. McMaster, D. L. Villeneuve and M. R. Embry, Alternatives to in vivo tests to detect endocrine disrupting chemicals (EDCs) in fish and amphibians-screening for estrogen, androgen and thyroid hormone disruption, *Crit. Rev. Toxicol.*, 2013, **43**, 45.

33. Nuclear Receptor Signalling Atlas, website accessed 24 May, 2013. http://www.nursa.org/index.cfm.

34. E. Swedenborg, J. Pongratz and J. A. Gustafsson, Endocrine disruptors targeting ERbeta function, *Int. J. Androl.*, 2010, **33**, 288.

35. T. F. H. Bovee and M. G. Pikkemaat, Bioactivity-based screening of antibiotics and hormones, *J. Chromatogr. A*, 2009, **1216**, 8035.

CHAPTER 11
Carcinogenicity Testing

ERROL ZEIGER

Errol Zeiger Consulting, 800 Indian Springs Road, Chapel Hill, NC 27514, USA
Email: zeiger@nc.rr.com

11.1 Introduction

Cancers arise naturally in the body as the result of genetic predisposition, the normal aging process (including reduced capacity to repair DNA damage), and by what appears to be purely random events, or they can be induced by radiation and certain chemicals. It is a disease characterized by cellular malfunction, in that the cell does not behave in a "normal" fashion, but is released from the normal growth constraints and grows in an uncontrolled fashion to form a tumor. During metastasis, malignant cells travel through the body to seed tumors at distal sites. Carcinogenicity is a disease of the whole animal, and the cancers arise as an end result of a series of genetic and epigenetic changes.

Unlike other diseases that can be triggered by a single event, or are the cumulative result of repeated similar events, e.g. cell toxicity or interference with normal metabolic processes, cancer requires a series of independent (key) events over time, following a triggering event which is often a mutation in a critical gene (Table 11.1). The sequence of disease development is a factor in this process because each event must lead to a clone of genetically or physiologically altered cells that are then the template for the next event in the series. Because these events differ physiologically and genetically and require cellular replication before the next event can occur, the process can not be measured or addressed by a single *in vitro* test system. Complicating the issue is that

Issues in Toxicology No. 19
Reducing, Refining and Replacing the Use of Animals in Toxicity Testing
Edited by David G. Allen and Michael D. Waters
© The Royal Society of Chemistry 2014
Published by the Royal Society of Chemistry, www.rsc.org

Table 11.1 Key events leading to the formation of a cancer by a mutagenic MoA.[1]

1a	Exposure of target tissue (e.g., stem cells) to electrophilic species.
1b	Reactions with other non-DNA cellular targets that have impact on adduct fate (e.g., depletion of detoxication pathways critical to clearance).
2	Reaction with DNA in target cells to produce promutagenic damage.
3	Misreplication on damaged DNA template or misrepair of DNA damage.
4	Mutations in critical genes in replicating target cell.
5	Mutations in critical genes lead to enhanced DNA/cell replication.
6	New cell replication leads to clonal expansion of mutant cells.
7	DNA replication leads to further mutations or epigenetic events in critical genes.
8	Imbalanced and uncontrolled clonal growth of mutant cells leads to preneoplastic lesions.
9	Progression of preneoplastic cells results in emergence of neoplasms, solid tumors, or leukemia.
10	Additional mutations in critical genes in subpopulation of cells as a result of clonal expansion result in malignant behavior.

although some carcinogens may induce an initial event in a majority of organs throughout the body, the cancer produced will arise only in specific organs, and that specificity appears to be determined by the initiating substance and the propensity of the target organ to respond to it.

Unlike other tests that use cultured target cells as surrogates for whole animals to measure the toxicological effect of concern, e.g., liver toxicity in liver cells; neurotoxicity in neuronal cells; skin corrosion using *in vitro* skin models; etc., cancer testing *in vitro* uses other toxic or cellular effects as surrogate endpoints for tumor development. Thus, although *in vitro* tests of various sorts have been used to predict cancer they are incapable of replicating the *in vivo* processes that lead to the actual tumors.

Cancer is an organ-specific event that is the end result of a series of cellular changes over time,[2–5] so it is not surprising that a purely *in vitro* series of tests would not be able to accurately definitively identify rodent or human carcinogens. This is why, in this area, the term carcinogen/cancer "prediction" is used because positive responses in the *in vitro* genetic toxicity tests trigger the presumption that the substance is a carcinogen. This presumption cannot be overcome in the absence of a subsequent rodent cancer assay.

Because of the concern for the introduction of carcinogens into the environment (i.e., through the food supply, drugs, pesticides, industrial chemicals) carcinogenicity testing is required for many of these chemicals before they can be marketed or used on a wide scale. The current cancer testing guidelines require 2-year tests in male and female rats and mice. A positive (carcinogenic) response in any of these four groups is sufficient to classify a substance as a carcinogen.[6]

The most widely used *in vitro* test for identifying carcinogens is the *Salmonella* mutagenicity (Ames) test.[7] Approximately 25–30% of the substances positive in the Ames test do not cause cancer in rodents, and approximately

50% of the chemicals that have been shown to be rodent carcinogens, and a smaller proportion (25–30%) of human carcinogens, do not produce positive results in the Ames test. The high predictive accuracy of the Ames test was initially based on carcinogens identified prior to the 1980s; subsequent surveys using carcinogens identified using more comprehensive animal testing protocols have led to a reduced predictive accuracy (see, e.g., Table 11.2). Therefore, positive results in the test can only be used to estimate the probability that the substance will be a rodent carcinogen, whereas negative results in the test are not informative for non-carcinogenicity. Other tests (e.g., *in vivo* mammalian cell genetic effects; *in vitro* mammalian cell transformation; quantitative structure–activity relationships [QSAR]) have been used to refine the predictions of carcinogenicity of the substances positive in the Ames test, and efforts have been underway to use these and other tests to identify the carcinogens that are not mutagenic in the Ames test.

With respect to mutagenicity tests, this chapter will focus on the use of the Ames test and other *in vitro* tests for identifying carcinogens. As has been shown previously,[13–15] although addition of *in vitro* mammalian cell mutation and/or chromosome aberration tests to the test battery will increase the number of carcinogens identified, they also increase the numbers of "false positives". As a result, many substances may be discarded as potentially carcinogenic and many cancer tests may be performed on substances that should have been considered to have a high probability of not being carcinogenic. Many believe that this increase in the numbers of mutagens that are not carcinogenic weakens the predictive effectiveness of the mutagenicity tests, leading to less confidence in their predictivity.

Although mutagenicity is used as a surrogate for carcinogenicity, a mutation is not equivalent to a cancer. Mutation is an initial step in the initiation of a cancer and/or may be an intermediate step in the progression of a precancerous cell. Therefore, it is not surprising that many carcinogens are not mutagenic, and not all mutagenic substances cause cancer.

Because cancer causing agents act on only one or a few stages of tumor initiation and development, tests for more than one effect/endpoint can, in theory, be more predictive than tests that measure only a single biological effect. As a consequence of the multi-stage nature of carcinogenesis,[1–5] there are no *in vitro* assays that serve as a surrogate for carcinogenicity in mammals and therefore, as desirable as it may be, there currently is no way to completely eliminate animal testing from carcinogen identification and characterization. However, a number of *in vitro*, *in vivo*, and *in silico* practices developed over the past 30–40 years have been able to reduce the numbers of animals used to identify potential human carcinogens, and eliminate potential carcinogens from subsequent *in vivo* testing. In addition, a number of newly developed *in vitro* and *in vivo* tests have been proposed that have shown limited success in predicting rodent carcinogenicity. Because of the current limitations of these tests (i.e., few substances from a limited range of chemicals tested by the test developer; testing only substances that are known to produce liver tumors; absence of replication in independent laboratories) they are not addressed here.

As a consequence, the focus of this chapter will be on the most commonly used *in silico* and *in vitro* tests.

11.2   The Rodent Carcinogenicity Assay

Before addressing *in vitro* and other tests to replace the rodent cancer assay, it is helpful to have an overview of that assay. As it is currently required by regulatory authorities,[6] the standard carcinogenicity assay consists of administration of the test substance to young, male and female rats and mice for up to 2 years by the route of exposure most appropriate for human experience (i.e., oral, dermal, or inhalation). In the basic protocol, there is a control group (no test substance) plus three dose groups; the highest dose being the maximum tolerated dose (MTD) which does not cause any chronic symptoms or decrement in weight gain. In the absence of other data, the MTD is determined by a preliminary 90-day study in rats and mice, which is also used to presumptively identify target organs or systemic toxicity of the test substance. Typically, 50 animals are used per dose group in the definitive study, which results in the use of 1600 animals for the 2-year study, i.e., control plus three dose levels in male and female rats and mice.[6,16] Approximately 30 different tissue or organ samples per animal are evaluated for tumor formation. Modifications may be made in the protocol if they are justified by the test substance or on prior information about the specific substance or class of substances.

The combination of 90-day and 2-year studies are used, in addition to determining the substance's carcinogenic properties, to identify the target organs, the rapidity of tumor onset, which is a measure of the potency of the response, and the dose–response relationship. The test data also allows the identification of a no observed adverse effect level or point of departure for establishment of a benchmark dose (BMD; the dose needed to increase the background tumor incidence by a certain increment, typically 10%), which are used to extrapolate the carcinogenic potency to low-dose human exposure levels.

A significant increase in tumors at any one tissue or organ in any one of the four test gender/species categories (i.e., male or female, rats or mice) is sufficient to label the substance as a rodent carcinogen and a likely human carcinogen. It is possible to overcome the presumption of human carcinogenicity by showing that the tumors in the rats or mice were produced by a molecular mechanism that is not relevant for humans, e.g., tumors in the male rat kidney associated with the accumulation of α2μ-globulin which does not occur in mice or humans.

The inability of *in vitro* genetic toxicity tests to identify large proportions of carcinogenic chemicals, including chemicals clustering in specific chemical classes, has led to the distinction of genotoxic versus non-genotoxic carcinogens. The carcinogens that are not responsive in the current genetic toxicity tests are classified as non-genotoxic, meaning that their carcinogenic response is most likely initiated by a non-DNA reactive mechanism (e.g., via a hormone receptor mediated mechanism). At present, there is no clear definition of the distinction between genotoxic and non-genotoxic carcinogens. A range of criteria for this distinction have been used, the most common being positive responses in any of the *in vitro* or *in vivo* mutagenicity or chromosome damage tests, unless that response has been shown to be the result of a procedural artifact.

11.3   Tests Used

A wide variety of *in vitro* and short-term *in vivo* tests have been used, or are proposed for use to identify carcinogens. These range from computerized QSAR to tests *in vitro* for mutagenicity in bacteria, yeast, and mammalian cells, cell transformation in mammalian cells, and changes in gene expression in mammalian cells. *In vivo* animal studies that have been proposed for genetic effects and cancer tests use mammalian species, typically rats or mice, or fewer animals than the standard rodent cancer assay, and have a shorter duration.

11.3.1   *In Silico* Tests

Computer-driven QSAR tests are based on the premise that, because biological activity derives from chemical structure, the structure can be used to predict biological activity.[17–22] These tests are described elsewhere in this volume (see Chapter 5) and will not be discussed in detail here.

Depending on the size of the learning set of chemicals, and the biology of the test system, the QSAR approach will have different levels of effectiveness for different chemical classes. There are a number of commercially available systems; some have been developed and are used by academic researchers, and some pharmaceutical and chemical companies have their own, proprietary, systems.

These systems are partly based on the studies of the Millers[23,24] and expanded by Ashby,[25] who identified a number of chemical moieties that were electrophilic, meaning that they were capable of binding to DNA nucleotides, and likely to be mutagenic and, therefore, more likely to be carcinogenic. Prediction of mutagenicity and carcinogenicity has been facilitated by the large numbers of chemicals tested in the *in vitro* tests and the chemicals tested in rodents for cancer by the National Toxicology Program (NTP),[26] and those identified from other sources.[27,28] One aspect of QSAR systems for predicting carcinogenicity that has been receiving much attention is their inability to predict non-genotoxic carcinogens (i.e., those that do not have electrophilic moieties and are not capable of binding to DNA). These include hormonally active substances, immunotoxins, among others, that produce cancers by diverse, non-DNA-reactive mechanisms.

QSAR systems are widely used in industry as an initial screen for potential carcinogens, by regulatory authorities to predict mutagenicity or carcinogenicity in the absence of test data, or to assist in determining when to acquire such data. There are a number of different commercial QSAR systems as well as in-house proprietary systems used by many large pharmaceutical and chemical companies.

Because these tests are developed and validated against chemicals with well-defined biological effects, they are most effective in predicting chemicals that are structurally related to the chemicals in the training set, and less effective in predicting chemical structures that are not well represented in their training sets. As would be expected, QSAR systems cannot predict the biological effects of chemicals whose structures have not been previously seen by the system (i.e., not part of its training set).

Another limitation is that QSAR is not effective against mixtures or chemically undefined substances, partly because the biological effects of the mixture can vary according to the relative proportions of its different components. They have also not been shown to be effective for inorganic substances or organometallics.

QSAR systems are often used as an initial screen before *in vitro* testing is performed. As noted, they are most effective when the chemical in question is a member of a larger set of structurally related chemicals with a rich *in vitro* or *in vivo* test database. QSAR is also used to help clarify questionable *in vitro* data, or in an attempt to predict the mutagenicity or carcinogenicity of presumptive metabolites of the substance of interest, or of substances not available in sufficient quantities for biological testing.

11.3.2 Mutagenicity Tests

Mutagenicity is often the initial insult that leads to the development of a tumor,[1,4,5,29] so that mutagenic chemicals are considered to be potential carcinogens. As a result, *in vitro* mutagenicity tests are often treated as presumptive carcinogenicity tests, although many mutagens have been shown to be non-carcinogenic in animals, and many chemicals that are negative in mutagenicity tests do induce cancers in test animals and humans. Although a positive response in these tests is highly predictive for rodent carcinogenicity, it does not inform the types or locations of the tumors that may be produced, or the potency of the cancer response.[30,31]

A historical limitation to the use of *in vitro* bacterial and mammalian cell tests to identify mutagens and carcinogens is that (with the exception of a few mammalian cell lines, and some primary mammalian cell cultures) the cells lack the metabolic capability of the intact animals. This is crucial because many chemicals are not DNA reactive, per se, but require metabolism to an active form. The mammalian liver is the primary site for such metabolism, but other organs, including the intestinal flora, also have metabolic capability. In order to overcome this discrepancy, a rodent liver homogenate is added to approximate *in vivo* metabolism. This liver homogenate, usually designated as S9 (i.e., a $9000 \times g$ supernatant of whole liver) is typically prepared from livers of male rats pretreated with the PCB Aroclor 1254, or a combination of β-naphtho-flavone and phenobarbital, to induce higher levels of the cytochrome enzymes responsible for the metabolism.[7,32] The S9 is mixed with a number of cofactors to support the enzyme activity before being added to the bacterial or mammalian cell cultures. Although rat liver is most commonly used, S9 can be prepared from other tissues and animal species according to the research or testing questions being asked. Human S9 has also been used, but has not been shown to confer any advantage over rodent S9 for identifying mutagens and potential carcinogens.[33–35] It has the disadvantage of being more difficult to obtain, and to be more variable from batch to batch than rodent S9.

In addition to the tests described below, a large number of other mutagenicity and chromosome damage tests have been proposed or used for presumptive carcinogen identification. For the most part, the performances and inter-laboratory reproducibility of these other tests have not been well studied, or the tests have been shown to be redundant or less effective than the tests currently in use.

11.3.2.1 Salmonella (Ames) test

Since 1975 when Ames, and then others,[8–10] showed that the *in vitro Salmonella* microsome mutagenicity test was able to correctly identify carcinogens and discriminate them from non-carcinogens approximately 90% of the time, the Ames test became the predominant *in vitro* screen for carcinogens. A positive result in the test is considered sufficient to label the substance as a potential carcinogen.

The test consists of a number of *Salmonella typhimurium* "tester" strains isolated by Ames and his colleagues.[7,36] The strains carry single mutations in different histidine (his) synthesis genes which renders them unable to grow in the absence of the amino acid, histidine. They can only divide and form colonies if histidine is added to their growth media or if they undergo a subsequent mutation that restores their ability to synthesize the amino acid. The individual tester strains respond selectively to mutagenic events so that substances producing various gene mutations and small deletions can be differentially identified. Ames also engineered a number of additional changes into the tester strains that renders them more permeable to the test chemicals, and more sensitive to mutation induction.[7,36]

Following the initial demonstrations of effectiveness of the *in vitro* genetic toxicity tests for predicting carcinogens, compilations using different sets of cancer data showed that the positive predictivity of the Ames test was in the 50–60% range, and approximately 50% of the carcinogens were not positive in the test (Table 11.2).[12–15] The mammalian cell tests (see below) identified more mutagenic chemicals than the Ames test, but also, because of their sensitivity to artifacts inherent in the test, misidentified a larger proportion of non-carcinogens.[12–15,37,38] One advantage to the Ames test is that, unlike the mammalian cell tests (below), it does not appear to be subject to producing artifactual positives. The most widely used protocols for the bacterial mutagenicity tests are in OECD Test Guideline 471.[39]

11.3.2.2 In Vitro Mammalian Cell Chromosome Damage

Chromosome aberrations have long been associated with cancer because cancer cells generally have an abnormal chromosome complement or contain damaged

Table 11.2   Prediction of carcinogenicity using *Salmonella* Mutant Tester Strains.*

Chems.	Sensitivity	Specificity	+Predictivity	−Predictivity	Concordance	Ref.
224	0.88	0.75	0.90	0.71	0.84	8
139	0.93	0.74	0.89	0.82	0.87	9
120	0.92	0.93	0.93	0.92	0.93	10
60	0.76	0.59	0.69	0.67	0.68	11
73	0.45	0.86	0.83	0.51	0.62	12
114	0.48	0.91	0.89	0.55	0.66	13
363	0.54	0.79	0.77	0.57	0.65	14
717	0.59	0.74	0.87	0.37	0.62	15

*The compilations from 1985 through 2005 contain many of the same test results produced by the US National Toxicology Program.
Chems., the number of chemicals evaluated; Sensitivity, the proportion of carcinogens positive in the test; Specificity, the proportion of non-carcinogens negative in the test; +Positive predictivity, the proportion of positives in the test that are carcinogens; −Predictivity, the proportion of negatives in the test that are not carcinogens; Concordance, the overall proportion of correct predictions.

chromosomes. Additionally, chromosome morphology is easily examined using optical microscopes without the need for molecular techniques.

When a chemical damages DNA or the chromosome structure it can cause a break in the chromosome. Alternatively, a chromosome break can be produced as the cell tries to repair DNA damage. These double-strand breaks are visible with a light microscope during mitosis and are used to identify chromosome damaging agents. Some cells may try to repair the breaks by rejoining the chromosome ends. If this repair does not restore the chromosomes to their normal configuration, it may lead to chromosome rearrangements, which are also visible during mitosis. Chromosome breaks are caused by damage in cells that are in their resting or DNA synthesis phase. When the cell enters mitosis, the individual chromosomes form visible structures, which allows the breaks or rearrangements to be seen. Most of the cells with chromosome breaks can successfully enter mitosis but cannot form viable daughter cells because of the loss of vital chromosomal material, or because chromosome rearrangements that may form will interfere with the final stages of mitosis. Therefore, unlike in gene mutation studies, mutant cells that can progress to carcinogenesis are produced in only a small fraction of the total damaged cells. As a consequence, the measure of total chromosome damage is used as a surrogate for the damage that will allow the cell to survive and successfully reproduce.

During the past few years, the chromosome aberration assay has been supplanted by the micronucleus (MN) test. This test also identifies cells with damaged or rearranged chromosomes, but is easier to perform and is less time consuming to score, and may result in fewer artifactual positives because, unlike the chromosome aberration test, the cells must be sufficiently viable to complete nucleus formation and mitosis. An added advantage to the MN test is that it allows the identification of aneuploid cells (i.e., cells with whole chromosome loss).

While any cell with a stable chromosome number can be used, the cell lines most commonly used for chromosome aberration and MN tests are the

Chinese hamster ovary (CHO) and lung (CHL) cells, and primary human lymphocytes (HuLy). Because of its aberrant karyotype, the L5178Y mouse lymphoma cell line is not useful for chromosome aberration studies, but can be used for MN studies. The cells are treated with and without S9 because most cell lines do not have the metabolic capability needed to metabolize the test substances.

One limitation of chromosome aberration assays is that chromosome breakage can occur as a secondary effect of cell toxicity or excursions in the pH of the test medium or its osmolality.[40–42] It is generally not possible to distinguish these secondary effects (artifactual positives) from chromosome damage caused by the direct interaction of the test chemical with the chromosome, although transcriptomic methods using human cells have been proposed.[43] As a result, these assays have responded to many chemicals that are not carcinogenic in animal tests. This has led some to attach the appellation "irrelevant positives" to these responses, to distinguish them from the "false positives" which are true chromosome breaking agents *in vitro* but, for one reason or another, are not effective *in vivo* or are not carcinogens.

These "irrelevant positives" are being addressed by new test guidelines under development which will require that cell toxicity be kept below 50–60%, as compared to previous guidelines which did not specify a maximum toxicity, or allowed up to 90% cell killing. Such limits will reduce the incidence of chromosome damage as a secondary effect to cell killing rather than via a direct interaction with the DNA or chromosome structure. It has not been shown that such "irrelevant positives" can be produced in the MN test because the cells have to be sufficiently viable to complete a round of replication for the MN to be formed.

Values for the effectiveness of *in vitro* chromosome aberration tests in CHO cells are in shown in Table 11.3. Insufficient data are available to compare the relative effectiveness of the other cell lines. Although there are less data available on the performance of the MN test for identifying carcinogens, the mechanism of MN formation, and comparisons of chromosome aberration and MN tests in the same cell lines support the use of MN tests in lieu of chromosome aberration tests. The most widely used protocols for chromosome aberration tests and the *in vitro* micronucleus test are in OECD Test Guideline 473,[44] and Test Guideline 487,[45] respectively.

Table 11.3 Predictivity of carcinogenicity using *in vitro* chromosome aberration tests with CHO cells.

Chems.	Sensitivity	Specificity	+Predictivity	−Predictivity	Concordance	Ref.
73*	0.55	0.69	0.73	0.50	0.60	12
114*	0.52	0.72	0.73	0.52	0.61	13
218*	0.52	0.68	0.73	0.46	0.58	14
460*	0.66	0.45	0.79	0.36	0.60	15

Footnotes as in Table 11.2.

11.3.2.3 *In Vitro Mammalian Cell Gene Mutation*

A number of tests are routinely used, the most common being mouse lymphoma L5178Y cells, although CHO cells and human fibroblast cell lines also provide acceptable results. Transgenic mouse or rat cells have also been used. In these tests, the cells are treated with and without S9, unless the cells used are known to possess sufficient cytochrome metabolic activity, and gene mutation is measured by the ability of the cells to grow under selective conditions. The mutation endpoint studied usually involves the loss of a specific enzyme activity which will then allow the cell to grow in the presence of a selective pressure (an antimetabolite or other substance) that would be toxic to the unmutated (wild-type) cell. For example, the mouse lymphoma cell assay (MLA) measures mutation at the thymidine kinase (TK) locus which enables the cell to survive in the presence of the pyrimidine analog, trifluorothymidine. Mutation in CHO and human fibroblast cells measure loss of function of the hypoxanthine-guanine phosphoribosyl transferase (HPRT) or xanthine-guanine phosphoribosyl transferase (XPRT) enzymes, which confer resistance to 6-thioguanine or 8-azaguanine.

Gene mutation tests take longer to perform than chromosome aberration or MN tests because the mutation must be fixed in the viable cells and then the cells may have to go through a number of generations to allow expression of the mutant phenotype. This delay allows the non-mutated enzyme to be cleared from the cell so that it would not be killed by the mutation-specific selective pressure.

One proposed advantage to the MLA is its ability to respond to both gene mutation and chromosome damage effects. The two effects have been distinguished by the induction of "small" mutant colonies resulting from gross chromosome damage, and "large" colonies which are the result of point mutations.[38,46] However, it has been shown that extremes of pH of the growth medium, or high osmolality of the test chemical, can induce positive responses in the MLA,[40,41] and it is assumed that the same effects would be produced in other mammalian cell assays. Similarly to chromosome aberrations, there is concern that high levels of toxicity could also generate positive responses that are not the result of the chemical's interaction with DNA or chromosomes. The effectiveness of the MLA for identifying presumptive carcinogens is described in Table 11.4. Detailed current protocols for *in vitro* gene mutation tests are included in OECD Test Guideline 476.[46]

11.3.3 *In Vitro* Cell Transformation Tests

Cell transformation tests measure the conversion of a non-malignant or pre-malignant mammalian cell in culture to a malignant form. The test chemical's potential malignancy is indicated by its ability to induce phenotypically transformed colonies in culture and, where follow-up testing is done, the ability of the transformed cells to produce tumors when injected into animals. The exact nature of the chemically induced cell damage leading to the transformed

Table 11.4 Predictivity of carcinogenicity using *in vitro* mammalian cell mutagenicity tests with mouse lymphoma cells.

Chems.	Sensitivity	Specificity	+Predictivity	−Predictivity	Concordance	Ref.
49*	0.88	0.11	0.81	0.17	0.73	37
126*	0.94	1.00	1.00	0.46	0.94	38
73*	0.70	0.45	0.66	0.50	0.60	12
114*	0.72	0.400	0.63	0.50	0.59	13
191*	0.74	0.32	0.61	0.46	0.57	14
332*	0.73	0.39	0.77	0.42	0.63	15

Footnotes as in Table 11.2.

phenotype, and the specific mechanisms of cell transformation to a malignant phenotype, are not known. These are not considered mutation tests, although mutagens and non-mutagens are able to transform the cells.

The majority of studies to determine the cancer predictivity of the assay have been performed with three cell types, SHE (Syrian hamster embryo), Balb/c (originally derived from an Balb/c mouse whole embryo culture), and C3H-10T1/2 (originally derived from embryonic fibroblast cells of C3H mice).[47–51] SHE cells, which are a primary culture, have been the most extensively studied and evaluated. The other cells most commonly used, (Balb/c; C3H-10T1/2) are partially transformed and grow continuously in cell culture, and are believed to have already experienced a number of the key events (see Table 11.1) needed to become tumor cells. Treatment with carcinogens or tumor promoters is believed to induce the final stages in the malignant progression of these cells. In practice, the cells are treated with the chemical being studied, and their progression to malignancy is monitored by the phenotypic appearance of the transformed colonies (i.e., from an orderly two-dimensional array of cells to a random orientation of cells) which tend to "pile up."

There have been numerous reports of the effectiveness of these tests in identifying carcinogens, including carcinogens that are negative in the *in vitro* gene mutation and chromosome aberration tests (i.e., the so-called non-genotoxic carcinogens).[49,52,53] As a consequence, cell transformation tests had been promoted as being effective test systems for identifying non-genotoxic carcinogens (i.e., those that are negative in the *in vitro* gene mutation and chromosome damage tests). However, surveys of the published studies using the assays have shown that the majority of carcinogens tested were previously shown to be mutagenic, and that many non-mutagenic chemicals were tested because they were already known to be non-carcinogenic. Validation studies have been underway since the early part of this century to examine the intra-lab reproducibility and between-laboratory transferability of the assays.[49–56]

At the time of this writing, there are no regulatory requirements for performing cell transformation assays, although positive results from such assays are considered as indicating potential carcinogenicity. It remains to be seen if acceptably validated cell transformation assays will be used to supplement the currently required assays, or as a replacement.

11.4   *In Vivo* Genetic Toxicity Assays

A number of *in vivo* assays are being used, or developed, for identifying carcinogens through their genetic activity, or their patterns of effects on genes and/or enzymes. Given the lower sensitivity of the *in vivo* genetic toxicity assays, a negative response in these assays for a chemical positive *in vitro* may not be sufficient to rule out its potential carcinogenicity, and there have been relatively few carcinogens that are not mutagenic or clastogenic *in vitro* that have been shown to be mutagenic or clastogenic *in vivo*. The more simple *in vivo* genetic toxicity assays have been used primarily to confirm *in vitro* positive results or to obtain additional information about potential cancer mechanisms. Other *in vivo* assays (e.g., transgenic rodents[57] and gene mutation in peripheral lymphocytes)[58,59] are not used for routine screening of chemicals for carcinogenicity, primarily because of their cost and/or the level of effort involved in performing them. Their value lies in their use to measure biomarkers of genetic damage or for investigating the mode of action of chemicals already known to be carcinogenic. These uses will not be addressed here.

Reductions in animal use have been achieved by using the same animals for genetic toxicity studies as were being used for short-, and intermediate-, term toxicity studies. These assays are described below in less detail than the *in vitro* assays (above).

11.4.1   *In Vivo* Mammalian Cell Chromosome Damage

Rodent bone marrow cells have been extensively used for assays of chromosome damage. Chromosome aberrations and MN are measured in immature bone marrow red blood cells, and MN can also be measured in mature bone marrow and peripheral blood erythrocytes. These assays have been used more to confirm or supplement the *in vitro* genetic toxicity results, rather than as a carcinogen predictor because of their relatively low sensitivity.[14] OECD Guidelines for the bone marrow and peripheral blood chromosome aberration and MN tests have been published.[60,61]

11.4.2   Toxicogenomics

Toxicogenomics encompasses the measurement of gene expression changes in the test animal in response to chemical exposures.[62] As such, it is an indicator of the initial cell responses to the exposure, rather than a measure of mutation, cancer, or other effect. Predictive toxicogenomics seeks to develop profiles of tissue responses to identify gene expression signatures and gene expression pathways that can be used to predict toxicological outcomes such as carcinogenicity. Low, human-exposure doses can be used, relatively few animals are needed, and exposures are generally short-term (i.e., ≤ 90 days).

Studies using known carcinogens and non-carcinogens have been performed by multiple laboratories in an attempt to identify patterns of response that will distinguish carcinogens from non-carcinogens and, also, to distinguish genotoxic carcinogens (those that can be identified using *in vitro* genetic toxicology tests), from non-genotoxic carcinogens (those that initiate tumor induction via a non-mutagenic mechanism and are not detected by the standard genetic toxicity tests).[63–68] It is hoped that these procedures will not only be able to identify all types of carcinogens, but also identify the tissues at risk for tumor development. An additional advantage to this system is that selected tissues in humans can be sampled in situations of human exposure, and the gene expression patterns can be compared to the animal's expression patterns or be used directly as an indicator of potential cancer risk or susceptibility.

11.4.3 Short-term Cancer Assays in Transgenic Rodents

A number of mouse strains have been developed that are highly sensitive to tumor development. This sensitivity is usually achieved by mutations that make them more susceptible to mutation, or mutations or other changes that make them more susceptible to tumor formation (e.g., p53; Tg.Ac; Hras).[69–72]

One advantage to the transgenic models is that, because of their engineered genetic disposition to develop tumors, the resultant tumors generally develop within a 6–12 month period, rather than the 1–2 years in traditional rodent models. Also, because of the sensitivity of these animals, the tests are typically performed using 15–25 animals per sex, instead of the 50 used in the traditional 2-year assay.

11.5   Summary

Cancer is a disease that is the result of a progression of cell and tissue changes leading to a tumor. In the animal, the progression of the tumor and the tissue it appears in will differ according to the specific carcinogen and the animal in which it is being tested. Not all tissues that are affected by a carcinogen develop tumors; tissue-specific and immunological factors will affect the progression and survival of the altered cells. As a consequence, no single *in vitro* test system will mimic the entire cancer process.

When commercial organizations screen chemicals under development for potential carcinogenicity, a positive response in any one of the screening tests (usually the Ames test or a chromosome damage test) may be sufficient to remove a substance from further development. The alternative would be to perform a definitive cancer assay in the hopes that it would not be carcinogenic, and this is done only if the perceived commercial return from the chemical is substantial enough to justify the risk. Although numbers are not publically available, many chemicals are declared to be presumptive carcinogens, and not tested further in animals, based on the *in silico* and *in vitro* test results.

As noted above, none of the *in vitro* tests measure cancer induction, but positive results in any of the tests are predictive for cancer, albeit with varying degrees of accuracy. Unfortunately, negative results in these tests are not predictive of non-carcinogenicity. Many testing schemes favored by regulatory authorities require two or three *in vitro* tests that measure different genetic endpoints (e.g., gene mutation; chromosome aberrations). In addition, tests that measure DNA damage (e.g., comet assay) and cell transformation are increasingly being used. One challenge associated with these multiple geno-toxicity tests is that each test has its unique false positives. Therefore, as more tests are added to the battery, the proportion of both true positives and false positives will increase. If a positive response in any of the test systems is sufficient to label a substance as potential carcinogen, the use of multiple test systems could have the effect of discarding a high proportion of non-carcinogenic substances, or, alternatively lead to additional *in vivo* tests being performed in an attempt to distinguish the true positives from false positives.

Current approaches are moving towards integrating the results of QSAR analysis, genetic toxicity tests, and tests for other precarcinogenic mechanisms (i.e., genetic recombination; cell transformation, in addition to toxicogenomics) to refine the estimation of the probability of carcinogenicity or non-carcinogenicity.

References

1. A. M. Jarabek, L. H. Pottenger, L. S. Andrews, D. Casciano, M. R. Embry, J. H. Kim, R. J. Preston, M. V. Reddy, R. Schoeny, D. Shuker, J. Skare, J. Swenberg, G. M. Williams and E. Zeiger, Creating context for the use of DNA adduct data in cancer risk assessment: I. Data organization, *Crit. Rev. Toxicol.*, 2009, **39**, 659.
2. E. R. Fearon and B. Vogelstein, A genetic model for colorectal tumori-genesis, *Cell*, 1990, **61**, 759–767.
3. B. Vogelstein and K. H. Kinzler, Cancer genes and the pathways they control, *Nat. Med.*, 2004, **10**, 789.
4. D. Hanahan and R. A. Weinberg, The hallmarks of cancer, *Cell*, 2000, **100**, 57.
5. D. Hanahan and R. A. Weinberg, Hallmarks of cancer: the next gener-ation, *Cell*, 2011, **144**, 646.
6. Organization for Economic Co-operation and Development (OECD), OECD Guideline for the Testing of Chemicals. No. 451. Carcinogenicity Studies. Adopted 7 September 2009.
7. K. Mortelmans and E. Zeiger, The Ames Salmonella/microsome muta-genicity assay, *Mutat. Res.*, 2000, **455**, 29.
8. J. McCann, E. Yamasaki and B. N. Ames, Detection of carcinogens in the Salmonella/microsome test: Assay of 300 chemicals, *Proc. Natl. Acad. Sci. USA*, 1975, **72**, 5135–5139.
9. T. Sugimura, S. Sato, M. Nagao, T. Yahagi, T. Matsushima, H. Seino, M. Takeuchi and T. Kawachi, Overlapping of carcinogens and mutagens, in *Fundamentals of Cancer Prevention*, ed. P. N. Magee, S. Takayama, T. Sugimura and T. Matsushima, University Park Press, Baltimore, 1976, p. 191.

10. I. F. H. Purchase, E. Longstaff, J. Ashby, J. A. Styles, D. Anderson, P. A. Lefevre and F. R. Westwood, An evaluation of 6 short-term tests for detecting organic chemical carcinogens, *Br. J. Cancer*, 1978, **37**, 873.

11. V. C. Dunkel, E. Zeiger, D. Brusick, E. McCoy, D. McGregor, K. Mortelmans, H. S. Rosenkranz and V. F. Simmon, Reproducibility of microbial mutagenicity assays: II. Testing of carcinogens and non-carcinogens in *Salmonella typhimurium* and *Escherichia coli*. *Environ. Mutagen.*, 1985, **7** (Suppl. 5), 1.

12. R. W. Tennant, B. H. Margolin, M. D. Shelby, E. Zeiger, J. K. Haseman, J. Spalding, M. Resnick, S. Stasiewicz, B. Anderson and R. Minor, Prediction of chemical carcinogenicity in rodents from in vitro genetic toxicity assays, *Science*, 1987, **236**, 933.

13. E. Zeiger, J. K. Haseman, M. D. Shelby, B. H. Margolin and R. W. Tennant, Evaluation of four in vitro genetic toxicity tests for predicting rodent carcinogenicity: Confirmation of earlier results with 41 additional chemicals, *Environ. Mol. Mutagen.*, 1990, **16**(Suppl. 18), 1.

14. E. Zeiger, Identification of rodent carcinogens and noncarcinogens using genetic toxicity tests: premises, promises, and performance, *Regul. Toxicol. Pharmacol.*, 1998, **28**, 85.

15. D. Kirkland, M. Aardema, L. Henderson and L. Muller, Evaluation of the ability of a battery of three in vitro genotoxicity tests to discriminate rodent carcinogens and non-carcinogens. I. Sensitivity, specificity and relative predictivity, *Mutat. Res.*, 2005, **584**, 1.

16. OECD, Guidance Document No. 116 on the Conduct and Design of Chronic Toxicity and Carcinogenicity Studies, Supporting Test Guidelines 451, 452 and 453. 2nd Edition. Adopted 13 April 2012.

17. Y. T. Woo, D. Y. Lai, M. F. Argus and J. C. Arcos, Development of structure-activity relationship rules for predicting carcinogenic potential of chemicals, *Toxicol. Lett.*, 1995, **7**, 219.

18. Y. T. Woo, Mechanisms of action of chemical carcinogens, and their role in Structure-Activity Relationships (SAR) analysis and risk assessment. In: R. Benigni, (ed.), *Quantitative Structure-Activity Relationship (QSAR) Models of Mutagens and Carcinogens*. Boca Raton: CRC Press, 2003, p. 41.

19. R. Benigni, C. Bossa, O. Tcheremenskaia and A. Giuliani, Alternatives to the carcinogenicity bioassay: in silico methods, and the in vitro and in vivo mutagenicity assays, *Exp. Opin. Drug Metab. Toxicol.*, 2010, **6**, 1.

20. J. Devillers, E. Mombelli and R. Samsera, Structural alerts for estimating the carcinogenicity of pesticides and biocides, *SAR QSAR Environ. Res.*, 2011, **22**, 89.

21. R. Benigni and C. Bossa, Alternative strategies for carcinogenicity assessment: an efficient and simplified approach based on in vitro mutagenicity and cell transformation assays, *Mutagenesis*, 2011, **26**, 455.

22. R. Benigni, C. Bossa, S. Alivernini and M. Colafranceschi, Assessment and validation of US EPA's OncoLogic® expert system and analysis of its modulating factors for structural alerts, *J. Environ. Sci. Health. C. Environ. Carcinog. Ecotoxicol. Rev.*, 2012, **30**, 152.

23. E. C. Miller and J. A. Miller, mechanisms of chemical carcinogenesis: nature of proximate carcinogens and interactions with macromolecules, *Pharmacol. Rev.*, 1966, **18**, 805.

24. J. A. Miller, Carcinogenesis by chemicals: an overview – G. H. A. Clowes memorial lecture, *Cancer Res.*, 1970, **30**, 559.

25. J. Ashby and R. W. Tennant, Definitive relationships among chemical structure, carcinogenicity and mutagenicity for 301 chemicals tested by the U.S. NTP, *Mutat. Res.*, 1991, **257**, 229.

26. NTP (National Toxicology Program) available on-line at: http:// ntp.niehs.nih.gov/index.cfm?objectid = 7DA86165-BDB5-82F8- F7E4FB36737253D5.

27. L. S. Gold and E. Zeiger, *Handbook of Carcinogenic Potency and Genotoxicity Databases*. CRC Press, Boca Raton, FL, 1997.

28. IARC (International Agency for Research on Cancer) available on-line at: http://www.iarc.fr/en/publications/list/monographs/index.php.

29. R. J. Preston and G. M. Williams, DNA-reactive carcinogens: Mode of action and human cancer hazard. *Crit. Rev. Toxicol.*, 2005, **35**, 673.

30. B. A. Fetterman, B. S. Kim, B. H. Margolin, J. S. Schildcrout, M. G. Smith, S. M. Wagner and E. Zeiger, Predicting rodent carcinogenicity from mutagenic potency measured in the Ames Salmonella assay, *Environ. Mol. Mutagen.*, 1997, **29**, 312.

31. J. S. Schildcrout, B. H. Margolin and E. Zeiger, Predicting rodent carcinogenicity using potency measures of the in vitro sister chromatid exchange and chromosome aberration assays, *Environ. Mol. Mutagen.*, 1999, **33**, 59.

32. W. W. Ku, A. Bigger, G. Brambilla, H. Glatt, E. Gocke, P. J. Guzzie, A. Hakura, M. Honma, H. J. Martus, R. S. Obach and S. Roberts, Strategy for genotoxicity testing – metabolic considerations, *Mutation Res.*, 2007, **627**, 59.

33. P. Beaune, R. Lemestre-Cornet, P. Kremers, A. Albert and J. Gielen, The Salmonella/microsome mutagenicity test: Comparison of human and rat livers as activating systems, *Mutation Res.*, 1985, **156**, 139.

34. A. Hakura, S. Suzuki and T. Satoh, Advantage of the use of human liver S9 in the Ames test, *Mutation Res.*, 1999, **438**, 29.

35. A. Hakura, H. Shimada, M. Nakajima, H. Sui, S. Kitamoto, S. Suzuki and T. Satoh, Salmonella/human S9 mutagenicity test: a collaborative study with 58 compounds, *Mutagenesis*, 2005, **20**, 217.

36. D. M. Maron and B. N. Ames, Revised methods for the Salmonella mutagenicity test, *Mutation Res.*, 1983, **113**, 173.

37. W. J. Caspary, D. S. Daston, B. C. Myhr, A. D. Mitchell, C. J. Rudd and P. S. Lee, Evaluation of the L5178Y mouse lymphoma cell mutagenesis assay: Interlaboratory reproducibility and assessment, *Environ. Mol. Mutagen.*, 1988, **12**(Suppl. 13), 195.

38. A. D. Mitchell, A. E. Auletta, C. Clive, P. E. Kirby, M. M. Moore and B. C. Myhr, The L5178Y/tk$^{+/-}$ mouse lymphoma specific gene and

chromosomal mutation assay. A phase III report of the U.S. Environmental Protection Agency Gene-Tox Program, *Mutation Res.*, 1997, **394**, 177.

39. Organization for Economic Co-operation and Development (OECD), OECD Guideline for the Testing of Chemicals. No. 471. Bacterial Reverse Mutation Test. Adopted 21 July 1997.

40. D. Brusick, Genotoxic effects in cultured mammalian cells produced by low pH treatment conditions and increased ion concentrations, *Environ. Mutagen.*, 1986, **8**, 789.

41. D. Scott, S. M. Galloway, R. R. Marshall, M. Ishidate Jr., D. Brusick, J. Ashby and B. C. Myhr, International Commission for Protection Against Environmental Mutagens and Carcinogens. Genotoxicity under extreme culture conditions. A report from ICPEMC Task Group 9, *Mutation Res.*, 1991, **257**, 147.

42. S. Galloway, Cytotoxicity and chromosome aberrations in vitro: Experience in industry and the case for an upper limit on toxicity in the aberration assay, *Environ. Mol. Mutagen.*, 2000, **35**, 191.

43. H. Ellinger-Ziegelbauer, J. M. Fostel, C. Aruga, D. Bauer, E. Boitier, S. Deng, D. Dickinson, A.-C. Le Fevre, A. J. Fornace Jr., O. Grenet, Y. Gu, J.-C. Hoflack, M. Shiiyama, R. Smith, R. D. Snyder, C. Spire, G. Tanaka and J. Aubrecht, Characterization and interlaboratory comparison of a gene expression signature for differentiating genotoxic mechanisms, *Toxicol. Sci.*, 2009, **110**, 341.

44. Organization for Economic Co-operation and Development (OECD), OECD Guideline for the Testing of Chemicals. No. 473. In Vitro Mammalian Chromosome Aberration Test. Adopted 21 July 1997.

45. Organization for Economic Co-operation and Development (OECD), OECD Guideline for the Testing of Chemicals. No. 487. In Vitro Mammalian Cell Micronucleus Test. Adopted 22 July 2010.

46. Organization for Economic Co-operation and Development (OECD), OECD Guideline for the Testing of Chemicals. No. 476. In Vitro Mammalian Cell Gene Mutation Test. Adopted 21 July 1997.

47. IARC/NCI/EPA Working Group, Cellular and molecular mechanisms of cell Transformation and standardization of transformation assays of established cell lines for the prediction of carcinogenic chemicals: Overview and recommended protocols. *Cancer Res.*, 1985, **45**, 2395.

48. R. Combes, M. Balls, R. Curren, M. Fischbach, N. Fusenig, D. Kirkland, C. Lasne, J. Landolph, R. LeBoeuf, H. Marquardt, J. McCormick, L. Muller, E. Rivedal, E. Sabbioni, N. Tanaka, P. Vasseur and H. Yamasaki, Cell transformation assays as predictors of human carcinogenicity. The report and recommendations of ECVAM Workshop XX, *ATLA*, 1999, **27**, 745.

49. OECD (Organisation for Economic Co-operation and Development), Detailed review on cell transformation assays for detection of chemical carcinogens. OECD Environment, Health and Safety Publications, Series on Testing and Assessment. 2007, No. 31.

50. P. Vanparys, R. Corvi, M. Aardema, L. Gribaldo, M. Hayashi, S. Hoffmann and L. Schechtman, ECVAM prevalidation of three cell transformation assays, *ALTEX*, 2010, **27**, 267.

51. S. Creton, M. J. Aardema, P. L. Carmichael, J. S. Harvey, F. L. Martin, R. F. Newbold, M. R. O'Donovan, K. Pant, A. Poth, A. Sakai, K. Sasaki, A. D. Scott, L. M. Schechtman, R. R. Shen, N. Tanaka and H. Yasaei, Cell transformation assays for prediction of carcinogenic potential: state of the science and future research needs, *Mutagenesis*, 2012, **27**, 93.

52. E. J. Matthews, J. W. Spalding and R. W. Tennant, Transformation responses of 168 chemicals compared with mutagenicity in Salmonella and carcinogenicity in rodent bioassays, *Environ. Health Perspect.*, 1993, **101**(Suppl. 2), 347.

53. R. J. Isfort and R. A. LeBoeuf, The Syrian hamster embryo (SHE) cell transformation system: a biologically relevant in vitro model--with carcinogen predicting capabilities--of in vivo multistage neoplastic transformation, *Crit. Rev. Oncogen.*, 1995, **6**, 251.

54. V. C. Dunkel, R. J. Pienta, A. Sivak and K. A. Traul, Comparative neoplastic transformation responses of BALB/c 3T3 cells, Syrian hamster embryo cells, Rauscher murine leukemia virus-infected Fischer 344 rat embryo cells to chemical carcinogens, *JNCI*, 1981, **67**, 1303.

55. V. C. Dunkel, L. M. Schechtman, A. S. Tu, A. Sivak, R. A. Lubet and T. P. Cameron, Interlaboratory evaluation of the C3H 10T1/2 transformation assay, *Environ. Mol. Mutagen.*, 1988, **12**, 21.

56. C. A. Jones, E. Huberman, M. F. Callaham, A. Tu, W. Hallowell, S. Pallota, A. Sivak, R. A. Lubet, M. D. Avery and R. E. Kouri, An interlaboratory evaluation of the Syrian hamster embryo cell transformation assay using eighteen coded chemicals, *Toxicol. In Vitro*, 1988, **2**, 103.

57. I. B. Lambert, T. M. Singer, S. E. Boucher and G. R. Douglas, Detailed review of transgenic rodent mutation assays, *Mutation Res.*, 2005, **590**, 1.

58. F. J. van Dam, A. T. Natarajan and A. D. Tates, Use of a T-lymphocyte clonal assay for determining HPRT mutant frequencies in individual rats, *Mutation Res.*, 1992, **271**, 231.

59. A. D. Tates, F. J. van Dam, F. A. de Zwart, C. M. van Teylingen and A. T. Natarajan, Development of a cloning assay with high cloning efficiency to detect induction of 6-thioguanine-resistant lymphocytes in spleen of adult mice following in vivo inhalation exposure to 1,3-butadiene, *Mutation Res.*, 1994, **309**, 299.

60. Organization for Economic Co-operation and Development (OECD), OECD Guideline for the Testing of Chemicals. No. 474. Mammalian Erythrocyte Micronucleus Test. Adopted 21 July 1997.

61. Organization for Economic Co-operation and Development (OECD), OECD Guideline for the Testing of Chemicals. No. 475. Mammalian Bone Marrow Chromosome Aberration Test. Adopted 21 July 1997.

62. M. D. Waters and J. M. Fostel, Toxicogenomics and systems toxicology: aims and prospects, *Nature Rev. Genet.*, 2004, **5**, 936.

63. H. Ellinger-Ziegelbauer, B. Stuart, B. Wahle, W. Bomann and H. J. Ahr, Comparison of the expression profiles induced by genotoxic and non-genotoxic carcinogens in rat liver, *Mutation Res.*, 2005, **575**, 61.

64. K. Nakayama, Y. Kawano, Y. Kawakami, N. Moriwaki, M. Sekijima, M. Otsuka, Y. Yakabe, H. Miyaura, K. Saito, K. Sumida and T. Shirai, Differences in gene expression profiles in the liver between carcinogenic and non-carcinogenic isomers of compounds given to rats in a 28-day repeat-dose toxicity study, *Toxicol. Appl. Pharmacol.*, 2006, **217**, 299.

65. M. R. Fielden, R. Brennan and J. Gollub, A gene expression biomarker provides early prediction and mechanistic assessment of hepatic tumor induction by nongenotoxic chemicals, *Toxicol. Sci.*, 2007, **99**, 90.

66. K. Z. Guyton, A. D. Kyle, J. Aubrecht, V. J. Cogliano, D. A. Eastmond, M. Jackson, N. Keshava, M. S. Sandy, B. Sonawane, L. Zhang, M. D. Waters and M. T. Smith, Improving prediction of chemical carcinogenicity by considering multiple mechanisms and applying toxicogenomic approaches, *Mutation Res.*, 2009, **681**, 230.

67. S. S. Auerbach, R. R. Shah, D. Mav, C. S. Smith, N. J. Walker, M. K. Vallant, G. A. Boorman and R. D. Irwin, Predicting the hepatocarcinogenic potential of alkenylbenzene flavoring agents using toxicogenomics and machine learning, *Toxicol. Appl. Pharmacol.*, 2010, **243**, 300.

68. M. D. Waters, M. A. Jackson and I. A. Lea, Characterizing and predicting carcinogenicity and mode of action using conventional and toxicogenomics methods, *Mutation Res.*, 2010, **705**, 184.

69. J. E. French, J. W. Spalding, J. K. Dunnick, R. R. Tice, M. Furedi-Machacek and R. W. Tennant, The use of transgenic animals in cancer testing, *Inhal. Toxicol.*, 1999, **11**, 541.

70. D. Gulezian, D. Jacobson-Kram, C. B. McCullough, H. Olson, L. Recio, D. Robinson, R. Storer, R. Tennant, J. M. Ward and D. A. Neumann, Use of transgenic animals for carcinogenicity testing: considerations and implications for risk assessment, *Toxicol. Pathol.*, 2000, **28**, 482.

71. J. W. Spalding, J. E. French, S. Stasiewicz, M. Furedi-Machacek, F. Conner, R. R. Tice and R. W. Tennant, Responses of transgenic mouse lines p53$^{(+/-)}$ and Tg.AC to agents tested in conventional carcinogenicity bioassays, *Toxicol. Sci.*, 2000, **53**, 213.

72. R. D. Storer, J. E. French, L. A. Donehower, D. Gulezian, K. Mitsumori, L. Recio, R. H. Schiestl, F. D. Sistare, N. Tamaoki, T. Usui, H. van Steeg and IWGT Working Group, Transgenic tumor models for carcinogen identification: the heterozygous Trp53-deficient and RasH2 mouse lines, *Mutation Res.*, 2003, **540**, 165.

Reproductive and Developmental Toxicity Testing: Issues for 3Rs Implementation

STEFANO LORENZETTI* AND ALBERTO MANTOVANI

Istituto Superiore di Sanità – ISS, Dpt. of Food Safety and Veterinary Public Health, Food and Veterinary Toxicology Unit, viale Regina Elena 299, 00161 Rome, Italy
*Email: stefano.lorenzetti@iss.it

12.1 Reproductive and Developmental Toxicity Testing: Risk Assessment and REACH Regulation

Toxicity testing is generally performed with the clear aim to determine the safety of a chemical by identifying its effects on a living organisms and patterns (amount, duration, ways of exposure) at which such effects may occur. In a few cases, like well-known environmental pollutants (e.g. arsenic, lead) or nutrients that may be toxic at excessive intakes (e.g. vitamin A, selenium), the available human studies (bio-monitoring, epidemiological and observational studies) provide major scientific evidence. However, for the vast majority of chemicals, the characterization of toxicological hazards for humans relies mostly, or solely, on studies on laboratory animals and, to an increasing extent, on *in vitro* systems. This leads, unavoidably, to uncertainties to be dealt with, related to the relevance to humans of effects observed in experimental systems, from both qualitative (biological relevance) and quantitative (are the effects observable at realistic exposure levels?) standpoints.

Issues in Toxicology No. 19
Reducing, Refining and Replacing the Use of Animals in Toxicity Testing
Edited by David G. Allen and Michael D. Waters
© The Royal Society of Chemistry 2014
Published by the Royal Society of Chemistry, www.rsc.org

The above considerations become even more relevant in the modern toxicological risk assessment, where attention is given to differential chemical hazards in relation to life-stages and gender. Under this respect, reproductive and developmental toxicity features most prominently, as they involve both genders and the mother/conceptus grouping: the pre- and peri-natal development are themselves a timed sequence of events, encompassing windows with different susceptibility, from organogenesis through to fetal and neonatal stages. For instance, true teratogenic effects (i.e. the induction of structural congenital anomalies) can generally be elicited only during the organogenesis phase, that in rodents corresponds to gestation day 6/7 (implantation) up to day 16/17 (palate closure), whereas in humans corresponds to the first three months of pregnancy. For agents inducing a specific malformation (like the drug thalidomide inducing limb reduction defects in humans) the most susceptible phase may be when the specific organogenesis process does occur. Conversely, even a short-term exposure in the specific susceptible developmental window might elicit an adverse effect in the conceptus.[1–4]

It would be unsound to strictly separate reproductive and developmental toxicity since they both affect the potential to produce a healthy new generation as well as involve similar biological targets, such as endocrine homeostasis and the proliferation/differentiation balance. The European Union (EU) approach considers both of them under the general term 'toxic to reproduction', where:

- *reproductive toxicity* includes effects on reproductive organs/tissues and their functions, fertility/fecundity up to implantation, and
- *developmental toxicity* includes effects relevant to life and development during prenatal and neonatal stages; an up-to-date concept of developmental toxicity could also include the impairment of developmental processes (immunity, skeleton, neuro-behavioural and sexual maturation) occurring during the post-natal life up to puberty, which may have serious consequences at medium- and long-term.[1–4]

Testing for reproductive and developmental toxicity is always considered, and often is mandatory, within all regulatory frameworks for chemicals, including pharmaceuticals, industrial chemicals, biocides, agricultural pesticides, compounds used in farm animal production and food additives. Whereas many research data show the unique sensitivity of the reproductive cycle to certain chemicals, these regulatory requirements rise from past disasters originating from the unchecked exposure to developmental toxicants in food (e.g. the developmental neurotoxicity caused by methyl mercury in seafood in Minamata, Japan)[5] or in pharmaceuticals, such as the thalidomide and diethylstilbestrol episodes: the latter, a synthetic estrogen used to prevent spontaneous abortion, was a particularly interesting case since it did not cause apparent birth defects but elicited an increased risk of reproductive cancer in women after puberty and reproductive malfunctioning in both sexes.[6] More recently, the increasing rate of male infertility and testicular cancer in many industrialized countries has increased the attention by toxicologists and regulators towards the reproductive effects of endocrine disrupters (EDs).[7]

Currently, the demand for reproductive and developmental toxicity testing is a critical requirement of new regulations such as the EU Registration, Evaluation, Authorization and Restriction of Chemicals (REACH, EC no.1907/ 2006; http://ec.europa.eu/environment/chemicals/reach/). New and existing chemicals identified as toxic to reproduction, together with those identified as carcinogenic or mutagenic, are pointed out as Substances of Very High Concerns (SVHC, REACH Article 57, EC no.1907/2006); moreover, SVHCs can be also identified, on a case-by-case basis, from scientific evidence as causing an equivalent level of concern as those above, such as EDs. Indeed, EDs are currently the major global issue as regards reproductive toxicity from the regulatory standpoint. The identification of the EDs, i.e. substances causing adverse health effects by altering function(s) of the endocrine system (WHO/ IPCS 2002),[8] is still a matter of debate, also because the definition of the adversity of an effect (WHO/IPCS 2004)[9] is not unequivocally accepted as regards endocrine endpoints. Although essentially all tissues of a mammalian organism are regulated from the endocrine system, the reproductive tissues and functions are considered of a fundamental importance due to the critical action of steroids and other hormones for their development and homeostasis. Hence, and notwithstanding many data gaps, reproductive and developmental endpoints are considered critical for the risk assessment of EDs (Mantovani, Ann NYAS, 2006).[3] An ED is characterized primarily by its endocrine-mediated mode of action rather than by its effects – which might vary according to sex and life-stage: the decision whether a substance would have to be regarded as an ED in a regulatory sense require a stepwise approach, as proposed in the OECD toolbox (OECD2005, http://www.oecd.org/env/ehs/testing/oecdconceptualframeworkforthetestingandassessmentofendocrinedisruptingchemicals.htm)[10] to define whether the endocrine mode of action is the leading mechanism underlying the critical effects for risk assessment of human health effects.[11,12]

In general, reproductive toxicology is particularly challenging for risk assessment, due to both complexity and duration of the mammalian reproductive cycle, which covers at least two generations and possibly three generations (the parents, the offspring till sexual maturity and the gametes within sexually mature offspring). Indeed, to assess the potential reproductive interference of chemicals, test methods in experimental animals (see next paragraph) must cover the following summarized steps:

1 growth and maturation of sperm and oocyte;
2 fertilization and first zygotic divisions;
3 implantation, placentation, intrauterine development and birth;
4 post-natal development throughout the period of lactation; and
5 development of the offspring to fertile adult animals able to re-start the reproductive cycle producing a second generation.

Furthermore, adverse effects may also occur in the next generation/offspring, as evidenced by the activity of vinclozolin, an antiandrogenic fungicide, on epigenetic inheritance.[13] Effects on the homeostasis and functionality of other tissues may also induce adverse effects on reproduction [e.g. neurobehavioural impairment reducing sexual function, liver functional impairment leading to altered metabolism/clearance of hormones or chronic malabsorption impairing the uptake of nutrients (e.g. vitamin A, zinc) essential for fertility or development]. These latter effects situate in a sort of 'grey area' that need expert judgment on case-by-case basis: indeed, a primary criterion to identify a substance as 'toxic to reproduction' is the ability to elicit reproductive or developmental effects that are not secondary to general toxicity (i.e. the reproductive cycle should be a chemical's target).

The toxicity of the mammalian reproductive cycle can be investigated in a holistic way through costly and time-consuming long-term *in vivo* tests such as the two-generation test (OECD TG 416). A different approach, exploiting *in vitro* assays, requires breakdown into essential steps such as female and male fertility, placenta, organogenesis and so on.[14] Such an approach was implemented within the EU Integrated Project ReProTect (LSHB-CT-2004-503257; http://www.reprotect.eu/), whose aim was to set a series of *in silico* and *in vitro* assays as building blocks to be further developed in an Integrated Testing Strategy (ITS; see below). ReProTect has finally delivered a feasibility study to show the capability of the selected *in vitro* test methods to correctly detect reproductive toxicants in a double-blind ring trial.[15]

Political pressures within the EU led the European Commission to emphasize the principle of 3Rs not only within the REACH regulation but also in other legislative tools such as the new Cosmetics Regulation (EC no. 1223/2009) and the Regulation on Classification, Labelling and Packaging/CLP of substances and mixtures (EC no 1272/2008). In all of them, the 3Rs are part of the new toxicological requirements, in particular the processes of screening and prioritization of chemicals. Under this respect, reproductive and developmental toxicology represent a major challenge for 3Rs implementation, due to the many endpoints, mechanisms and targets involved. In fact, regulatory toxicology has been hesitant to embrace with enthusiasm the new *in vitro* (and *in silico*) methods and approaches to complement or substitute traditional animal tests due to, among others, the slow rate of formal validation of the proposed alternative methods and the difficulties to individuate markers of effects that unequivocally represent a gold standard, especially for endocrine endpoints. Alternative tests in the field of reproductive and developmental toxicity might be of great interest since they could provide timely and cost-effective outputs, but a testing strategy and associated framework to interpret and use results are still lagging behind. The US is adopting a different approach, which could be called 'reductionist' since it is based on mechanisms at subcellular level to extrapolate information for hazard characterization at organism level. This approach pivots on a mechanistic toxicology based on the biotechnological and bioinformatic revolutions derived by the high-throughput screening allowed by omics and data mining tools.[16,17] Such a reductionist approach led to the development of the so-called 'toxicology for the 21st century' (Tox21c)[18] and to a toxicity testing strategy based on several human cell cultures to individuate and characterize specific pathways of toxicity (PoT) underlying individual chemical adverse effects (see below).

12.2 Existing *In Vivo* Tests: An Overview

The current *in vivo* tests are the ones collected by the Organisation for Economic Co-operation and Development (OECD) as OECD Test Guidelines (TGs) and considered as the most relevant internationally agreed test methods used by government, industry and independent laboratories to determine the safety of chemicals and chemical preparations for human health and the environment. OECD TGs in the field of reproductive toxicity include *in vivo* 'screening' reproductive tests for high-production volume substances for which no data are available (OECD TG 421 and 422) and short-term *in vivo* assays to detect EDs (uterotrophic assay for (anti)-estrogenic substances, OECD TG 440; Hershberger assay for (anti)-androgenic substances, OECD TG 441). However, the core tests are:

1 OECD TG 414 for prenatal developmental toxicity;
2 OECD TG 426 for developmental neurotoxicity upon pre- and early post-natal exposure, a regulatory requirement mainly triggered by pesticides;
3 the two-generation reproductive toxicity study (OECD TG 416) which is considered the core test for hazard characterization of EDs and it is the only test where an organism (the F1 animal) is exposed from the gamete stage through to sexual maturity and production of a next generation, so covering effects during the whole reproductive cycle. Within REACH, OECD TG 416 is a default requirement for substances produced or imported at 1000 tonnes per year and may be a requirement for substances produced or imported at 100 tonnes per year or more;
4 the 'extended one-generation reproductive toxicity study' (EOGRTS; OECD TG 443) which is a new (2011) development of the now outdated 'one-generation study' (OECD TG 415).

OECD TG 443 is quite interesting since it aims at replacing OECD TG 416 by providing at least as much information while reducing by 50% the number of animals employed as well as the study duration. The rationale of EOGRTS is that adding a second generation (as in OECD TG 416) does not contribute significant new information over and above that found in the first generation.[19,20] Moreover, many critical effects for EDs (and possibly other toxicants, like those affecting nervous, renal or immune systems) are the 'delayed developmental effects'[21] (i.e. mechanisms triggered during early development with effects detectable at puberty or adulthood). Therefore, the EOGRTS aims at a refined and more-extensive assessment of various parameters at F1 adulthood. In order not to make EOGRTS an organizational nightmare in the attempt to test many possible effects, the test should be integrated in a tiered testing strategy and properly aimed to the relevant parameters by prior testing steps. However, the current database on EOGRTS is very limited and does not allow a direct comparison with the OECD TG 416. Finally, one can mention the current debate at OECD level about the development of a TG for testing specific effects on post-natal development, which could contribute significantly to the risk assessment of chemicals relevant to children's products such as food contact materials, toys and textiles.[4]

In addition to those TGs above, since 2011 three OECD TGs have been adopted to deal with the identification of ED effects *in vitro*, namely:

1. the H295R Steroidogenesis Assay (OECD TG 450);
2. the 'Performance-Based Test Guideline for Stably Transfected Transactivation *In Vitro* Assays to Detect Estrogen Receptor Agonists' (OECD TG 455); and
3. the 'BG1Luc Estrogen Receptor Transactivation Test Method for Identifying Estrogen Receptor Agonists and Antagonists' (OECD TG 457).

The importance of having these tests included among the other OECD TGs should not be overlooked: the assays exploit recombinant technologies, identify mechanisms rather than effects and have been developed upon performance assessment and protocol validation. These three TGs cover a very narrow, albeit important range of mechanisms (steroidogenesis and estrogen receptor α anti-agonism); this has to be taken into account when considering the inclusion of the TGs in a consistent and robust tiered testing strategy for EDs. This is surely not meant to state that *in vitro* assays are currently of no use in risk assessment: clarification of mechanisms underlying *in vivo* effects by *ad hoc*-developed *in vitro* assays is important for the classification and characterization of chemicals under scrutiny, especially when a weight of evidence approach is adopted.[22,23] However, to implement the role of *in vitro* assays as first tier tools for screening and hazard identification would definitely require thinking in terms of a testing battery, and possibly to turn towards ITS, rather than of single, stand-alone tests.

12.3 Gaps of Current Testing

Despite the apparent large availability of available *in vivo* tools, an appraisal of the list of OECD TGs could lead to a 'counter-list' of potential gaps of current testing in terms of missing targets and/or effects that are not easily detectable or not fully covered. A few examples will be presented here. Overall, with regard to the endocrine system, a weak point of current OECD TGs is the main focus on very few nuclear receptor (NRs)-mediated pathways that somehow neglects the existence of a total number of 48 and 60 NRs, in humans and rodents respectively.[24,25] Hence, even NRs having a recognized ligand (e.g. ERβ) are currently overlooked by toxicological testing; the pathophysiological role of many ligand–NR interactions[24,25] is not covered by any validated test and so are unknown. Therefore, several ligand–NR interactions are not linked to a toxicological, endocrine-related effect although some NRs are broadly involved in xenobiotic sensing and metabolism, such as the pregnane X receptor (PXR) and the constitutive androstane receptor (CAR).[26,27]

Within the endocrine system, the role of the neuroendocrine axes is also not fully implemented. In particular, the impact and mode of action (MoA) of EDs at the level of the hypothalamic/pituitary/adrenal axis and their consequences on reproduction, growth onset, puberty, homeostatic weight control and/or neuro-behaviour should be taken into account considering the demonstrated neuro-endocrine effects of pesticides such as chlorpyrifos.[28–30]

The neglected outputs of the ligand–NR interactions is also well represented by the disorders connected to the metabolic syndrome and the above-mentioned homeostatic weight control. Long-term effects of metabolic re-programming due to pre-natal and developmental exposures are not considered by any testing guidelines, although the worldwide major endocrine diseases are currently obesity and diabetes and many experimental and epidemiological data hint to a relationship between the perinatal exposure to obesogens (i.e. EDs such as organotins, phthalates, BPA, TCDD, PCBs, As and others) and the incidence of metabolic syndrome.[28,31–34] Indeed, no systematic investigation or consolidated *in vitro* test methods considering the interaction of such obe-sogens with NRs involved in lipid metabolism (such as peroxisome proliferator-activated receptors/PPARs, liver X receptors/LXRs) and/or their modulation of adipokines and growth factors (e.g. insulin, insulin-like growth factors/ IGFs) have been so far developed. As an example, the prenatal exposure to di-(2-ethylhexyl)phthalate (DEHP) affects liver morphology and metabolism in post-natal CD-1 mice (at doses corresponding to the no- or low-observed adverse effect level [NOAEL/LOAEL] for altered liver maturation) has been recognized as an inducer of hepatosteatosis in the F1 generation causing an overall altered glycogen storage and lipid metabolism.[35]

Finally, some important reproductive tissues are overlooked by current toxicological testing (e.g. prostate, placenta) that were investigated within ReProTect.[36–39]

The prostate gland is critical for male fertility and its homeostasis and function strictly depend on androgen regulation. Within ReProTect, a novel *in vitro* approach was implemented to investigate prostate-mediated effects on male reproduction and to detect EDs affecting androgen-regulated pathways: the assay was based on the secretion of prostate-specific antigen (PSA) in the human prostate epithelium cell line LNCaP.[36,37] PSA was selected since it is a major clinical biomarker of prostate function.

The placenta has been long investigated in developmental toxicology from the standpoint of embryo/fetal exposure; in fact, the placenta can also be a direct target of toxicants impairing its major roles in fetal nutrition and in the endocrine and immune regulation of pregnancy. Placental development pre-sents many differences in humans from rodents. For ethical reasons, studies in human placenta can be performed only in tissues obtained after natural or elective termination of pregnancy. Therefore, *in vitro* models need to be de-veloped to screen chemicals affecting placenta formation and development. Within ReProTect, two *in vitro* models have been developed, namely the trophoblast-derived choriocarcinoma cell line BeWo and the primary cultures of chorionic villous explants from human placenta.[38,39] The two models were used in sequential steps: first, the BeWo cells to evaluate direct toxicity on trophoblast-like cells, and then the human chorionic villous explants to assess dose responses of functional effects at non-toxic concentrations.[38]

12.4   *In Vitro* Alternatives: Testing Batteries and Integrated Testing Strategies

As mentioned above, animal testing in reproductive toxicology has been challenged by the search for 'alternative' test methods trying to implement the 3Rs concept. Such challenges have been supported by worldwide resource investments; many interesting scientific outputs have been produced but it has not been very successful in proposing alternatives acceptable at regulatory level. However, success is badly needed.

Within the EU REACH regulation, requirements for chemical testing include reproductive and developmental toxicity testing in experimental animals for those chemicals produced at 10 tonnes per year or more. Almost 10 years ago, it was estimated that within this decade, approximately 30 000 chemicals may need to be tested for safety, and under current OECD TGs such testing would require the use of approximately 7.2 million laboratory animals.[40] In a recent re-evaluation of costs and animal use, the new estimates for the number of substances falling under REACH range from 68 000 to 101 000 chemicals and the most demanding studies are in the area of reproductive toxicity testing with about 90% of all animal use and 70% of the required costs for registration. Even using the lower estimate of 68 000 chemicals, the current testing requirements led to estimate an overall demand of 54 million vertebrate animals and testing costs of 9.5 billion euro.[41] This is practically unfeasible: even with the availability of adequate human and material resources (which would be unlikely), the time needed to obtain and analyse data would result in a prolonged and unchecked exposure of EU citizens to a score of chemicals potentially hazardous for fertility and/or the next generation. However, under REACH, the testing on vertebrate animals may not be performed without permission and the use of alternative methods of filling data gaps on the toxicological properties of chemicals is encouraged. These alternatives might include *in vitro* and structure–activity relationship studies, but the REACH technical guidance indicates that these kinds of studies are not adequate to replace reproductive and developmental toxicity testing in whole animals. Therefore, the current situation presents quite a few noticeable contradictions between the need to protect the public and the environment by sound (as well as time- and cost-effective) tools and the availability of such tools. Indeed, despite extensive research, the variety of alternative methods developed for reproductive toxicity testing and the recent adoption of few of them as OECD TGs, not a single one has reached so far regulatory acceptance within a testing framework; this is due mainly to limitations in their predictability and applicability domains.[42]

Hence, the most practical opportunities for the avoidance of whole animal reproductive and developmental toxicity testing currently remain the approaches exploiting available information, namely: (i) the 'read-across', a process in which gaps are filled using data from related compounds; and (ii) the 'weight of evidence' method, based on the integration of existing data from regulatory as well as non-regulatory studies and using factors such as chemical structure and anticipated exposure.[43,44] Finally, also an approach based on 'thresholds of toxicological concerns' (TTC) might be accepted under REACH or also in other frameworks, as a method to reduce vertebrate animal testing through prioritization; however, application of TTC require a robust toxicological database with reduced uncertainties about dose–response relationships, thus, it may be difficult to apply TTC to somewhat controversial (yet major) issues like EDs.[43,44]

As mentioned above, due to the complexity of the mammalian reproductive cycle, it is clear that any reproductive toxicity testing strategy that includes *in vitro* assays will require a test battery consisting of the integration of different, complementary assay systems. A test battery is usually formed by a series of stand-alone tests designed to complement the other tests and generally to measure a different component of a multi-factorial toxic effect (OECD, 2005).[10] Such an approach has been used within the EU Integrated Project ReProTect, in which more than 20 different *in vitro* assays have been developed or improved aiming to predict adverse effects of chemicals on male and female fertility, on implantation, or on embryonic development. The final 'core battery' consisted of 14 different assays which covered a variety of reproductive and/or endocrine disruption endpoints and were used in a double-blinded ring-trial, the so-called 'feasibility study', in order to evaluate their ability to predict toxicity of ten chemicals whose *in vivo* toxicological profiles where already known.[15]

The 'core battery' ReProTect tests listed below were the most promising, since they identified the toxicologically relevant properties of the ten tested chemicals with relatively high accuracy in the majority of cases. To assess endocrine disruption, the (anti)-androgenic and the (anti)-estrogenic activities were analysed by the following six gene-reporter assays:

- the AR-binding assay (ARBA):[45] binding of a radiolabelled ligand to the androgen receptor AR;
- the ER-binding assay (ERBA):[46] binding of a radiolabelled ligand to the estrogen receptor ERα;
- the AR chemically activated luciferase expression assay (AR CALUX):[47,48] luciferase activity of an androgen-response element (ARE)-promoter driven reporter plasmid;
- the ER chemically activated luciferase expression assay (ER CALUX):[47,49] luciferase activity of an estrogen-response element (ERE)-promoter driven reporter plasmid;
- the PC-3-androgen receptor-luciferase-MMTV assay (PALM):[50,51] luciferase activity of an ARE-promoter driven reporter plasmid but in a different human cell line than the AR CALUX;

- the MCF-7 cells (ER+) stably transfected cells (with ERE-ßGlob-Luc-SVNeo) MELN:[52,53] luciferase activity of an ERE-promoter driven reporter plasmid.

To investigate effects on fertility, five tests have been used in the ReProTect 'feasibility study', two of them focusing on the same part of the reproductive cycle, folliculogenesis and oogenesis, but one in mouse and the other in bovine, whereas the others addressed fertilization (bovine) and implantation (mouse and humans), namely:

- the mouse follicle bioassay FBA;[53–55]
- the bovine *in vitro* maturation assay bIVM;[56,57]
- the bovine *in vitro* fertilization assay bIVF;[56]
- the mouse embryonic peri-implantation assay MEPA;[58,59]
- the Ishikawa cell test.[60]

Finally, three ReProTect tests assessed embryonic development by:

- the whole embryo culture (WEC),[15,61] detecting growth and morphology of rat embryos;
- the embryonic stem cell test (EST),[15,62,63] detecting inhibition of differentiation into beating cardiomyocytes of mouse embryonic stem cells;
- the ReProGlo assay:[15,64] luciferase activity of Tcf/Lef-promoter (Wnt signalling) driven reporter plasmid in mouse embryonic stem cells.

The ReProTect 'feasibility study' results demonstrated, as a proof of principle, that complementary *in vitro* tests performed in parallel, provided predictive toxicological profiles that can be used in hazard assessment although extrapolation to humans need to incorporate at least kinetic studies for a proper risk assessment.

The ReProTect 'core battery' did not include specific tests for male fertility or placental function.[15] Some tests, however, did not enter the 'feasibility study' simply due to time constraints, namely the PSA secretion assay in human prostate epithelium cells LNCaP and the β-hCG release in human placenta cells BeWo. These assays were successfully conducted with promising results following the same criteria of the 'feasibility study'.[36–39] These two last mentioned *in vitro* tests highlight one interesting aspect of the use of *in vitro* methods, namely the potential 'phenotypic anchoring' properties of the functional tests applied as markers of exposure/effect (see below), a process by which changes in gene expression are linked to changes in phenotype and, in particular, to the ones used as clinical biomarkers.[65–67] Indeed, both PSA secretion assay and β-hCG release possess such features and can be successfully used to anchor toxicogenomics to a cell/tissue function.[67–69]

Overall, a battery of tests is the simplest possible strategy to replace or substitute for traditional animal tests. To mimic the complexity and responsiveness of a living organism or, at least, of one of its organs and to model the respective kinetics, ITS has been proposed as a solution: 'in the context of safety assessment, integrated testing strategy is a methodology which integrates information for toxicological evaluation from more than one source, thus fa-cilitating decision-making. This should be achieved whilst taking into con-sideration the principles of 3Rs'.[70] The perspectives and current limitations of the ITS have been recently extensively reviewed.[71] Here, it is important to highlight how ITS has to be considered a tool for chemical hazard and risk assessment that, unfortunately, lacks a clear definition as well as a methodo-logical basis for a multi-parameter evaluation bringing together existing in-formation with new data from different sources; last but not least, ITS still awaits a consistent approach for the analysis of test results within and across testing stages in order to combine the different building blocks in an efficient and effective decision-making process.[70,72] Furthermore, a guidance on ITS has to be developed to determine which tools and methods have to be integrated to fulfil the regulatory endpoints, independently of the current legislative re-quirements. Accordingly, the current ECVAM modular approach to validate stand-alone tests has to be adapted to the whole ITS as well as to each of its components and ITS should progressively include as more as possible prob-abilistic pharmacokinetics and exposure modelling, quantitative structure–activity relationship (QSAR) data and read-across, high-throughputs screening and 'omics' data.[71] Thus, ITS is a most promising concept that still needs appropriate instruments to be implemented without 'drowning into complexity'.

12.5 Mechanistic Understanding and Systems Biology in Risk Assessment

Systems biology is an integrative discipline that, through a reductionist ap-proach mainly based on novel molecular biology and genetic methodologies such as the so-called 'omics', seeks to reveal the properties and behaviour of an organism, or any complex biological systems, in terms of an integrated and interacting network of genes, proteins and biochemical reactions as ultimately responsible for an organism's form and functions. The limitations and per-spectives of systems biology have been discussed frequently in recent years and some authors stated that, from the perspective of integrative physiology, sys-tems biology appears to be a collection of tools in search of questions versus a collection of hypothesis-driven questions searching for an answer.[73] Indeed, Joyner and Pedersen suggested that 'the challenge is to use integrative phy-siology to incorporate the findings from 'omics' into models with relevance to patient groups and/or populations, to facilitate the study of gene-physiology, environmental and cultural interactions'.[73]

The novel US approach to toxicity testing (Tox21c)[16,18,74] pointed to mechanisms and precisely to the pathways of toxicity (PoT) in order to breakdown chemical hazards into their MoA making use of 'omics' (e.g. tox-icogenomics, toxicometabolomics). The basic assumption is that there are thousands of potentially toxic chemicals, whereas the PoT influencing cell/tissue functions are in a relatively limited number. Indeed, to combine different technological advances in informatics, high-throughput screening (HTS) technologies and systems biology, the US Environmental Protection Agency (EPA) developed computational tools to be applied to chemicals and mixtures in the so-called computational toxicology in order to support research on the evaluation of chemical toxicity.[75] In this view, next-generation risk assessment will take advantage from quantitative high-throughput data that should provide a much higher capacity for assessing chemical toxicity than is currently available. The identification of PoT requires 'omics' technologies; once a PoT is identified through its 'omics' pattern, a high-throughput test battery can be constructed. In turn, the ensemble of identified PoT can be used to map the human toxome allowing the identification of both (non)toxic chemicals, depending on which relevant PoT is triggered, and (non)toxic concentrations of chemicals, depending at which concentrations the PoT is triggered.[74]

Whereas the 'omics' data set is rapidly growing, the fundamental question is 'which is the functional meaning?' Such a question might remain without an answer if toxicogenomics is not consistently integrated by proteomics and/or metabolomics (i.e. the sources of functional biomarkers that may link gene expression to cellular functions). The above-mentioned concept of 'phenotypic anchoring' could provide an answer if clinical biomarkers can be developed starting from proteins, metabolites or even RNA (such as circulating miRNA) that are positively associated ('functional validation') to changes in patho-physiological conditions. Thus, the development of 'phenotypic anchoring' may be rather critical to the characterization of PoT.

12.6   Conclusions

The implementation of 3Rs in reproductive and developmental toxicity needs to consider the complex physiology of the target systems, including endocrine regulation, and the consequent multiplicity of possible targets and PoT. Any *in vitro* strategy in this field has first to break down the mammalian reproductive cycle into building blocks and then to integrate different assays into a robust and cost-effective strategy. The assessment of EDs as a cutting-edge topic in toxicology can, indeed, spearhead this process. EDs are primarily defined by their MoAs: moreover, endocrine adverse effects should be characterized considering physiological life-stage changes. Therefore, EDs appear as a priority field to apply the PoT-based Tox21c. However promising, the increase of toxicogenomics data should be viewed with caution. It would be rather unsound to replace current *in vivo* testing, with all their limitations, by molecular-based tools whose predictive capacity has not been properly assessed. Thus, toxicogenomics needs to be integrated with a parallel development of 'phenotypic anchoring' in order to support the functional relevance of gene expression pattern changes.

References

1. S. Bremer, R. Cortvrindt, G. Daston, B. Eletti, A. Mantovani, F. Maranghi, O. Pelkonen, I. Ruhdel and H. Spielmann, Reproductive and developmental toxicity, *Altern. Lab. Anim.*, 2005, **33**(Suppl. 1), 183–209.
2. A. Mantovani and F. Maranghi, Risk assessment of chemicals potentially affecting male fertility, *Contraception*, 2005, **72**(4), 308–313.
3. A. Mantovani, Risk assessment of endocrine disrupters. The role of toxicological studies, *Ann. NY Acad. Sci.*, 2006, **1076**, 239–252.
4. F. Maranghi and A. Mantovani, Targeted toxicological testing to investigate the role of endocrine disrupters in puberty disorders, *Reprod. Toxicol.*, 2012, **33**(3), 290–296.
5. M. Harada, Minamata disease: methylmercury poisoning in Japan caused by environmental pollution, *Crit. Rev. Toxicol.*, 1995, **25**(1), 1–24.
6. R. R. Newbold, E. Padilla-Banks and W. N. Jefferson, Adverse effects of the model environmental estrogen diethylstilbestrol are transmitted to subsequent generations, *Endocrinology*, 2006, **147**(Suppl. 6), S11–17.
7. R. M. Sharpe and N. E. Skakkebaek, Testicular dysgenesis syndrome: mechanistic insights and potential new downstream effects, *Fertil. Steril.*, 2008, **89**(Suppl. 2), e33–38.
8. WHO/IPCS, *Global assessment of the state-of-the-science of endocrine disruptors*, World Health Organization, Geneva, 2002.
9. WHO/IPCS, IPCS Risk Assessment Terminology, *IPCS/OECD Key Generic Terms used in Chemical Hazard/Risk Assessment*, World Health Organization, Geneva, 2004. http://www.who.int/ipcs/methods/harmonization/areas/ipcsterminologyparts1and2.pdf.
10. OECD, Guidance document on the validation and international acceptance of new or updated test methods for hazard assessment, ENV/JM/MONO(2005)14, OECD Series on Testing and Assessment 34, Organisation for Economic Co-operation and Development, Paris, France, 2005.
11. A. R. Boobis, S. M. Cohen, V. Dellarco, D. McGregor, M. E. Meek, C. Vickers, D. Willcocks and W. Farland, IPCS framework for analyzing the relevance of a cancer mode of action for humans, *Crit. Rev. Toxicol.*, 2006, **36**(10), 781–792.
12. A. R. Boobis, J. E. Doe and B. Heinrich-Hirsch, IPCS framework for analysing the relevance of a non-cancer mode of action for humans, *Crit. Rev. Toxicol.*, 2008, **38**(2), 87–96.
13. M. K. Skinner and M. D. Anway, Epigenetic transgenerational actions of vinclozolin on the development of disease and cancer, *Crit. Rev. Oncog.*, 2007, **13**(1), 75–78.
14. H. Spielmann, The way forward in reproductive/developmental toxicity testing, *ATLA*, 2009, **37**, 641–656.
15. B. Schenk, M. Weimer, S. Bremer, B. van der Burg, R. Cortvrindt, A. Freyberger, G. Lazzari, C. Pellizzer, A. Piersma, W. R. Schäfer, A. Seiler, H. Witters and M. Schwarz, The ReProTect Feasibility Study, a

novel comprehensive in vitro approach to detect reproductive toxicants, *Reprod. Toxicol.*, 2010, **30**, 200–218.

16. T. Hartung, From alternative methods to a new toxicology, *Eur. J. Pharm. Biopharm.*, 2001, **77**(3), 338–349.

17. M. Bouvier d'Yvoire, S. Bremer, S. Casati, M. Ceridono, S. Coecke, R. Corvi, C. Eskes, L. Gribaldo, C. Griesinger, H. Knaut, J. P. Linge, A. Roi and V. Zuang, ECVAM and new technologies for toxicity testing, *Adv. Exp. Med. Biol.*, 2012, **745**, 154–180.

18. T. Hartung, Lessons learned from alternative methods and their validation for a new toxicology in the 21st century, *J. Toxicol. Environ. Health B. Crit. Rev.*, 2010, **13**(2–4), 277–290.

19. G. Janer, B. C. Hakkert, W. Slob, T. Vermeire and A. H. Piersma, A retrospective analysis of the two-generation study: What is the added value of the second generation?, *Reprod. Toxicol.*, 2007, **24**, 97–102.

20. M. E. Reaves, J. Kidwell, E. Mendez, V. Dellarco, D. Dix, M. Martin, R. Cooper and T. Stoker, *Retrospective Analysis of 350 Multi-Generation Reproductive Toxicity Rat Studies to Support the Proposed Extended One-Generation Reproductive Toxicity Test Guideline*, US EPA Office of Pesticide Programs, Arlington, VA, USA, 2008.

21. A. Mantovani and G. Calamandrei, Delayed developmental effects following prenatal exposure to drugs, *Curr. Pharm. Des.*, 2001, 7(9), 859–880.

22. F. Maranghi, M. Rescia, C. Macrì, E. Di Consiglio, G. De Angelis, E. Testai, D. Farini, M. De Felici, S. Lorenzetti and A. Mantovani, Lindane may modulate the female reproductive development through the interaction with ER-beta: an in vivo-in vitro approach, *Chem. Biol. Interact.*, 2007, **169**(1), 1–14.

23. F. Maranghi, R. Tassinari, D. Marcoccia, I. Altieri, T. Catone, G. De Angelis, E. Testai, S. Mastrangelo, M. G. Evandri, P. Bolle and S. Lorenzetti, The food contaminant semicarbazide acts as an endocrine disrupter: Evidence from an integrated in vivo/in vitro approach, *Chem. Biol. Interact.*, 2010, **183**(1), 40–48.

24. F. M. Sladek, What are nuclear receptor ligands?, *Mol. Cell. Endocrinol.*, 2011, **334**(1–2), 3–13.

25. S. Lorenzetti and L. Narciso, Nuclear receptors: connecting human health to the environment, in *Computational Approaches to Nuclear Receptors*, ed. by P. Cozzini and G. E. Kellogg, Drug Discovery series of Royal Society of Chemistry, Cambridge, UK, pp. 1–22, 2012. DOI: 10.1039/9781849735353.

26. J. Gao and W. Xie, Targeting xenobiotic receptors PXR and CAR for metabolic diseases, *Trends Pharmacol. Sci.*, 2012, **33**(10), 552–558.

27. Y. M. Wang, S. S. Ong, S. C. Chai and T. Chen, Role of CAR and PXR in xenobiotic sensing and metabolism, *Expert Opin. Drug Metab. Toxicol.*, 2012, **8**(7), 803–817.

28. F. Grün and B. Blumberg, Endocrine disrupters as obesogens, *Mol. Cell. Endocrinol.*, 2009, **304**(1–2), 19–29.

29. S. Tait, L. Ricceri, A. Venerosi, F. Maranghi, A. Mantovani and G. Calamandrei, Long-term effects on hypothalamic neuropeptides after

developmental exposure to chlorpyrifos in mice, *Environ. Health Perspect.*, 2009, **117**(1), 112–116.

30. A. Venerosi, L. Ricceri, S. Tait and G. Calamandrei, Sex dimorphic behaviors as markers of neuroendocrine disruption by environmental chemicals: the case of chlorpyrifos, *Neurotoxicology*, 2012, **33**(6), 1420–1426.

31. S. Decherf and B. A. Demeneix, The obesogen hypothesis: a shift of focus from the periphery to the hypothalamus, *J. Toxicol. Environ. Health B. Crit. Rev.*, 2011, **14**(5–7), 423–448.

32. A. Janesick and B. Blumberg, Endocrine disrupting chemicals and the developmental programming of adipogenesis and obesity, *Birth Defects Res. C Embryo Today*, 2011, **93**(1), 34–50.

33. A. Janesick and B. Blumberg, Obesogens, stem cells and the developmental programming of obesity, *Int. J. Androl.*, 2012, **35**(3), 437–448.

34. W. Holtcamp, Obesogens: an environmental link to obesity, *Environ. Health Perspect.*, 2012, **120**(2), a62–68.

35. F. Maranghi, S. Lorenzetti, R. Tassinari, G. Moracci, V. Tassinari, D. Marcoccia, A. Di Virgilio, A. Eusepi, A. Romeo, A. Magrelli, M. Salvatore, F. Tosto, M. Viganotti, A. Antoccia, A. Di Masi, G. Azzalin, C. Tanzarella, G. Macino, D. Taruscio and A. Mantovani, In utero exposure to di-(2-ethylhexyl)phthalate affects liver morphology and metabolism in post-natal CD-1 mice, *Reprod. Toxicol.*, 2010, **29**(4), 427–432.

36. S. Lorenzetti, I. Altieri, S. Arabi, D. Balduzzi, N. Bechi, E. Cordelli, C. Galli, F. Ietta, S. C. Modina, L. Narciso, F. Pacchierotti, P. Villani, A. Galli, G. Lazzari, A. M. Luciano, L. Paulesu, M. Spanò and A. Mantovani, Innovative non-animal testing strategies for reproductive toxicology: the contribution of Italian partners within the EU project ReProTect, *Ann. 1st Super Sanita.*, 2011, **47**(4), 429–444.

37. S. Lorenzetti, D. Marcoccia, L. Narciso and A. Mantovani, Cell viability and PSA secretion assays in LNCaP cells: a tiered in vitro approach to screen chemicals with a prostate-mediated effect on male reproduction within the ReProTect project, *Reprod. Toxicol.*, 2010, **30**(1), 25–35.

38. T. J. Mørck, G. Sorda, N. Bechi, B. S. Rasmussen, J. B. Nielsen, F. Ietta, E. Rytting, L. Mathiesen, L. Paulesu and L. E. Knudsen, Placental transport and in vitro effects of Bisphenol A, *Reprod. Toxicol.*, 2010, **30**(1), 131–137.

39. N. Bechi, G. Sorda, A. Spagnoletti, J. Bhattacharjee and E. A. Vieira Ferro, B. de Freitas Barbosa, M. Frosini, M. Valoti, G. Sgaragli, L. Paulesu and F. Ietta, Toxicity assessment on trophoblast cells for some environment polluting chemicals and 17β-estradiol, *Toxicol. In Vitro*, 2013, **27**(3), 995–1000.

40. T. Höfer, I. Gerner, U. Gundert-Remy, M. Liebsch, A. Schulte, H. Spielmann, R. Vogel and K. Wettig, Animal testing and alternative approaches for the human health risk assessment under the proposed new European chemicals regulation, *Arch. Toxicol.*, 2004, **78**(10), 549–564.

41. C. Rovida and T. Hartung, Re-evaluation of animal numbers and costs for in vivo tests to accomplish REACH legislation requirements for chemicals – a report by the transatlantic think tank for toxicology (t(4)), *ALTEX*, 2009, **26**(3), 187–208.

42. B. van der Burg, E. Kroese and A. H. Piersma, Towards a pragmatic alternative testing strategy for the detection of reproductive toxicants, *Reprod. Toxicol.*, 2011, **31**(4), 558–561.

43. A. R. Scialli and A. J. Guikema, REACH and reproductive and developmental toxicology: still questions, *Syst. Biol. Reprod. Med.*, 2012, **58**(1), 63–69.

44. I. Tluczkiewicz, M. Batke, D. Kroese, H. Buist, T. Aldenberg, E. Pauné, H. Grimm, R. Kühne, G. Schüürmann, I. Mangelsdorf and S. E. Escher, *The OSIRIS weight of evidence approach: ITS for the endpoints repeated-dose toxicity (RepDose ITS)*, Regul. Toxicol. Pharmacol., 2013. http://dx.doi.org/10.1016/j.yrtph.2013.02.004.

45. A. Freyberger, M. Weimer, H. S. Tran and H. J. Ahr, Assessment of a recombinant androgen receptor binding assay initial steps towards validation, *Reprod. Toxicol.*, 2010, **30**(1), 2–8.

46. A. Freyberger, V. Wilson, M. Weimer, S. Tan, H. S. Tran and H. J. Ahr, Assessment of a robust model protocol with accelerated throughput for a human recombinant full length estrogen receptor-alpha binding assay: protocol optimization and intralaboratory assay performance as initial steps towards validation, *Reprod. Toxicol.*, 2010, **30**(1), 50–59.

47. E. Sonneveld, H. J. Jansen, J. A. Riteco, A. Brouwer and B. van der Burg, Development of androgen- and estrogen-responsive bioassays, members of a panel of human cell line-based highly selective steroid-responsive bioassays, *Toxicol. Sci.*, 2005, **83**(1), 136–148.

48. B. van der Burg, R. Winter, H.-Y. Man, C. Vangenechten, M. Weimer, P. Berckmans, *et al.*, Optimization and prevalidation of the in vitro AR CALUX method to test androgenic and antiandrogenic activity of compounds, *Reprod. Toxicol.*, 2010, **30**(1), 18–24.

49. B. van der Burg, R. Winter, M. Weimer, P. Berckmans, G. Suzuki, L. Gijsbers, *et al.*, Optimization and prevalidation of the in vitro ERα CALUX method to test estrogenic and antiestrogenic activity of compounds, *Reprod. Toxicol.*, 2010, **30**(1), 73–80.

50. A. Freyberger, H. Witters, M. Weimer, W. Lofink, P. Berckmans and H. J. Ahr, Screening for (anti)androgenic properties using a standard operation protocol based on the human stably transfected androgen sensitive PALM cell line. First steps towards validation, *Reprod. Toxicol.*, 2010, **30**(1), 9–17.

51. P. Berckmans, H. Leppens, C. Vangenechten and H. Witters, Screening of endocrine disrupting chemicals with MELN cells, an ER-transactivation assay combined with cytotoxicity assessment, *Toxicol. In Vitro*, 2007, **21**(7), 1262–1267.

52. H. Witters, A. Freyberger, K. Smits, C. Vangenechten, W. Lofink, M. Weimer, *et al.*, The assessment of estrogenic or anti-estrogenic activity

of chemicals by the human stably transfected estrogen sensitive MELN cell line: results of test performance and transferability, *Reprod. Toxicol.*, 2010, **30**(1), 60–72.

53. R. G. Cortvrindt and J. E. Smitz, Follicle culture in reproductive toxicology: a tool for in vitro testing of ovarian function?, *Hum. Reprod. Update*, 2002, **8**(3), 243–254.

54. K. Van Wemmel, E. Gobbers, U. Eichenlaub-Ritter, J. Smitz and R. Cortvrindt, Ovarian follicle bioassay reveals adverse effects of diazepam exposure upon follicle development and oocyte quality, *Reprod. Toxicol.*, 2005, **20**(2), 183–193.

55. P. J. Devine and R. Cortvrindt, *In Vitro Ovarian Model Systems*, 2nd edn, Elsevier Ltd, 2010.

56. G. Lazzari, I. Tessaro, G. Crotti, C. Galli, S. Hoffmann, S. Bremer, *et al.*, Development of an in vitro test battery for assessing chemical effects on bovine germ cells under the ReProTect umbrella, *Toxicol. Appl. Pharmacol.*, 2008, **233**(3), 360–370.

57. A. M. Luciano, F. Franciosi, V. Lodde, D. Corbani, G. Lazzari, G. Crotti, *et al.*, Transferability and inter-laboratory variability assessment of the in vitro bovine oocyte maturation (IVM) test within ReProTect, *Reprod. Toxicol.*, 2010, **30**(1), 81–88.

58. K. Lemeire, V. Van Merris and R. Cortvrindt, The antibiotic streptomycin assessed in a battery of in vitro tests for reproductive toxicology, *Toxicol. In Vitro*, 2007, **21**(7), 1348–1353.

59. V. Van Merris, K. Van Wemmel and R. Cortvrindt, In vitro effects of dexamethasone on mouse ovarian function and pre-implantation embryo development, *Reprod. Toxicol.*, 2007, **23**(1), 32–41.

60. W. R. Schaefer, L. Fischer, W. R. Deppert, A. L. S. Hanjalic-Beck, M. Weimer, *et al.*, In vitro Ishikawa cell test for assessing tissue-specific chemical effects on human endometrium, *Reprod. Toxicol.*, 2010, **30**(1), 89–93.

61. A. H. Piersma, E. Genschow, A. Verhoef, M. Q. Spanjersberg, N. A. Brown, M. Brady, *et al.*, Validation of the postimplantation rat whole-embryo culture test in the international ECVAM validation study on three in vitro embryotoxicity tests, *Altern. Lab. Anim.*, 2004, **32**(3), 275–307.

62. A. E. Seiler, R. Buesen, A. Visan and H. Spielmann, Use of murine embryonic stem cells in embryotoxicity assays: the embryonic stem cell test, *Methods Mol. Biol.*, 2006, **329**, 371–395.

63. E. Genschow, H. Spielmann, G. Scholz, I. Pohl, A. Seiler, N. Clemann, *et al.*, Validation of the embryonic stem cell test in the international ECVAM validation study on three in vitro embryotoxicity tests, *Altern. Lab. Anim.*, 2004, **32**(3), 209–244.

64. F. Uibel, A. Muhleisen, C. Kohle, M. Weimer, T. C. Stummann, S. Bremer, *et al.*, ReProGlo: a new stem cell based reporter assay aimed to predict embryotoxic potential of drugs and chemicals, *Reprod. Toxicol.*, 2010, **30**(1), 103–112.

65. M. L. Cunningham, R. Irwin and G. Boorman, Tox/path team takes on differential gene expression, *Environ. Health Perspect.*, 2003, **111**, A814–A815.
66. R. Paules, Phenotypic anchoring: linking cause and effect, *Environ. Health Perspect.*, 2003, **111**, A338–A339.
67. J. G. Moggs, Molecular responses to xenoestrogens: mechanistic insights from toxicogenomics, *Toxicology*, 2005, **213**(3), 177–193.
68. G. P. Daston, Gene expression, dose-response, and phenotypic anchoring: applications for toxicogenomics in risk assessment, *Toxicol. Sci.*, 2008, **105**(2), 233–234.
69. S. Lorenzetti, V. Lagatta, D. Marcoccia, F. Aureli, F. Cubadda, E. Aricò, I. Canini, L. Castiello, S. Parlato, L. Gabriele, F. Maranghi and A. Mantovani, Functional assays, integrated with gene expression signatures, as predictive toxicological biomarkers: from toxicogenomics to phenotypic anchoring, *Toxicol. Lett.*, 2008, **180S**(S1), S123–S124.
70. A. Kinsner-Ovaskainen, Z. Akkan, S. Casati, S. Coecke, R. Corvi, G. Dal Negro, J. De Bruijn, O. De Silva, L. Gribaldo, C. Griesinger, J. Jaworska, J. Kreysa, G. Maxwell, P. McNamee, A. Price, P. Prieto, R. Schubert, L. Tosti, A. Worth and V. Zuang, Overcoming barriers to validation of non-animal partial replacement methods/Integrated Testing Strategies: the report of an EPAA-ECVAM workshop, *Altern. Lab. Anim.*, 2009, **37**(4), 437–444.
71. T. Hartung, T. Luechtefeld, A. Maertens, A. Kleensang and J. Jaworska, Integrated testing strategies for safety assessments, *ALTEX*, 2013, **30**(1), 3–18.
72. J. Jaworska, S. Gabbert and T. Aldenberg, Towards optimization of chemical testing under REACH: a Bayesian network approach to Integrated Testing Strategies, *Regul. Toxicol. Pharmacol.*, 2010, **57**(2–3), 157–167.
73. M. J. Joyner and B. K. Pedersen, Ten questions about systems biology, *J. Physiol.*, 2011, **589**(Pt. 5), 1017–1030.
74. T. Hartung and M. McBride, Food for Thought . . . on mapping the human toxome, *ALTEX*, 2011, **28**(2), 83–93.
75. R. Kavlock and D. Dix, Computational toxicology as implemented by the U.S. EPA: providing high throughput decision support tools for screening and assessing chemical exposure, hazard and risk, *J. Toxicol. Environ. Health B Crit. Rev.*, 2010, **13**(2–4), 197–217.

Subject Index

Illustrations, figures, and tables are in **bold**.